European Studies in Philosophy of Science

Volume 9

This new series results from the synergy of EPSA - European Philosophy of Science Association - and PSE - Philosophy of Science in a European Perspective: ESF Networking Programme (2008–2013). It continues the aims of the Springer series "The Philosophy of Science in a European Perspective" and is meant to give a new impetus to European research in the philosophy of science. The main purpose of the series is to provide a publication platform to young researchers working in Europe, who will thus be encouraged to publish in English and make their work internationally known and available. In addition, the series will host the EPSA conference proceedings, selected papers coming from workshops, edited volumes on specific issues in the philosophy of science, monographs and outstanding Ph.D. dissertations. There will be a special emphasis on philosophy of science originating from Europe. In all cases there will be a commitment to high standards of quality. The Editors will be assisted by an Editorial Board of renowned scholars, who will advise on the selection of manuscripts to be considered for publication.

More information about this series at http://www.springer.com/series/13909

Alexander Christian • David Hommen
Nina Retzlaff • Gerhard Schurz
Editors

Philosophy of Science

Between the Natural Sciences, the Social Sciences, and the Humanities

Springer

Editors
Alexander Christian
Düsseldorf Center for Logic and Philosophy of Science
Heinrich Heine University Düsseldorf
Düsseldorf, Germany

David Hommen
Düsseldorf Center for Logic and Philosophy of Science
Heinrich Heine University Düsseldorf
Düsseldorf, Germany

Nina Retzlaff
Düsseldorf Center for Logic and Philosophy of Science
Heinrich Heine University Düsseldorf
Düsseldorf, Germany

Gerhard Schurz
Düsseldorf Center for Logic and Philosophy of Science
Heinrich Heine University Düsseldorf
Düsseldorf, Germany

ISSN 2365-4228 ISSN 2365-4236 (electronic)
European Studies in Philosophy of Science
ISBN 978-3-030-10229-6 ISBN 978-3-319-72577-2 (eBook)
https://doi.org/10.1007/978-3-319-72577-2

Softcover re-print of the Hardcover 1st edition 2018

Printed on acid-free paper

This Springer imprint is published by the registered company Springer International Publishing AG part of Springer Nature.
The registered company address is: Gewerbestrasse 11, 6330 Cham, Switzerland

Acknowledgment

The editors would like to thank Dennis Dieks, Maria Carla Galavotti, and Wenceslao J. Gonzalez for the inclusion of this volume in the European Studies in Philosophy of Science series.

We would like to thank our contributors for their competence, cooperativeness, and patience. Furthermore, we would like to express our gratitude to Paul Näger, Arne Weber, Ludger Jansen, Oliver Scholz, Andreas Hüttemann, Corina Strößner, Holger Lyre, Florian Boge, Alexander Gebharter, Christian Feldbacher-Escamilla, Georg Toepfer, Martin Carrier, and Susanne Hahn for reviewing the collected contributions.

Düsseldorf Center for Logic and Philosophy of Science
Heinrich Heine University Düsseldorf
Düsseldorf, Germany

Alexander Christian
David Hommen
Nina Retzlaff
Gerhard Schurz

Contents

Contributors

Antonio Piccolomini d'Aragona is a PhD student at Aix-Marseille University and "La Sapienza" University of Rome and a lecturer in logic at the Department of Philosophy of Aix-Marseille University. He works on Dag Prawitz's recent theory of grounds, under the conjoint direction of Prof. Gabriella Crocco, Aix-Marseille University, and Prof. Cesare Cozzo, "La Sapienza" University of Rome; his research areas are broadly mathematical logic and philosophy of logic, with the main focus on proof-theoretic semantics, verificationist theories of meaning, inferentialism, and type theories.

Martin Carrier is professor of philosophy at Bielefeld University and director of the Institute for Interdisciplinary Studies of Science (I2SoS). He earned his PhD at the University of Münster, spent his postdoc period at the University of Konstanz, and became professor of philosophy at the University of Heidelberg. Since 1998, Carrier has worked in Bielefeld. His chief area of research is the philosophy of science, in particular historical changes in science and scientific method, theory-ladenness and empirical testability, and presently the relationship between science and values and science operating at the interface with society. In this latter field, he addresses methodological changes imposed on science by the pressure of practice. He is a member of various German and European Academies of Science. He was awarded the Leibniz Prize of the German Research Association (DFG) for 2008, the Blaise Pascal Medal in Social Sciences and Humanities by the European Academy of Sciences for 2015, and the John G. Diefenbaker Award by the Canada Council for the Arts for 2016.

Roger Deulofeu is a PhD student at Logos Research Group, University of Barcelona. He is working on the notion of scientific laws and explanation in science, with a particular emphasis on the biological domain, arguing that explanation in biology does not uniquely work by describing mechanisms but by identifying regularities that make biological phenomena expectable.

Axel Gelfert received his PhD in History and Philosophy of Science from the University of Cambridge. He has held fellowships and appointments in Budapest, Edinburgh, and Singapore, and is currently Professor of Philosophy at the Technical University of Berlin. He is the author of *A Critical Introduction to Testimony* (Bloomsbury 2014) and *How to Do Science With Models: A Philosophical Primer* (Springer 2016).

Jens Harbecke is a professor of theoretical philosophy and philosophy of social sciences at Witten/Herdecke University, Germany. He is also the project coordinator of a European research project (www.insosci.eu) on the philosophy of social science and neuroscience, and he collaborates as a principal investigator within a philosophical research project funded by the German-Israeli Foundation (www.philosophy-cognitive-science.com/) on causation and computation in neuroscience. His research focuses on constitutive explanations in economics and neurosciences. He also works on questions about causality in the metaphysics of mind and on counterfactual and regularity theories of causation. His recent publications include "The regularity theory of mechanistic constitution and a methodology for constitutive inference," published in *Studies in History and Philosophy of Science Part C: Studies in History and Philosophy of Biological and Biomedical Sciences* (2015); "Regularity constitution and the location of mechanistic levels," published in *Foundations of Science* (2015); and "The role of supervenience and constitution in neuroscientific research," published in *Synthese* (2014).

Paul Hoyningen-Huene is a philosopher of science with a PhD in theoretical physics. In 2014, he retired from his professorship for theoretical philosophy, especially philosophy of science, at the Institute of Philosophy of Leibniz University of Hannover, Germany. He teaches now philosophy of economics at the Department of Economics of the University of Zurich, Switzerland. He is best known for his books *Reconstructing Scientific Revolutions: Thomas S. Kuhn's Philosophy of Science* (1993), *Formal Logic: A Philosophical Approach* (2004), and *Systematicity: The Nature of Science* (2013).

Lara Huber lectures in research ethics and philosophy of science at Kiel University. Her current research focuses on norms in the sciences from the perspective of epistemology and action theory. She has published on modelling and further topics in the philosophy of the life sciences (e.g., *Philosophie der biomedizinischen Wissenschaften*, in: Grundriss Wissenschaftsphilosophie, Die Philosophien der Einzelwissenschaften, ed. by Simon Lohse and Thomas Reydon, pp. 287–318, with Lara K. Keuck, Hamburg: Meiner Verlag 2017). She is the co-editor of *Standardization in Measurement: Philosophical, Historical and Sociological Issues* with Oliver Schlaudt (London: Routledge / Pickering & Chatto 2015), and recently has completed a monograph on scientific concepts and societal perceptions of normality: *Normal* (Hamburg: Textem, forthcoming).

Peter Hucklenbroich Dr. med., Dr. phil., is professor of philosophy and history of medicine at the Medical Faculty and University Hospital in Münster, F.R.G. From 1995 to 2015, he has been director of the Institute of Ethics, History, and Theory of Medicine in Münster. He has been chairman and cochairman of the Center for Philosophy of Science, the Clinical Ethics Committee, and the Institutional Review Board at the University of Münster. He has published numerous books and papers about philosophy of natural science, philosophy of medicine, and the medical concept of disease. His most recognized scientific contributions are the book *Wissenschaftstheoretische Aspekte des Krankheitsbegriffs* (*Philosophical Aspects of the Concept of Disease*, Münster 2013) and the paper "'Disease entity' as the key theoretical concept of medicine" (Journ Med Phil 39, 2014).

Ludger Jansen teaches philosophy at the Ruhr University Bochum and the University of Rostock. He has a strong research interest in the metaphysics of science, including topics like dispositions, functions, and causation. Together with Barry Smith, he has published the first introduction to applied ontology in German.

Elizaveta Kostrova is a research fellow at Sociology of Religion Research Seminar at St Tikhon's Orthodox University, Moscow. Her research focuses on philosophical interpretations of the Other and intersubjectivity, as well as their possible connections to social sciences and religion. She is interested in how social interaction (especially "unselfish" and "disinterested" kind of it) can be conceptualized and grounded philosophically.

Keizo Matsubara has received two PhD degrees from Uppsala University, one in theoretical physics (2004) and one in theoretical philosophy (2013). He is currently a postdoc working at the University of Illinois at Chicago within the project *Space and Time After Quantum Gravity.*

Anne Sophie Meincke is a research fellow at the Centre for the Study of Life Sciences (Egenis) at the University of Exeter. She works at the intersection of metaphysics and the philosophy of biology, focusing in particular on the implications of a process ontological concept of the organism for identity (biological and personal), agency, and free will. Meincke's master's and PhD studies were funded by the German Academic Scholarship Foundation. In 2014, Meincke was awarded the annual Prize for Scientific Research of the City of Innsbruck. She also won, together with John Dupré, the 2015/2016 annual conference grant of the Institute of Philosophy, UCL London, which led to an interdisciplinary conference on "Biological Identity" in summer 2016. Meincke's recent and forthcoming publications include *Auf dem Kampfplatz der Metaphysik: Kritische Studien zur transtemporalen Identität von Personen* (*On the Battlefield of Metaphysics: Critical Studies on the Transtemporal Identity of Persons*) (2015); "Potentialität und Disposition in der Diskussion über den Status des menschlichen Embryos: Zur Ontologie des Potentialitätsarguments" ("Potentiality and disposition in the debate on the status of the human embryo: On the ontology of the argument from potentiality") (2015); *Dispositionalism: Perspectives from Metaphysics and the Philosophy of Science* (as editor) (forthcoming); "How to stay the same while changing: Personal identity as a test case for reconciling 'analytic' and 'continental' philosophy through process ontology," in *Analytic-Bridge-Continental + (ABC+) Process Philosophy*, edited by R. Booth and Berlin et al. (forthcoming); and "Persons as Biological Processes. A Bio-Processual Way-Out of the Personal Identity Dilemma," in *Everything Flows: Towards a Process Philosophy of Biology*, edited by D. Nicholson and J. Dupré (forthcoming).

Andrea Reichenberger is currently working as research assistant at the Center for History of Women Philosophers and Scientists, Faculty of Arts and Humanities, Department of Philosophy, Paderborn University. Her research interests focus on the history and philosophy of physics, especially on women's studies. Awards include Certificate Colloquium Logicum 2006, DVMLG, and Certifico Institutio de Física de Líquidos y Systemas Biológicos La Plata 2010. Recent publications include *Émilie Du Châtelet's Institutions physiques: Über die Rolle von Prinzipien und Hypothesen in der Physik* (Springer 2016).

Kian Salimkhani is a PhD student and research associate at the Institute for Philosophy at the University of Bonn and a member of the DFG-funded research unit Inductive Metaphysics. His research interests include philosophy of physics (especially spacetime theories and quantum field theory), general philosophy of science, and metaphysics. In his PhD project, he investigates the issue of fundamentality of spacetime. He studied theoretical physics and philosophy in Bonn.

Javier Suárez is a PhD student at Egenis, the Centre for the Study of Life Sciences, University of Exeter. He is working on the intersection between philosophy of science and philosophy of biology with particular emphasis on the implications of symbiosis research for evolutionary theory, as well as for traditional topics in philosophy of science (scientific explanation, scientific representation, and the use of model organisms in science).

Introduction

This volume contains a selection of papers delivered at the Second International Conference of the German Society for Philosophy of Science (Gesellschaft für Wissenschaftsphilosophie, GWP) which took place at the Heinrich Heine University in Düsseldorf, Germany, from March 8 to 11, 2016, and was hosted by the Düsseldorf Center for Logic and Philosophy of Science (DCLPS). GWP.2016 was sponsored by the Heinrich Heine University Düsseldorf and the Düsseldorf Center for Logic and Philosophy of Science, the German Research Foundation (Deutsche Forschungsgemeinschaft, DFG), and the *Journal for General Philosophy of Science* (Springer). The GWP organizers were Holger Lyre (Magdeburg), Ulrich Krohs (Münster), Thomas Reydon (Hanover), and Uljana Feest (Hanover). The Local Organization Committee consisted of Gerhard Schurz (chair), Alexander Christian, Christian J. Feldbacher-Escamilla, Alexander Gebharter, David Hommen, Nina Retzlaff, and Paul Thorn.

The aim of GWP.2016 was to enable philosophers of science from Germany and other countries to meet and engage in fruitful discussions on current research topics in philosophy of science and to strengthen the international philosophy of science community. It was also intended to bring together philosophers of science working in different fields of philosophy of science; accordingly, the organizers decided to entitle GWP.2016 "Philosophy of Science: Between the Natural Sciences, the Social Sciences, and the Humanities." Since GWP.2016 comprised a number of outstanding contributions, the organizers decided to publish this volume, which is included in the Springer book series of the European Philosophy of Science Association, besides a special issue of the *Journal for General Philosophy of Science* (JGPS) devoted to GWP.2016.

GWP.2016 had more than 150 participants (approx. one-third were women and about one-fifth were students or graduate students), who came from 16 European and 6 non-European countries. There were 6 plenary lectures given by invited speakers, 62 contributed papers, and 7 contributed symposia (with 19 symposia talks). All in all, GWP.2016 featured 87 talks. The plenary lectures were given by Rainer Hegselmann (Bayreuth), Paul Hoyningen-Huene (Hanover), Michela Massimi (Edinburgh), Stathis Psillos (Athens), Alexander Rosenberg (Duke), and

Gila Sher (San Diego). The conference featured contributed papers and symposia covering all subfields of philosophy of science. The main sections were general philosophy of science (approx. 30%), philosophy of life sciences (approx. 20%), philosophy of natural sciences (approx. 15%), and philosophy of social sciences and humanities (approx. 10%). There were also sections on other fields of philosophy of science and also on more specific topics (all in all approx. 25%). In particular, these were causality, confirmation, history of philosophy of science, mechanisms, philosophy of mathematics, and values in science. The seven symposia dealt with absences in biological and medical explanations, constitution, genetics and culture, philosophy of science and engineering, and quantum gravity.[1]

The list of authors who agreed to contribute to this collection includes renowned experts from several fields in philosophy of science who contributed talks to GWP.2016, including one invited talk of GWP.2016, for which we are particularly thankful. Moreover, the collection presents research of young scientists and has a comparably high share of female authors (one-third).

The essays in this volume are divided into four parts: (1) philosophy of physics, (2) philosophy of life sciences, (3) philosophy of social sciences and values in science, and (4) philosophy of mathematics and formal modeling. We hope that the collection provide insights into a number of ongoing discussions in important subfields of philosophy of science and it will therefore be interesting for an interdisciplinary and multidisciplinary readership.

Philosophy of physics: This part includes papers on unification in high energy physics, cosmology, and causation in physics, including contributions about core arguments in favor of scientific realism, the unification of fundamental forces in physics, testability of multiverse theories, and causal determination in spacetime theories.

In his contribution, **Paul Hoyningen-Huene** addresses two famous arguments in favor of scientific realism. He first discusses a peculiarity of the realism-antirealism debate. Some authors defending antirealist positions in a philosophical discussion seem to be inconsistent with what they do when treating scientific subjects. In the latter situation, they behave as realists. Hoyningen-Huene argues that this tension can be dissolved by distinguishing different discourses belonging to different levels of philosophical radicality. Depending on the respective level, certain presuppositions are either granted or questioned. The author then turns to a discussion of the miracle argument by discussing a simple example of curve fitting. In the example, multiple use-novel predictions are possible without indicating the truth of the fitting curve. It is argued that because this situation has similarities with real scientific cases, it sheds serious doubt upon the miracle argument. Next, Hoyningen-Huene discusses the strategy of selective realism, especially its

[1]For more information about GWP.2016, please see Christian, A., Feldbacher-Escamilla, C. J., and Gebharter, A. (2016). The Second International Conference of the German Society for Philosophy of Science (GWP.2016), March 8–11, 2016. *Journal for General Philosophy of Science*, 1–3. http://doi.org/10.1007/s10838-016-9358-4.

additional crucial component, the continuity argument. The continuity of some X in a series of theories, with X being responsible for the theories' use-novel predictions, is taken to be a reliable indicator for the reality of X. However, the continuity of X could as well be due to the similarity of the theories in the series with an empirically very successful theory embodying X, without X being real. Thus, the author concludes that the two main arguments for scientific realism show severe weaknesses.

Kian Salimkhani's contribution deals with the central challenge of fundamental physics to develop a unified theory of quantum gravity (QG): the combination of general relativity and quantum mechanics. The common conviction is that the quest for QG is not only fueled but generated by *external* principles and hence driven, first and foremost, by reasoning involving philosophical assumptions. Against this, Salimkhani claims that it is exactly the particle physics stance – taken, e.g., by Weinberg and others – that reveals the issue of QG as a genuine physical problem arising within the framework of quantum field theory (QFT). Salimkhani argues that the quest for QG sets an important and often misconceived example of physics' *internal* unificatory practice. Physics' internal strategies – e.g., exploiting the explanatory capacities of an established theory – suffice to explain the search for a theory of quantum gravity. To set the stage for his argument, the author recaps what the research program of QG is about and what remarks suspecting a "dogma of unification" amount to. Subsequently, two important consequences for our understanding of general relativity (GR) and the issue of QG are briefly discussed: First, it is suggested that we should not take GR as a fundamental theory because it can be reduced to QFT. Second, the investigation serves as a clarification of what the problem with QG actually is. Afterward, some objections against the advocated picture are mentioned and very briefly replied to, before the author revisits the opening question concerning the alleged "dogma of unification."

Keizo Matsubara discusses predictions and explanations in multiverse scenarios. Many researchers in contemporary physics take the possibility that our universe is just one of many in a multiverse seriously. In the current debate, however, speculations about multiverses are often connected to arguments using the controversial anthropic principle, which many critics find to be untestable and unscientific. In his contribution, Matsubara suggests criteria that need to be satisfied before a multiverse theory should be considered scientifically respectable. While presently proposed multiverse scenarios do not yet live up to criteria strong enough to be counted as part of well-established science, the author argues that one could in principle find good scientific reasons for accepting a theory entailing that we live in a multiverse. Multiverse theories, if sufficiently developed, can have testable predictions. Accordingly, Matsubara is interested in the question how we in principle can test *specific* multiverse theories, as opposed to evaluating the generic idea that we live in a multiverse. For this, Matsubara focuses on string theory and its multiple stable solutions, which for Matsubara represent a landscape of possible multiverses. In some cases, a multiverse theory can be testable; however, to properly test a multiverse theory, it is important to distinguish new predictions from explanations based on the multiverse.

Andrea Reichenberger devotes her contribution to the work of mathematician and physician Luise Lange (1891–1978). In her articles on the clock paradox and the relativity of time, Lange defends the theory of relativity against philosophical refutations. The clock paradox concerns the phenomenon of time dilation, which is a direct consequence of special relativity: if there are two synchronous clocks at the same inertial reference frame and one of them is moved along a closed curve with constant velocity until it has returned after some time to its point of departure, this clock will lag on its arrival behind the clock that has not been moved. This effect seems to be paradoxical because, in relativity, it appears that either clock could "regard" the other as the traveler, in which case each should find the other delayed – a logical contradiction. Lange shows, however, that the apparent clock paradox is not a paradox but merely conflicts with common sense and is based on a misunderstanding of the theory. Reichenberger's study explores, contextualizes, and analyzes Lange's clear and sophisticated contribution to the debate for the first time.

Philosophy of life sciences: This part begins with a contribution by **Anne Sophie Meincke** about recent developments in the philosophy of biology toward a biologically grounded concept of agency. Herein, agency is described as bio-agency: the intrinsically normative adaptive behavior of human and nonhuman organisms, arising from their biological autonomy. Meincke's contribution assesses the bio-agency approach by examining criticism recently directed by its proponents against the project of embodied robotics. Defenders of the bio-agency approach have claimed that embodied robots do not, and for fundamental reasons cannot, qualify as artificial agents because they do not fully realize biological autonomy. More particularly, it has been claimed that embodied robots fail to be agents because agency essentially requires metabolism. Meincke argues that this criticism, while being valuable in bringing to the fore important differences between bio-agents and existing embodied robots, nevertheless is too strong. It relies on inferences from agency-as-we-know-it to agency-as-it-could-be which are justified neither empirically nor conceptually.

Roger Deulofeu and **Javier Suárez** focus on their contribution on the common appeal to mechanistic explanations in contemporary philosophy of science. Mechanists argue that an explanation of a phenomenon consists of citing the mechanism that brings the phenomenon about. In their contribution, the authors present an argument that challenges the universality of mechanistic explanation: in explanations of the contemporary features of the eukaryotic cell, biologists appeal to its symbiogenetic origin. Therefore, the notion of symbiogenesis plays the main explanatory role. Deulofeu and Suárez defend the notion that symbiogenesis is non-mechanistic in nature and that any attempt to explain some of the contemporary features of the eukaryotic cell mechanistically turns out to be at least insufficient and sometimes fails to address the question that is asked. Finally, the authors suggest that symbiogenesis is better understood as a pragmatic scientific law and present an alternative non-mechanistic model of scientific explanation. In the model they present, the use of scientific laws is supposed to be a minimal requirement of all scientific explanations, since the purpose of a scientific explanation is to make phenomena expectable. Therefore, this model would help to understand biologists' appeal to the notion of symbiosis and thus is shown to be better, for the case under examination, than the mechanistic alternative.

Ludger Jansen's contribution is concerned with functional explanations, which interestingly apply not only in cases of normal functioning but also in the case of malfunctioning. According to a straightforward analysis, a bearer of the function to F is malfunctioning if and only if it does not F although it should do so. This makes malfunctions and malfunctionings analogous to negative causation and thus problematic, because they seem to involve absent dispositions and absent processes. This analysis seems also to require that the function to F cannot be identical with the disposition to F. Thus, we seem to be trapped in a dilemma: If the realm of functions is separated from the realm of dispositions, then it seems that functions cannot be causally efficacious. Alternatively, functions are considered to be identical with dispositions, but then malfunctioning seems to be conceptually impossible. Jansen's contribution defends and further develops the thesis of Röhl and Jansen that functions are not a special type of dispositions. For this purpose, it first reviews different varieties of malfunction and malfunctioning and suggests definitions of both malfunction and malfunctioning. The author discusses the special-disposition account of the basic formal ontology (BFO), which Spear et al. have defended by suggesting various strategies on how a special-disposition account can deal with malfunctions. On the one hand, Jansen's contribution evaluates these strategies and indicates several problems arising from them. On the other hand, it describes how to account for the non-optionality and the causal efficacy of functions, if functions are not dispositions. While function types are not identical to disposition types, there are important interrelations between functions and dispositions, namely, (1) heuristically, (2) from a design perspective for artifact functions, and (3) from an evolutionary perspective for types of biological functions.

Peter Hucklenbroich's contribution deals with disease entities and the naturalness of disease classifications in medical pathology. In the twentieth- and twenty-first-century medicine, the concept of a disease entity has proven to be of key importance for pathology and the theory of diseases. Disease entities are kinds of complex clinical and etiopathogenetic processes that are triggered by specific primary causes and develop on anatomical, physiological, clinical, and subjectively experienced levels. They are distinguished from healthy states of life by definite criteria of pathologicity. Hucklenbroich sketches the prehistory as well as the central features of the current paradigm of disease entities. Since the 1970s, philosophical theories of disease tend to ignore or, at best, reject this concept. By examining the well-respected theories of H. Tristram Engelhardt, Jr., and Caroline Whitbeck, it is shown that this defensive attitude results from a philosophical misconception of the concept. Engelhardt criticizes the concept of disease entity because he erroneously assumes, as Hucklenbroich argues, that explanations using this concept are inconsistent with explanations by laws of physiology. On the other hand, Whitbeck correctly refers to the modern, scientific version of the concept. But in her opinion, the concept "cause of disease" is defined according to certain "instrumental interests" that may differ between subjects and is, thus, neither objective nor unique and unequivocal. Hence, the concept of disease entity is ambiguous and not suited for establishing a unique, unambiguous, and unequivocal *natural classification* of diseases. Hucklenbroich shows that Whitbeck's objections rest upon misconceptions concerning the concept of "primary cause," i.e., "etiological factor," and of the so-called "multi-factorial" causation. By reference to a careful, medically and philosophically correct reconstruction of these concepts, he aims to show that her objections do not apply.

Philosophy of social sciences and values in science: This part starts with a contribution by **Martin Carrier** who addresses matters of agnotology, a research field decisively influenced by Robert Proctor, who introduced the notion in 1992. Agnotology refers to the active creation and preservation of confusion and ignorance. Focusing on his contribution to the intentional production of misleading information or the deliberate creation of epistemically detrimental dissent, however, Carrier recognizes several nontrivial epistemological problems requiring clarification. First, the purpose of generating confusion is typically difficult to ascertain. Accordingly, identifying a publicly accessible mistake would be helpful for pinpointing agnotological ploys. Second, the idea underlying Proctor's notion is that sociopolitical motives have trumped or outplayed the quest for knowledge. However, implementing this idea demands the distinction between epistemic and non-epistemic values. The former appreciate knowledge and understanding, while the latter refer to sociopolitical interests and utility. Many philosophers of science do not acknowledge an in-principle distinction between the two. At the same time, they are committed to scientific pluralism. Both considerations come together in raising the problem which methodological standards are violated in the production and maintenance of ignorance. Carrier proposes to identify agnotological ploys by the discrepancy between the conclusions suggested by the design of a study and the conclusions actually drawn or indicated. This mechanism of "false advertising" serves to implement agnotological ploys and helps to identify them without having to invoke the intentions of the relevant agents. The author discusses three agnotological cases, i.e., studies on bisphenol A, Bt-maize/Roundup, and Gardermoen's airport in Oslo. Pinpointing agnotological endeavors is a means for weeding out approaches that look fitting at first glance but which are, in fact, blatantly inappropriate. Identifying such endeavors serves to reduce the range of studies under consideration and thus helps to manage pluralist diversity.

Elizaveta Kostrova investigates in her contribution the "ought" dimension in value theory and John Dewey's notion of the desirable from a philosophical as well as a sociological standpoint. The concept of "value" is widely used in various fields, and it has recently become the subject of empirical research. However, there is no common understanding of what it is. From the very start, the scope of value has been part of the opposition of what "is" to what "ought to be," and the fact that value judgments contained a normative element seemed to make the exclusion of value from the area of scientific analysis inevitable. As Kostrova shows in her contribution, John Dewey offers a different way of reasoning about values, which would allow scientists to keep the normativity in a way of saving the specificity of the concept. In order to do this, Dewey links the source of value with the evaluation process and introduces the concept of the "desirable" drawing the line between the "desirable" and the "desired." Clyde Kluckhohn later borrowed this concept from Dewey while formulating the concept of values within Parsons' theory of action. Thanks to him, the "desirable" has become a favorite part of value definition among different researchers. As a result of this development, the concept of "desirability" has been transformed: for example, in social psychology, the "desirable" has moved closer to the "important," and the significance of the normative aspect has diminished, evolving to a more descriptive understanding, while the social dimension, though present already in Dewey, has greatly increased. Kostrova's contribution considers the appearance of Dewey's notion of the desirable in the definition of value as well as its role in it and its further application in the study of values.

Lara Huber analyzes how standards shape scientific knowledge. Standards are said to provide trust in scientific methodology in general and measuring devices in particular. To standardize means to formalize and regulate scientific practices and to prioritize instrumental and methodological prerequisites of research: Standardization impacts on the design of experiments concern the reporting of outcomes and the assessment of research (e.g., peer review process). Studies in the history of science and technology have shown that standards contribute significantly to the evolution and validation of scientific practices. The philosophy of science is as yet only beginning to analyze systematic challenges posed by standardization. The main interest of Huber's contribution is to elaborate on the question how standards relate to ends that facilitate and/or allow for knowledge claims in experimental sciences in general. The author intends to inform about scientific practices in different fields of research that address given ends of standardization. First of all, Huber presents three examples of standards in science. Her contribution then focuses on three ends purported to serve epistemic needs in different fields of scientific inquiry: stability, homogeneity, and internal validity. She presents three case studies on standardization in different fields of scientific research, ranging from physics and measurement science to population-based trial design in psychology and medicine, in order to inquire into the reality of standards as being very specific tools with defined uses while sharing general suppositions about which ends they serve within the realm of science.

Philosophy of mathematics and formal modeling: This part starts with a contribution by **Jens Harbecke** who addresses a potential problem for his offered methodology of constitutive inference in the context of mechanistic explanation. According to the mechanistic approach, an adequate explanation demands an analysis of the mechanisms "underlying" an explanandum phenomenon at several levels. A central challenge for this approach consists in offering an account of how such mechanistic explanations can be established. As many authors have observed, the relationship between phenomena and their mechanisms cannot be a causal one, because a causal relationship is commonly considered to hold only between non-overlapping events, but a mechanism is believed to overlap with the phenomenon in space and time. Their noncausal and synchronous relation is usually referred to as "constitution." The problem seems to be that even when all causal relationships among mechanisms or parts of mechanism have been identified, it remains unclear whether all constitutive relationships among mechanisms and phenomena have been established thereby as well. Against this, Harbecke argues that it is possible to explicate a methodology for the establishment of constitutive explanations, although the latter differs substantially from methodologies establishing causal relationships. Harbecke's so-called methodology of constitutive inference is ultimately based on Mill's "method of difference," which requires a complete variation of factors in a given frame. In constitutive contexts, however, such a complete variation is often impossible. The author offers a solution to this problem that utilizes the notion of a "mechanism slice." In a first step, an example of a currently accepted explanation in neuroscience is reconstructed, which serves as a reference point of the subsequent discussion. It is argued that the proposed solution accommodates well all schematic situations in which the impossibility of varying all test factors could be expected either to lead to false inferences or to preclude the establishment of correct constitutive claims.

Antonio Piccolomini d'Aragona considers Dag Prawitz's recent theory of grounds. Since the 1970s, Prawitz has been interested in general proof theory. His normalization theorems play in natural deduction systems the role that Gentzen's cut-elimination plays in sequent calculi, a syntactic result which is extended to semantics through what Schroeder-Heister calls the "fundamental corollary of normalization theory," stating that every closed derivation in intuitionistic logic can be reduced to one using an introduction rule in its last step. The framework is inspired by Gentzen's notion that the introduction rules represent the definitions of the symbols concerned, and the elimination rules are no more than the consequences of these definitions. According to Prawitz, however, this is not the only possible approach to general proof theory, since one could also try to give a direct characterization of different kinds of proofs. From this standpoint, the influence of Gentzen and Dummett is accompanied by references to the Brouwer-Heyting-Kolmogorov (BHK) clauses. Already in 1977, Prawitz addressed the non-decidable character of the BHK proofs. In his more recent papers, Prawitz provides indications on how the ground-theoretic framework should be developed. However, the overall project still seems to be in an embryonic stage. In his contribution, Piccolomini d'Aragona addresses a threefold task. First, he analyzes the decidability problem within the BHK approach. Next, the author proposes a partial calculus for Prawitz's theory of grounds. After introducing a core calculus for Gentzen's introductions, he defines two expansions of it, one for full first-order minimal logic and another for a kind of "metalanguage" of grounds. These expansions help understand the final task, a ground-theoretic reformulation of the BHK decidability issue.

The final contribution by **Axel Gelfert** analyzes the concept and relevance of exploration in the context of scientific modeling. Traditional frameworks for evaluating scientific models have tended to downplay their exploratory function; instead they emphasize how models are inherently intended for specific phenomena and are to be judged by their ability to predict, reproduce, or explain empirical observations. By contrast, Gelfert argues that exploration should stand alongside explanation, prediction, and representation as a core function of scientific models.

Thus, models often serve as starting points for future inquiry, as proofs of principle, as sources of potential explanations, and as a tool for reassessing the suitability of the target system (and sometimes of whole research agendas). This is illustrated by a case study of the varied career of reaction-diffusion models in the study of biological pattern formation, which was initiated by Alan Turing in a classic 1952 paper. Initially regarded as mathematically elegant, but biologically irrelevant, demonstrations of how, in principle, spontaneous pattern formation could occur in an organism, such Turing models have only recently rebounded, thanks to advances in experimental techniques and computational methods. The long-delayed vindication of Turing's initial model, the author argues, is best explained by recognizing it as an exploratory tool (rather than as a purported representation of an actual target system).

Düsseldorf Center for Logic and Philosophy of Science
Heinrich Heine University Düsseldorf
Düsseldorf, Germany

Alexander Christian
David Hommen
Nina Retzlaff
Gerhard Schurz

Part I
Philosophy of Physics

Chapter 1
Are There Good Arguments Against Scientific Realism?

Paul Hoyningen-Huene

Abstract I will first discuss a peculiarity of the realism-antirealism debate. Some authors defending antirealist positions in a philosophical discussion seem to be inconsistent with what they do when treating scientific subjects. In the latter situation, they behave as realists. This tension can be dissolved by distinguishing different discourses belonging to different levels of philosophical radicality. Depending on the respective level, certain presuppositions are either granted or questioned. I will then turn to a discussion of the miracle argument by discussing a simple example of curve fitting. In the example, multiple use-novel predictions are possible without indicating the truth of the fitting curve. Because this situation has similarities with real scientific cases, it sheds serious doubt upon the miracle argument. Next, I discuss the strategy of selective realism, especially its additional crucial component, the continuity argument. The continuity of some X in a series of theories, with X being responsible for the theories' use-novel predictions, is taken to be a reliable indicator for the reality of X. However, the continuity of X could as well be due to the similarity of the theories in the series with an empirically very successful theory embodying X, without X being real. Thus, the two main arguments for scientific realism show severe weaknesses.

Keywords Miracle argument · Use-novel predictions · Continuity argument · Selective realism · Structural realism

---

P. Hoyningen-Huene (✉)
Institute of Philosophy, Leibniz University of Hanover, Hanover, Germany

Department of Economics, University of Zurich, Zurich, Switzerland
e-mail: hoyningen@ww.uni-hannover.de

A. Christian et al. (eds.), *Philosophy of Science*, European Studies in Philosophy of Science 9, https://doi.org/10.1007/978-3-319-72577-2_1

1.1 Introduction

There is a plausible *prima facie* answer to the title question whether there are good arguments against scientific realism, which simply is no! The source for this answer is the ubiquitous behavior of scientists, more specifically of physicists: they are usually straightforward realists when it comes to discussing scientific results. Good physicists have a solid education, are usually diligent, rational, intelligent, and self-critical people (at least as long as they talk science, not necessarily when they talk *about* science). Here is an example from recent, very topical science (suspect of earning some of its authors a Nobel Prize in physics). The upper half of Fig. 1.1 represents data that were measured on September 14, 2015 and published on Feb 11, 2016 (Abbott et al. 2016). The interpretation of these data is summarized in the conclusion of the paper:

> VIII. CONCLUSION
>
> The LIGO detectors have observed gravitational waves from the merger of two stellar-mass black holes. The detected waveform matches the predictions of general relativity for the inspiral and merger of a pair of black holes and the ringdown of the resulting single black hole. These observations demonstrate the existence of binary stellar-mass black hole systems. This is the first direct detection of gravitational waves and the first observation of a binary black hole merger.

The language of this conclusion (and of the whole body of the paper) is uncompromisingly realist: they "have observed gravitational waves", "the existence of binary stellar-mass black hole systems" is demonstrated, gravitational waves have been "directly" detected, and "a binary black hole merger" has been observed for the first time. There is no talk of or any argument for the given *realist* interpretation of the data: no other possibility is mentioned, let alone explicitly discarded based on some argument. Therefore, for the physicists involved – more than 1000 figure as authors

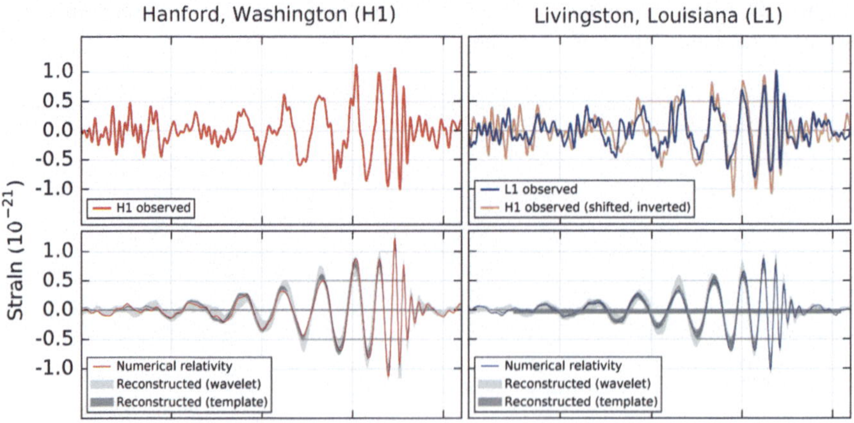

Fig. 1.1 Data for gravitational waves

of the paper – the case seems clear: they just detected really existing gravitational waves and observed the really existing merger of a pair of really existing black holes. Any argument for the evidently realist interpretation of the data is lacking. This suggests that the authors deem such an argument just totally unnecessary. If we stick to the hypothesis that this group of physicists is a bunch of fairly rational people, we must conclude that there simply are no serious arguments against the realist stance taken up in the paper, otherwise these arguments would have been confronted (and possibly disproved). Therefore, in the view of physics, as implicitly contained in the cited paper, the case seems clear: there are no serious arguments against scientific realism.

However, there seem to be serious dissenting voices: there are quite a few statements by (theoretical) physicists and chemists exactly to the contrary. A prominent example is Stephen Hawking:

> I take the positivist viewpoint that a physical theory is just a mathematical model and that it is meaningless to ask whether it corresponds to reality. All that one can ask is that its predictions should be in agreement with observation.[1]

This statement seems to be in blatant contradiction with the realist stance of the physicists who discovered gravitational waves. Is Hawking perhaps fundamentally different from these physicists? It may come as a surprise that he is not, at least as far as black hole physics and gravitational waves are concerned. In the context of the 1960s and 1970s discussion about the possible detection of gravitational waves, Hawking published a paper entitled "Gravitational radiation from colliding black holes" (Hawking 1971). Its abstract reads:

> It is shown that there is an upper bound to the energy of the gravitational radiation emitted when one collapsed object captures another. In the case of two objects with equal masses m and zero intrinsic angular momenta, this upper bound is $(2-\sqrt{2})\,m$.

Hawking refers to "gravitational radiation emitted" and "collapsed object[s]" (i.e., black holes), and there is no sign in the paper that these things are only calculational devices with no reality content, as one would expect from an instrumentalist. Instead, he speaks about them in the same language as one speaks about any ordinary real physical object. Hawking's stance in this paper is thus purely realist. However, what shall we make of this apparent contradiction between a purely realist and a radically instrumentalist stance occurring in the same author?

[1] Hawking (1996, 3–4). See also the very clear statement of the consequence of his positivism later in the book: "[Penrose] is worried that Schrödinger's cat is in a quantum state, where it is half alive and half dead. He feels that can't correspond to reality. But that doesn't bother me. I don't demand that a theory correspond to reality because I don't know what it is. Reality is not a quality you can test with a litmus paper. All I'm concerned with is that the theory should predict the results of measurements" (Hawking and Penrose 1996, 121). See also Hawking and Mlodinow (2010, esp. Chapter 3).

1.2 Levels of Philosophical Radicality

My suggestion is that we should distinguish, in somewhat fashionable terminology, different discourses, or: ways of reasonably discussing things (or "language games"). I shall describe the differences between these discourses as differences in the levels of their *philosophical radicality*. There is a ground level, or level zero, of philosophical radicality in which nothing is put into question for *philosophical* motives. On this level, nothing is doubted beyond what is doubted in normal scientific practice (or in everyday discourse, for that matter). For instance, in cutting edge scientific discourse about new hypothetical objects, many things are taken for granted, for instance realistically interpreted established theories and those parts of the experimental equipment that have been exhaustively tested. "Taken for granted" only means that these things are not questioned *in the given context* which does not, of course, exclude their being questioned in other contexts, be it scientific or philosophical contexts (more on the latter see below). For instance, in the recent discovery of gravitational waves and of inspiraling black holes, it was taken for granted (among many other things) that the theory on which the design of the lasers was based was correct, that the data that the two detectors produced were the result of optical interference, and that General Relativity Theory was the right theory to interpret the data (see Abbott et al. 2016).[2] The question in focus was the existence of gravitational waves and, for any particular case, their concrete sources. Clearly, this is a thoroughly realist stance: the pertinent scientific theories are interpreted realistically, and the question is whether gravitational waves really exist and what their sources are. Thus, the enterprise is a purely scientific one, devoid of any *additional* philosophical questioning.

In the given context, the first level of philosophical radicality is reached by questioning *in general* the step to a realist interpretation of scientific theories. This is what the standard philosophy of science discussion about scientific realism is all about. In this case, our knowledge of observable macroscopic objects as real objects is typically taken for granted. The question being asked is this: Are we in general justified to assume the existence and properties of those unobservable

[2] A referee of an earlier version of this paper objected to my description of level zero of philosophical radicality that "it is entirely legitimate for a scientist to question background theories in order to draw into doubt a conclusion like the detection of gravitational waves. Double checking and questioning scientific background assumptions fully plays out at the scientific level and constitutes an important element of scientific reasoning." No and yes. For instance, doubting the putative detection of gravitational waves on the basis that the use of Maxwell's equations should be questioned for the calculation of interference patterns would be far from being "entirely legitimate", as the referee has it. Although this doubt is not excluded as a matter of principle, in a series of steps of critically checking the experiment this particular step would come rather late. "Double checking and questioning scientific background assumptions" not referring to accepted fundamental theories, however, is a completely different matter. Of course, I never meant to deny the legitimacy of a critical scientific discussion of assumptions of this kind on level zero of philosophical radicality.

objects that our mature and well-confirmed theories about the pertinent domain postulate, based on our observations of macroscopic objects? A positive answer to this question is (roughly) the position of the scientific realist. Someone who denies the legitimacy of this step to a realist interpretation of well-confirmed mature theories is a scientific anti-realist, or instrumentalist. Clearly, the question about the *general* legitimacy of realist interpretations of well-confirmed mature theories is more radical than the zero level question about the legitimacy of the realist interpretation of a given *individual* theory. The former question is a philosophical question, the latter a scientific one. Clearly, on level zero, i.e., in the scientific context, the general legitimacy of realist interpretations of theories (under appropriate conditions) is taken for granted. In other words, the general *philosophical* doubt about realist interpretation articulated on level one does not come into play in the scientific practice on level zero.[3] The situation is similar to the situation we are confronted with by "the" problem of induction. Philosophers (since Hume) have asked the question of the legitimacy of inductive generalizations in general (level one). Scientists, by contrast, take the possibility of inductive generalization under appropriate conditions for granted and ask in any particular case, whether the conditions for a valid inductive generalization are met (level zero).[4]

One can push philosophical doubt even beyond level one of philosophical radicality, although this is much less fashionable in current philosophy of science. The main assumption of the first level of philosophical radicality is that we have knowledge of observable macroscopic objects. If one is a scientific anti-realist on the first level, one may extend one's doubt about the epistemic accessibility of unobservable objects to observable objects as well (this is of course but one route to this higher level of philosophical radicality). Thus, on this second level of philosophical radicality, the existence of and our epistemic access to macroscopic observable objects is questioned. Roughly, this is the level of philosophical radicality on which sense data based philosophies, Kant, perspectival realism, and model-dependent realism, among others, operate.[5] These philosophies question the

---

[3]I note in passing that in the history of philosophy, others have seen this difference also. For instance, Edmund Husserl denoted it as a difference between the "natural standpoint" and the "critical epistemological standpoint"; see Husserl (1967 [1922], §§ 27ff).

[4]With respect to the scientific realism debate, the above distinction between levels zero and one of philosophical radicality has been articulated somewhat differently in Magnus and Callender (2004). They distinguish "retail arguments for realism (arguments about specific kinds of things such as neutrinos, for instance) from wholesale arguments (arguments about all or most of the entities posited in our best scientific theories)" (321). Clearly, this distinction is very similar to the one proposed above. However, what is missing from my point of view in Magnus and Callender's version of the distinction is the explicit reference to the correlated difference of epistemic stances, here called different levels of philosophical radicality. Only the difference in the epistemic stances reveals the possibility to defend seemingly inconsistent positions at the different levels; see below.

[5]For perspectival realism, see, e.g., Giere (2006); for model-dependent realism, see, e.g., Hawking and Mlodinow (2010). It seems to me that these two positions are essentially identical.

givenness, or pure object-sidedness, of unitary observable macroscopic objects and propose to investigate the constitution of these objects, i.e., the contribution of subject-sided elements. Obviously, different philosophical positions may result from this questioning the apparently unquestionable pure object-sidedness of observable things. Pushing philosophical radicality even further, one may reach Cartesian skepticism (from which position it seems very difficult to move anywhere).

It should be noted that the different levels of philosophical radicality are not uniquely defined. Neither do I claim that a certain level structure of philosophical radicality in one area of philosophy can be immediately transferred to another area, say from philosophy of science to ethics. There, the levels of philosophical radicality may take on forms different from the ones in philosophy of science. The essential point is that at some level n, certain things are taken for granted, whereas at level n + 1, they are questioned. To move from one level to another, i.e. to participate in discourses situated at different levels, is not inconsistent.[6] Each level determines a certain discourse by fixing certain things as given and beyond dispute – for the sake of argument, or of conviction. A discourse determined in this way may be interesting or uninteresting, depending on one's goals and convictions. For instance, in order to understand certain every day or certain scientific practices, one should be aware of being at the zeroth level, whereas certain philosophical questions necessarily involve a move to a higher level of philosophical radicality. As I have illustrated above by the example of Steven Hawking, the same person can work at both levels as all antirealist philosophers do when it comes to normal everyday affairs, were they typically do not doubt the existence and cognizability of observable objects.

However, individual philosophers and scientists strongly differ in their willingness to engage with the various levels of philosophical radicality. The higher the degree of philosophical radicality, the further away from common sense one moves. If one uses the adherence to common sense as an argument against one's engagement with one of the levels beyond level zero, one should be conscious about this argument's persuasive force. It may be a convincing argument for those who think that with common sense one is epistemologically on a safer ground than with any mode of philosophical questioning that a particular higher level of philosophical radicality involves. However, for those defending the practice of philosophy on a higher level, the accusation of a deviation from common sense is certainly not persuasive. Quite on the contrary: for asking philosophical questions on a certain level of philosophical radicality above level zero is nothing but questioning certain common sense presuppositions. Thus for those philosophers, the refusal of engaging with that level of philosophical radicality is nothing but a refusal of philosophy itself.[7]

---

[6]Some defenders of common sense realism appear to assume the inconsistency of level zero and level one. See, for example, Richard Dawkins: "Show me a cultural relativist at thirty thousand feet and I'll show you a hypocrite": Dawkins (1995, 31–32).

[7]See, e.g., Rowbottom (2011) against the scientific realist philosopher Howard Sankey.

In the following, we will move beyond level zero. I shall investigate two arguments or strategies, respectively, which are standardly used in the defense of scientific and/or structural realism: the "miracle argument" and the "selective strategy".

1.3 The Miracle Argument

One of the most important arguments for scientific realism starts from an uncontroversial observation: science has been very successful repeatedly in producing novel predictions. The cases that are relevant for the argument must be described more carefully. "Predictions" in the given context are not necessarily predictions in the temporal sense, but are statements about observable putative facts that are derived from a certain hypothesis or theory. Typically, pertinent antecedent conditions and possibly other assumptions have to be included in the premises of the derivation. The "novelty" of the prediction means in the given context that the predicted data have not been used in the construction of the theory. For clarity, sometimes the expression "use-novel predictions" is used.[8] Here are two examples. In 1916, Einstein predicted the existence of gravitational waves as a consequence of his General Relativity Theory (GRT) (Einstein (1916), with corrections in Einstein (1918)). Gravitational waves were not used in the construction of GRT and were thus a use-novel prediction. Secondly, in 1927, Heitler and London derived from the newly developed quantum theory the existence of the covalent bond between hydrogen atoms (Heitler and London 1927). The covalent bond was well known at the time but unexplained in terms of physics, and it was not used in the construction of quantum mechanics. In this sense, the existence of the covalent bond was a use-novel prediction of quantum mechanics.

The question is, how are these use-novel predictions possible? How does a theory acquire the capability of predicting novel facts that the theory was not designed for? What exactly are the resources of a theory for such predictions? In other words, how can this particular success of theories be explained? Scientific realists have a plausible answer to these questions. Theories can produce correct use-novel predictions if they are approximately true, i.e., if their theoretical terms refer to real entities and if they get the properties of these entities at least approximately right. Thus, approximate truth of theories is *sufficient* for correct use-novel predictions. However, according to the scientific realist, approximate truth is also *necessary* for correct use-novel predictions. The reasoning is that without approximate truth of a theory, it is just incomprehensible how it could be capable of producing use-novel

[8]According to Schindler (2008, 266), the term "use-novelty" has been introduced by Deborah Mayo in Mayo (1991, 524). The concept of use-novel predictions, as opposed to temporally novel predictions, was apparently introduced by Zahar and Worrall in the 1970s and 1980s: see Worrall (1989, 148–149).

prediction: it just lacks the resources to do so. So, in Hilary Putnam's famous words, realism "is the only philosophy that does not make the success of science a miracle" (Putnam 1975, 73). The putative weight of this argument is well expressed by its denotation as the "ultimate argument" for realism (van Fraassen 1980, 39; Musgrave 1988). The point of the miracle argument is that it licenses the inference from the existence of correct use-novel predictions of a theory to its approximate truth.

Despite its undeniable plausibility, the question is whether the miracle argument is really correct. There has been an intense critical discussion of the miracle argument, focusing on different aspects of it.[9] Here, I am contributing to this discussion by providing an utterly simple model that may disclose a weakness of the argument.[10] I model the situation of finding an at least approximately true theory, given some empirical evidence, by a curve fitting exercise. Imagine a situation in which there is a true function that we want to identify by a number of measurements. This function itself is determined by 15 data points in five intervals but this is unknown to us ("only God knows" that the function has 15 free parameters). So God's point of view of the situation is depicted in Fig. 1.2.

Now imagine we earthlings have measured the first six data points because we want to determine the true function. Of course, based on these six data points there are many admissible functions and not only the true function, as is illustrated in Fig. 1.3.

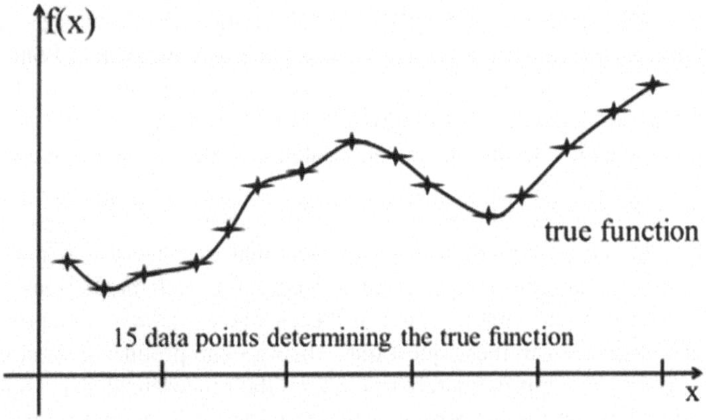

Fig. 1.2 Data points determining the true function

[9]See, e.g., Alai (2014), Hitchcock and Sober (2004), Howson (2013), Lee (2014), Lyons (2003), Lyons (2015), Magnus and Callender (2004), Menke (2014), Psillos (2006), and Saatsi and Vickers (2011).

[10]My first presentation of this model was very clumsy: see Hoyningen-Huene 2011, appendix. It was, to the best of my knowledge, ignored in the literature. The present presentation of the model got rid of much unnecessary mathematical baggage and is hopefully more palatable.

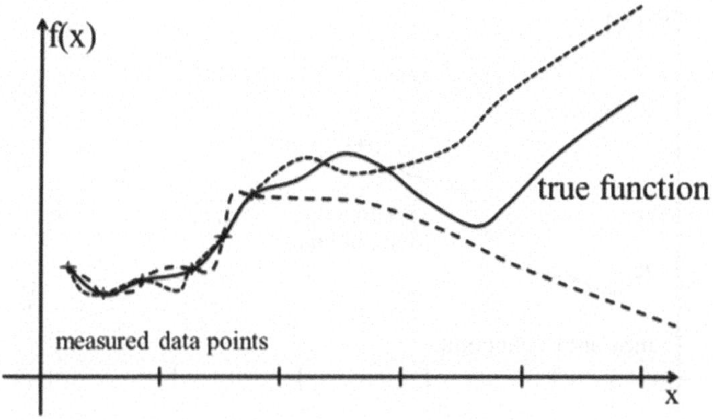

Fig. 1.3 Six measured data points with some fitting functions

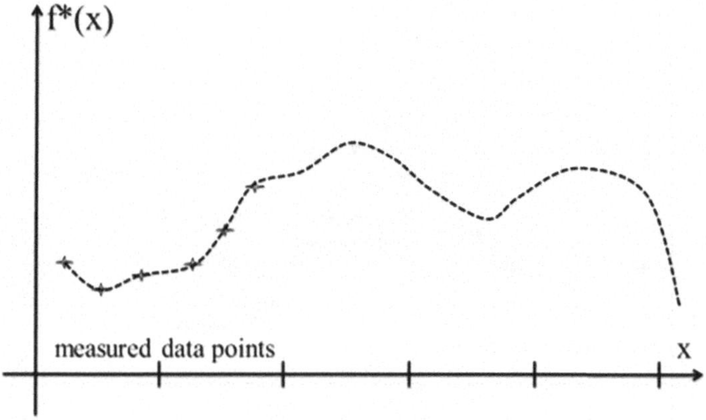

Fig. 1.4 Function f*(x) fitting six measured data points

Imagine now that in the course of our curve fitting exercise, we come up with a function f*(x) as depicted in Fig. 1.4.

As we know that there are many more functions than just the true function that fit the first six data points, we want to test the new candidate f*(x) by making predictions in the third interval of data points. Note that these predictions are strictly use-novel. The result is depicted in Fig. 1.5.

As it turns out, the predicted data points are empirically correct. In other words, we made a successful use-novel prediction based on our function f*(x). Now, overoptimistic realists will possibly exclaim that this proves that f*(x) is the true function! However, we may want to be more cautious because it could be just a lucky (though very improbable) accident that the predicted data points in the third

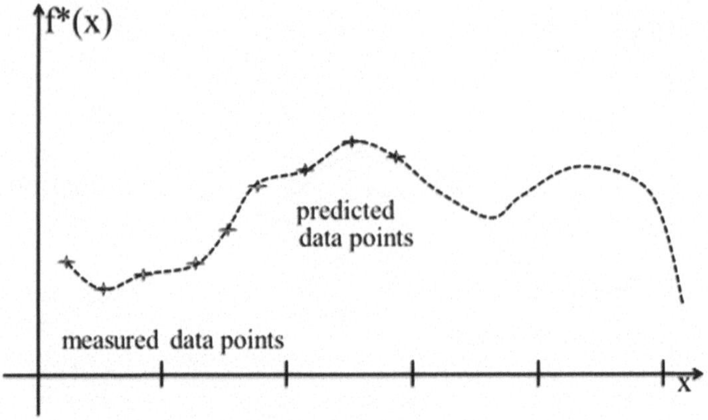

Fig. 1.5 Three data points, in the 3rd interval, correctly predicted by f*(x)

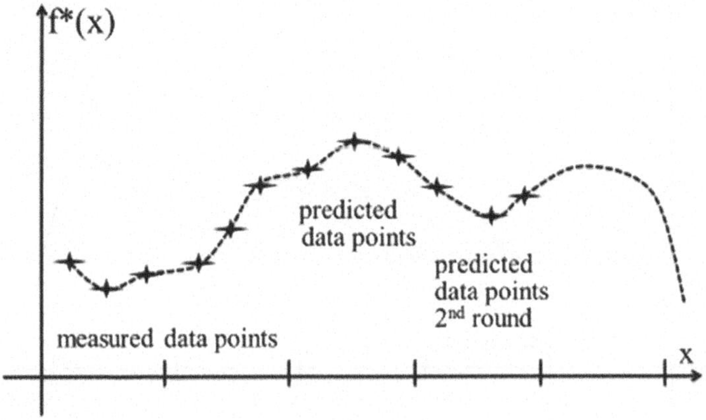

Fig. 1.6 Three more data points, in the 4th interval, correctly predicted by f*(x)

interval came out correctly. Therefore, we will make a second round of use-novel predictions with f*(x) in the fourth interval of data points: see Fig. 1.6.

As it turns out, the predicted data points are again empirically correct. In other words, we made a second round of successful use-novel predictions on the basis of our function f*(x). Now, the realist will have to say: f*(x) is the true function, everything else would be a miracle, or so the miracle argument goes! Unfortunately, in our model we cannot measure the data points in the fifth interval, but God has the full picture: see Fig. 1.7.

Unfortunately, f*(x) differs badly from the true function, in spite of its proven capability of two consecutive rounds of correct use-novel predictions.

How can the miracle argument mislead us so badly by suggesting that f*(x) must be the true function? The reason is that in the model situation, one must be very

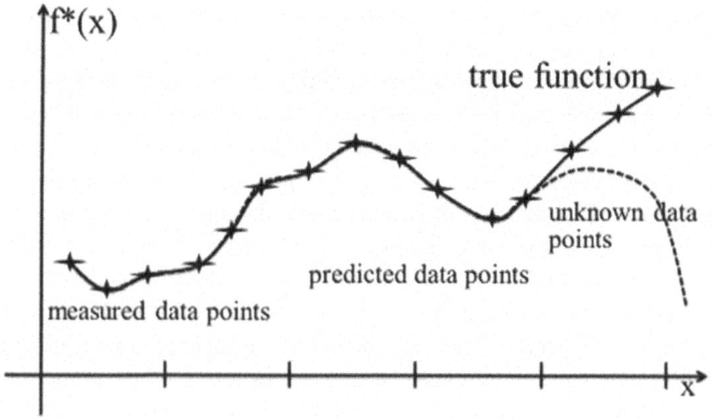

Fig. 1.7 Deviation of f*(x) from the true function

lucky to hit upon a function that produces correct use-novel predictions in the third and fourth interval. By sheer luck, this is exactly what happened when we stumbled upon f*(x). However, one must be even vastly luckier to hit, among the functions that make correct predictions in the third and fourth interval, upon the true function (or any approximately true function). There are many more fundamentally false functions producing correct use-novel predictions in the third and fourth interval than there are approximately true functions with the same predictive capability in these two intervals. In other words, predictive success of a function in the third and fourth data interval is no indicator for predictive success in the fifth interval, i.e., being the true (or an approximately true) function.

Based on this model, the essential weakness of the miracle argument can be presented as follows. The miracle argument is right in stating that it is very unlikely that a theory that has been constructed on the basis of a given set of data will be capable of making correct use-novel predictions. Because of this fact, the miracle argument concludes that there must be a reason for this unlikely possibility to be actual, and the only plausible reason seems to be – in most cases – the theory being (approximately) true. This is purely qualitative reasoning about probabilities. If one turns to comparative reasoning regarding the pertinent probabilities, the result reverts. Although the probability of finding a theory that makes correct use-novel predictions is indeed very low, the probability of finding one among them that is approximately true or even true is even much lower! The reason is that (approximately) true theories are only a small subset of the whole set of theories that are capable of making use-novel predictions. In other words, the miracle argument is fallacious: the capability of producing use-novel predictions is no reliable indicator for (approximate) truth.

However, one may object that the above reasoning is based upon a highly idealized model, and it is very questionable whether this model situation has anything to do with real science.[11] It seems to me that there are striking parallels of the model with real cases, regarding both its seductive power to realist conclusions and their failure. As is well known, classical physics was extremely successful for roughly 150 years, from the early to mid-eighteenth century (general acceptance of Newtonian physics in the 1740s) and the end of the nineteenth century (first doubts about classical physics arising). During this period, classical physics produced numerous use-novel predictions in various areas like the theory of gases, acoustics, gravitational physics (especially regarding the planetary system), electrodynamics, and so on. Up to the end of the nineteenth century, most physicists believed that Newton's theory and its extensions were the last word of physics because they were thought to be literally true.[12] This was indeed the miracle argument in action: how could the repeated success of classical physics in terms of its numerous correct use-novel predictions be explained except by classical physics being final and true? This era corresponds to the third and fourth interval in our highly idealized model. However, as is generally known the belief in the ultimate truth of classical physics has thoroughly dissolved due to the introduction of Special Relativity (1905), General Relativity (1915), and Quantum Mechanics (1925). The data these theories responded to were just not compatible with classical physics. Of course, this situation corresponds to the fifth data interval in our model in which the extrapolation of a hitherto extremely successful theory or function, respectively, proves an utter failure. I conclude that the analogy between the above model and historical reality is close enough to fundamentally doubt the validity of the miracle argument.

However, at this point I have to deal with an objection. One reviewer was not sure whether the paper was, as I had declared, "really dealing with a "level 1"-consideration only [. . .], but rather (at least elements of) radical level 3-skepticism. And to protect scientific realism from the threat of skepticism is not part of the

[11]For instance, a perceptive reviewer of an earlier version of this paper objected that the question of scientific realism is concerned with the introduction of scientific objects whereas the curve-fitting example is not; therefore, the curve fitting example cannot be of direct relevance to the issue of scientific realism. Yes, it is a fundamental presupposition of any analogy that there is a substantive difference between the things that are stipulated to be analogous. However, in the present case the point of comparison is the presumed truth of the function and the (approximate) truth of the theory in question. In both cases, truth is erroneously inferred from a limited amount of data that were in part even produced by correct use-novel predictions. In the curve fitting case, truth does not imply the existence of certain objects, whereas in the scientific case it does, but that does not invalidate the analogy. – For further illustration of the analogy, see the example of classical physics in the main text.

[12]Remember the well-known story by Max Planck about his Munich teacher Philipp von Jolly who in 1874 advised him not to study physics because Jolly portrayed "physics as a highly developed, almost fully matured science . . . Possibly in one or another nook there would perhaps be a dust particle or a small bubble to be examined and classified, but the system as a whole stood there fairly secured, and theoretical physics approached visibly that degree of perfection which, for example, geometry has had already for centuries" (Max Planck in 1924, printed in Planck 1933, cited after the translation in *Scientific American*, Feb. 1996, 10).

scientific realist's project." Let me take up the issue of skepticism. I take a skeptical doubt to consider and affirm a logical possibility that radically contradicts deeply engrained convictions (typically about reality) and that cannot be compellingly rejected. Take, for example, Descartes' "evil demon" who is supposedly systematically deceiving us about reality: it is a logical possibility that cannot be compellingly rejected. The weakness of such skeptical arguments is that we do not have the slightest positive hint that the logical possibility in question is indeed realized, and that there are infinitely many such logical possibilities. Surely, my objection to the miracle argument based on the curve-fitting example appears, at first sight, to articulate a logical possibility only. However, what distinguishes this objection from a purely skeptical one are the historical facts that I mentioned above. The possibility that even repeated use-novel predictions can falsely lead us to realistic conclusions, is not just a logical possibility, but was a (bitter) reality with respect to classical physics, as it turned out in the twentieth century. Unlike any skeptical doubt that is purely hypothetical, I presented empirical evidence that the miracle argument can be misleading. Thus, my objection does not, like a skeptical argument, articulate a purely logical possibility only (that can be eventually dismissed) but shows, by empirical example, that this possibility can indeed be real and has therefore to be taken seriously.

Do all these considerations answer the title question, "Are there good arguments against scientific realism?" Of course, they do not. They only show that what has perhaps imprudently been called "the *ultimate* argument for scientific realism" fails badly because it is fallacious. However, there may be other arguments for scientific realism besides the "ultimate" one, and there may be other forms of realism, relatives of scientific realism, that are supported by different arguments. This is indeed the case. I shall now investigate something that is not really a single argument in support of a particular kind of realism, but rather an argumentative strategy that singles out particular kinds of realism together with arguments supporting them. So we are dealing with a family of different realist positions that are united by a common type of supportive argument. The strategy is called "selective realism".

1.4 Selective Realism

Here is a rough sketch of the strategy of selective realism.[13] The starting point is a presupposition that is nowadays widely shared by philosophers of science of different persuasions (and by some fraction of scientists). This presupposition claims that most, if not all, of our accepted scientific theories, even the very best ones, are strictly speaking false. This implies that there are (at least) some parts of these theories that do not correspond to reality. Therefore, a realist interpretation

[13]For a very clear summary, see Chakravartty (2014), Sections 1.3 and 2.3. An alternative denomination of selective realism is "deployment realism"; see, e.g., Lyons (2016).

of these theories as wholes is inappropriate. However, some of these theories are remarkably successful, e.g., with regard to use-novel predictions. Now the first step in the strategy of selective realism is this: identify and select those elements of the theories in questions, which are responsible for the pertinent success. It is those elements that are potentially worthy of a realist commitment. Thus, the first step of the argument identifies, in the spirit of the miracle argument, the *candidates* for a realist commitment.

Why does this step only identify *candidates* for a realist commitment, why is an additional second step necessary? The reason is that it is known from the history of science that also theories relying on theoretical entities that were later shown not to exist, may have been persuasively successful, for instance with respect to use-novel predictions. This historical fact has been used as an argument against cruder forms of scientific realism: theoretical entities responsible for the empirical success of a theory may still be abandoned during the next scientific revolution and be replaced by something substantially different. Thus, for a robust realist commitment to some theoretical element it is not sufficient that this element is responsible for persuasive empirical success (whatever kind the favored persuasive empirical success is). In addition, this theoretical element must also survive serious theory change, i.e., revolutionary developments. This second step of the strategy of selective realism is also called "the continuity argument".

This is then the general strategy of selective realism: identify those elements of empirically successful scientific theories, which are (a) responsible for the particular empirical success and which (b) have survived revolutionary developments. Each of these conditions is necessary, and together they are seen as sufficient for the realist commitment to these elements of scientific theories. Sophisticated realists, however, do not claim that fulfilment of these two conditions *guarantees* the correctness of the realist commitment, because they are fallibilists. They only claim that under the two conditions it is reasonable to interpret these elements realistically.

Different versions of selective realism can now be distinguished according to what they take to be the most promising candidates for their realist commitments. *Scientific realism* claims that it is theoretical entities that are responsible for a theory's given empirical success. However, one has to show that in addition, these entities survive revolutionary breaks. Typically, this has been done by deflecting and containing the (undisputed) meaning change of key concepts involved in revolutionary breaks, i.e. incommensurability, to the intensions of the concepts in question, and thereby keeping their referents, i.e. their extensions, stable.[14] *Explanationism* leaves first open what the candidates for a realist commitment might be. It tries to identify them functionally by their role in producing the success of the respective theory, i.e., novel predictions. These functionally important parts of theories are called their "working posits", to which the realist should commit, in contrast to the "idle parts" of the theories, which do not contribute to the

[14]This presupposes a theory of meaning and of reference that supports this move. See among many others, e.g., Sankey (1994), especially Chapters 2 and 5.

pertinent empirical success. The working posits then explain the empirical success of these theories, which gives the position its name. As successor theories should be capable of reproducing the empirical success of their predecessors, the working posits of the predecessors should be present in them, too. Finally, *structural realism* singles out as the locus of realist commitments structural aspects of the theories in question, typically mathematical aspects of the theories.[15] The basic idea is that it is correspondence of the formalism of our best theories with physical structures that is responsible for the empirical successes of such theories. These structures may be preserved through scientific revolutions although the ontologies and the explanatory mechanisms of the theories divided by the revolutionary break may differ greatly. Again, not everything in these structures may be involved in the empirical success of a theory, so only those parts are singled out that represent "operative" relations.[16]

Clearly, in the first step of this strategy the miracle argument is involved, at least as a heuristic device, otherwise this step would not be plausible. However, for the sake of argument I will at this point not doubt the legitimacy of this first step, because only on the basis of its acceptance can the strength of the second step be assessed. Let us now turn to this second step of the strategy of selective realism, the continuity argument.

The precise formulation of the continuity argument merits careful discussion. Ioannis Votsis has devoted a full paper to this issue. His result is the following: "Preservation [through theory change] is a reliable guide to (approximate) truth"[17]. Votsis stated this in the context of his discussion of structural realism, but his statement can be generalized to the other positions we are also discussing here. This is because not only structures with a claim to reality must survive theory change, but also entities or working posits (or whatever) with a claim to reality. We can thus formulate the second step of the strategy of selective realism as follows.

Let a sequence of empirically progressive theories be given with "entity continuity", or "working posits continuity", or "structural continuity". At any given time, the respective theoretical entities, or working posits, or structures, are identified as responsible for the specific empirical success of those theories. Typically, this empirical success is spelled out as the capability of correct use-novel predictions, but it may also be another form of particularly persuasive empirical success. For the following, however, neither what is taken to be responsible for the empirical success (entities, structures ...), nor what is taken the persuasive empirical success to consist in (use-novel predictions ...), will play a role, so we can abstract from them. The (abstract) situation we are dealing with can thus be described as follows. Let a sequence theories be given with X continuity, and X is seen as at least partly responsible for their increasing, particularly persuasive empirical success. Then, so

---

[15]As is well-known, structural realism has gained much renewed attention in the last decades, mainly through John Worrall's work: Worrall (1996 [1989]). For a comprehensive reference to the current status of structural realism, see, e.g., French 2014.

[16]See Votsis (2011, 107–108).

[17]Votsis (2011, 116).

the continuity argument goes, X can very reliably be interpreted realistically. In this way, the continuity of X, i.e. its stability through theory change, is taken to be a (fallible) indicator of its reality.

Against the continuity argument, the following objection can be brought forward; I call it the "dead end objection".[18] For the sake of argument, let us make the (temporary) assumption that the given X continuity in the series of empirically increasingly successful theories is due to their convergence to a certain limit theory that will also display X, like the elements of the series. Thus, the elements in the series of empirically increasingly successful theories are successive approximations to the limit theory.

Under these suppositions, what do we know about this limit theory? First, the limit theory is empirically more powerful than any of the elements in the series. By "empirically more powerful", I mean that its predictions are more accurate than those of any of the elements of the series, and it may make additional correct predictions in comparison to the elements in the series. Imagine, for instance, that the predictions of the limit theory attain a relative accuracy of at least 10^{-30} for all observed quantities. However, from the given suppositions about the limit theory we cannot infer that the limit theory is also capable of producing correct use-novel predictions (or whatever the measure of the pertinent empirical success is). For instance, the potential to make use-novel predictions by means of the presence of X may have already been exhausted by the elements in the series, such that the limit theory cannot contribute additional use-novel predictions based on X.

Second, by construction, the empirical power of the limit theory is at least partly due to the presence of X. The decisive question is, are we allowed to interpret X in the limit theory realistically? The continuity argument would clearly give a positive answer to this question: the continual presence of X in all the elements of the series of theories and X's capability of producing persuasive empirical success licenses a realist interpretation of X in the elements of the series. If X can be interpreted realistically in the elements of the series, clearly the same X contained in the limit theory can also be realistically interpreted. This would be the answer of the continuity argument.

However, we should now put the continuity argument on hold because we want to investigate its validity. In order to do so, we have to investigate the properties of the limit theory more carefully. We will do that in two steps. First, we investigate the limit theory in isolation of the elements of the series of theories whose limit it is. Second, we put the limit theory back in its proper context, namely, as being the limit of the series of empirically increasingly successful theories.

[18]Following a paper that I gave in 2012 at the GAP conference in Constance, Ron Giere suggested to me the name "dead end objection", instead of "impasse objection" that I had used in Hoyningen-Huene (2013). I am grateful for this suggestion. Furthermore, the presentation of the objection in the present paper is much more careful than in the older paper. Specifically, I am now avoiding the objection that my argument is just skeptical and thus void of any argumentative force for the particular case in question.

When we consider the limit theory separately, i.e., independently of the series of theories whose limit it is, we can ask whether we have reasons to interpret its component X realistically. The situation is this: We have a theory that is empirically extremely accurate, and we can ascribe its predictive accuracy to some component X of it. However, we do not know whether the limit theory has those specific empirical virtues that we have identified earlier as the necessary criteria for a realist interpretation of X, like the capacity to produce correct use-novel predictions (or whatever). In other words, we cannot run the miracle argument on the limit theory because it does not fulfill its necessary empirical suppositions (capability to produce use-novel predictions, or whatever). Thus, when applied to the limit theory in isolation, the first step of the strategy of selective realism fails for *empirical* reasons: we just do not know whether the limit theory fulfills the necessary empirical suppositions for the application of the miracle argument. However, also the second step of the strategy of selective realism, the continuity argument, fails for the limit theory, this time for *conceptual* reason. As we are considering the limit theory in isolation, there simply are no other theories with which any continuity with the limit theory could exist, thus the continuity argument is inapplicable. In other words, the two arguments that are otherwise seen as supporting realist interpretations of (parts of) theories are not applicable to the limit theory. So, despite its undoubted empirical virtues (e.g., a relative empirical accuracy of at least 10^{-30}), the limit theory's epistemic status with respect to a realist interpretation is open at this point. Of course, there is the *possibility* of X deserving a realistic interpretation, thus explaining the fantastic predictive success of the limit theory. However, there is also the possibility that the limit theory is a model, whose component X that is responsible for immensely accurate predictions is nevertheless entirely unrealistic. So, considered independently of the series of theories whose limit it is, we have no reason to interpret the X contained in the limit theory realistically: the arguments usually adduced to support a realist interpretation of an X contained in a theory do not apply to the limit theory. Considered in separation of the series of theories whose limit it is, the epistemic status of X in the limit theory with regard to realism is open: we just do not know and have no arguments to assess it; it may be (approximately) true, or it may be radically false.[19]

How does the picture change when the limit theory is now put back in its proper context, namely, being the limit of a series of empirically increasingly successful theories? Any single theory in the series is capable of making use-novel predictions (or whatever the standard of persuasive empirical success is) due to the presence of X, and X can also be found in its successor theory. This is a fact that, according to

[19]A reviewer of an earlier version of this paper opined that "it is assumed that the limit theory is not the true theory or indeed an approximately true theory. But that's begging the question against the realist." Had I had made this assumption, the reviewer would of course be right. However, I am *not* assuming that the limit theory is not true. I leave it entirely open what the epistemic status of the limit theory is and ask what arguments we have to determine its epistemic status. Then I claim that *we have no arguments* to the effect that the X in the limit theory can be interpreted realistically, so its epistemic status *remains* open.

the continuity argument, is evidence for the legitimacy of a realist interpretation of X. Therefore, focusing on the theories in the series, we appear to have as many independent pieces of evidence for the realist interpretation of X as we have theory pairs. However, taking into account that there is a limit theory, the theories in the series are empirically successful because they participate (at least approximately) in the component X of the limit theory. Therefore, the X-continuity in the series of theories becomes a consequence of their convergence to the limit theory. Furthermore, also the fact that the theories in the series can produce correct novel predictions (or some other persuasive empirical success), is just a consequence of their (at least approximate) participation in the component X of the limit theory. It is the component X of the limit theory that the theories in the series increasingly exploit, and that increasing exploitation makes use-novel predictions (or some other persuasive empirical success) possible. In other words, interpreting the theories' capacity to produce use-novel predictions as indicators of the reality of X is a mistake; it is a consequence of the existence of a limit theory that contains X. Moreover, as there is no good reason to interpret the limit theory realistically with respect to X, there is no reason to interpret the presence of X realistically in any of the theories in the series either. Thus, instead of having as many pieces of evidence for the (approximate) reality of X as there are theory pairs in the series, we have no evidence for the (approximate) reality of X at all. Because the continuity argument suggests the opposite, it is fallacious.

In the above argument, I have used the assumption that the series of theories converges to a limit theory. This is, of course, a problematic assumption because to explicate what the convergence of theories exactly means is a highly nontrivial affair. However, it is easy to relax this condition. The series of theories does not have to converge to an empirically much more accurate theory. It is enough that there is such a theory in the background such that all the theories in the series derive their empirical potential from the kinship with regard to X to this theory. Again, the fact that this theory in the background fits many empirical facts to a very high accuracy is not by itself a sign for its verisimilitude with respect to aspect X. As the theories in the series do not in any sense go beyond the background theory, their continuously containing X cannot be interpreted realistically, as the background theory's containing X cannot be either.

1.5 Conclusion

If the given analysis is correct, I think that at the bottom level, both the continuity argument and the miracle argument share the same fundamental weakness. In the case of the miracle argument, from a theory's (possibly even repeated) use-novel predictions realist conclusions are drawn. In the case of the continuity argument, from continuity of some X through (possibly repeated) theory change realist conclusions are drawn. However, the realist conclusions are overhasty because the observed properties of the theory or theories, respectively, could equally well be due to an underlying fundamentally false theory, i.e., a theory that cannot be realistically interpreted in the pertinent respect. The highly idealized model that I presented suggests the following hypothesis. Certain astonishing properties of theories that prompt their realist interpretation are in fact due to a theory in the background that may be fundamentally false in the critical respects. Though in general, it is indeed very unlikely that a series of given theories has a false theory in the background that is responsible for their realism-suggesting properties, it is *even more unlikely* that these properties are due to a (approximately) true theory in the background.

Let me sum up. My title question was, are there good arguments *against* scientific realism? The answer is, perhaps there are. What I have argued in this paper is that the miracle argument and the continuity argument, which are supposed to be the best arguments *for* scientific and other forms of selective realism, are fallacious.

References

Abbott, B.P., et al. 2016. Observation of gravitational waves from a binary black hole merger. *Physical Review Letters* 116 (6): 061102.

Alai, M. 2014. Novel predictions and the no miracle argument. *Erkenntnis* 79 (2): 297–326.

Chakravartty, Anjan. 2014. Scientific realism. In *The Stanford encyclopedia of philosophy*, ed. E.N. Zalta. Spring 2014 ed. http://plato.stanford.edu/archives/spr2014/entries/scientific-realism/

Dawkins, R. 1995. *River out of Eden: A Darwinian view of life*. New York: Basic Books.

Einstein, A. 1916, Juni 29. Näherungsweise Integration der Feldgleichungen der Gravitation. *Königlich Preußische Akademie der Wissenschaften (Berlin). Sitzungsberichte*, 688–696.

———. 1918, February 21. Über Gravitationswellen. *Königlich Preußische Akademie der Wissenschaften (Berlin). Sitzungsberichte*, 154–167.

French, S. 2014. *The structure of the world: Metaphysics and representation*. Oxford: Oxford University Press.

Giere, R.N. 2006. *Scientific perspectivism*. Chicago: University of Chicago Press.

Hawking, S.W. 1971. Gravitational radiation from colliding black holes. *Physical Review Letters* 26 (21): 1344–1346.

Hawking, S.W. 1996. Classical theory. In *The nature of space and time*, ed. S.W. Hawking and R. Penrose. Princeton: Princeton University Press.

Hawking, S.W., and L. Mlodinow. 2010. *The grand design*. London: Bantam.

Hawking, S.W., and R. Penrose. 1996. *The nature of space and time*. Princeton: Princeton University Press.

Heitler, W., and F. London. 1927. Wechselwirkung neutraler Atome und homöopolare Bindung nach der Quantenmechanik. *Zeitschrift für Physik* 44: 455–472.

Hitchcock, C.H., and E. Sober. 2004. Prediction versus accommodation and the risk of overfitting. *The British Journal for the Philosophy of Science* 55 (1): 1–34.

Howson, C. 2013. Exhuming the no-miracles argument. *Analysis* 73 (2): 205–211.

Hoyningen-Huene, P. 2011. Reconsidering the miracle argument on the supposition of transient underdetermination. *Synthese* 180 (2): 173–187.

———. 2013. The ultimate argument against convergent realism and structural realism: The impasse objection. In *EPSA11 perspectives and foundational problems in philosophy of science*, ed. V. Karakostas and D. Dieks, 131–139. Cham: Springer.

Husserl, E. 1967 [1922]. *Ideas: General introduction to pure phenomenology*. Trans. W.R.G. Gibson. New York: Humanities Press.

Lee, W.-Y. 2014. Should the no-mircacle argument add to scientific evidence. *Philosophia* 42 (4): 999–1004.
Lyons, T.D. 2003. Explaining the success of a scientific theory. *Philosophy of Science* 70: 891–901.
———. 2015. Scientific Realism. In *The Oxford handbook of philosophy of science*, ed. P. Humphreys. New York: Oxford University Press.
———. 2016. Structural realism versus deployment realism: A comparative evaluation. *Studies in History and Philosophy of Science Part A* 59: 95–105.
Magnus, P.D., and C. Callender. 2004. Realist ennui and the base rate fallacy. *Philosophy of Science* 71 (3): 320–338.
Mayo, D.G. 1991. Novel evidence and severe tests. *Philosoophy of Science* 58 (4): 523–552.
Menke, C. 2014. Does the miracle argument embody a base rate fallacy? *Studies in History and Philosophy of Science* 45 (3): 103–108.
Musgrave, A. 1988. The ultimate argument for scientific realism. In *Relativism and realism in science*, ed. R. Nola. Dordrecht: Kluwer Academic.
Planck, M. 1933. *Wege zur physikalischen Erkenntnis: Reden und Vorträge*. Leipzig: S. Hirzel.
Psillos, S. 2006. Thinking about the ultimate argument for realism. In *Rationality and reality: Conversations with Alan Musgrave*, ed. C. Cheyne and J. Worrall, 133–156. Berlin: Springer.
Putnam, H. 1975. What is mathematical truth? In *Mathematics, matter and method. Philosophical papers*, vol. 1, 60–78. Cambridge: Cambridge University Press.
Rowbottom, D.P. 2011. What's at the bottom of scientific realism? *Studies in History and Philosophy of Science Part A* 42 (4): 625–628.
Saatsi, J.T., and P. Vickers. 2011. Miraculous success? Inconsistency and untruth in Kirchhoff's diffraction theorie. *British Journal for the Philosophy of Science* 62 (1): 29–46.
Sankey, H. 1994. *The incommensurability thesis*. Aldershot: Avebury.
Schindler, S. 2008. Use-novel predictions and Mendeleev's periodic table: Response to Scerri and Worrall (2001). *Studies in History and Philosophy of Science Part A* 39 (2): 265–269.
van Fraassen, B.C. 1980. *The scientific image*. Oxford: Clarendon.
Votsis, I. 2011. Structural realism: Continuity and its limits. In *Scientific structuralism*, ed. P. Bokulich and A. Bokulich, 105–117. Dordrecht: Springer. Available at: http://philsci-archive.pitt.edu/5233/1/VotsisStructuralRealismContinuityanditsLimits.pdf
Worrall, J. 1989. Fresnel, poisson, and the white spot: The role of successful predictions in the acceptance of scientific theories. In *The use of experiment. Studies in the natural sciences*, ed. D. Gooding, T. Pinch, and S. Schaffer, 135–157. Cambridge: Cambridge University Press.
———. 1996 [1989]. Structural realism: The best of both worlds?. In *The philosophy of science*, ed. D. Papineau, 139–165. Oxford: Oxford university press (originally in *Dialectica* 43, 99–124 (1989)).

Chapter 2
Quantum Gravity: A Dogma of Unification?

Kian Salimkhani

Abstract The quest for a theory of quantum gravity is usually understood to be driven by philosophical assumptions external to physics proper. It is suspected that specifically approaches in the context of particle physics are rather based on metaphysical premises than experimental data or physical arguments. I disagree. In this paper, I argue that the quest for a theory of quantum gravity sets an important example of physics' *internal* unificatory practice. It is exactly Weinberg's and others' particle physics stance that reveals the issue of quantum gravity as a genuine physical problem arising within the framework of quantum field theory.

Keywords Principle of equivalence · Unification · Quantum field theory · Quantum gravity · General relativity · Graviton/spin-2 particle · Lorentz-invariance

2.1 Introduction

To 'combine general relativity and quantum mechanics'—as the issue of quantum gravity (QG) is frequently summarized—is typically understood to be the central challenge for fundamental physics. The common conviction is that this quest for QG is not only fuelled, but generated by external principles (cf. Mattingly 2005; Wüthrich 2006, 2012). Accordingly, the research program of QG is believed to be driven, first and foremost, by reasoning involving philosophical assumptions.

I thank Andreas Bartels, Cord Friebe, Stefan Heidl, Niels Linnemann, James Read, Matthias Rolffs, Thorsten Schimannek, and Christian Wüthrich for helpful discussions and remarks. Furthermore, I thank the anonymous referee for pressing me to clarify some paragraphs, especially in the opening sections.

K. Salimkhani (✉)
Institute for Philosophy, University of Bonn, Am Hof 1, 53113 Bonn, Germany
e-mail: ksalimkhani@uni-bonn.de

A. Christian et al. (eds.), *Philosophy of Science*, European Studies in Philosophy of Science 9, https://doi.org/10.1007/978-3-319-72577-2_2

It is suspected that specifically approaches in the context of particle physics are essentially based on, for example, metaphysical premises rather than experimental data or physical arguments. I disagree. In fact, it is exactly Weinberg's and others' particle physics stance that reveals the issue of QG as a genuine physical problem arising *within* the framework of quantum field theory (QFT).

In this paper, I argue that the quest for QG sets an important and often misconceived example of physics' *internal* unificatory practice. Physics' internal strategies, e.g. exploiting the explanatory capacities of an established theory, suffice to explain the search for a theory of quantum gravity. To set the stage, I will first recap what the research program of QG is about and what remarks suspecting a 'dogma of unification' amount to. To support my claim, I will then investigate the spin-2 approach to the matter focussing on Weinberg's quantum field theoretic reconstruction of Einstein's principle of equivalence. Subsequently, two important consequences for our understanding of general relativity (GR) and the issue of QG are briefly discussed: First, it is suggested that we should not take GR as a fundamental theory because it can be reduced to QFT. Second, the investigation serves as a clarification of what the problem with QG actually is. Afterwards, some objections against the advocated picture are mentioned and very briefly replied to. Finally, I will revisit the opening question concerning the alleged 'dogma of unification'.

2.2 Some Remarks on Quantum Gravity

Fundamental physics is based on two theories, the standard model of particle physics (SM)[1] and the general theory of relativity (GR). While the first describes the electromagnetic, the weak and the strong interaction of subatomic matter particles as well as the Higgs mechanism in a quantum field theoretic framework, the latter addresses the fourth fundamental interaction, i.e. gravity, in terms of a classical field theory. To reconcile these basic pillars within one framework uncovering their common ground or, more specifically, finding a quantum theory of gravitation or an even richer all-encompassing theory comprising it (e.g. string theory), is frequently understood to be the central challenge for contemporary physics.

Such reconciliations do not necessarily result in a particularly substantial type of unification. One might simply be concerned with issues of consistency or some law-like connection. For instance, as a first attempt, one could merely try to merge or couple classical GR and QFT without further modifications to form so-called semi-classical theories (e.g. Carlip 2008). While such theories acknowledge that according to quantum theory the matter fields are fundamentally quantum theoretic

[1] I do not distinguish between 'theory' and 'model (of a theory)' here. More accurately, one would refer to the SM as a 'model (of QFT)'.

structures, they insist that gravitation, i.e. spacetime, is fundamentally classical ('non-quantum'). Accordingly, a simple semi-classical theory rewrites Einstein's equations as:

$$G_{ab} = 8\pi \langle T_{ab} \rangle \ . \tag{2.1}$$

Here, the matter fields are introduced by the expectation value of the stress energy tensor, $\langle T_{ab} \rangle$. However, despite some convenient properties according to the Ehrenfest theorem that links the quantum mechanical expectation value to Newton's classical equations of motion, the expectation value is not a fully classical object. Therefore, it gives rise to problematic discontinuities as many have pointed out (e.g. Eppley and Hannah 1977; see also Wald 1984; Kiefer 2007). As a result, most physicists typically do not seriously consider semi-classical theories. It should be noted though that some have criticized these objections against semi-classical theories as not compelling (cf. Huggett and Callender 2001a,b; Mattingly 2005, 2006; Wüthrich 2005). So, technically semi-classical theories may not be ruled out yet.

Still, what is typically understood by 'quantum gravity' is a more substantial reconciliation in terms of some sort of *quantization* of gravity (e.g. Huggett and Callender 2001a,b; Wüthrich 2005). To 'quantize' a classical theory means to construct a quantum theory whose classical limit agrees with the classical theory. Note that quantization does not necessarily imply discreteness. For instance, in the case of quantum mechanics (QM) some observables did become discrete after quantization, but others like position and momentum operators did not. Accordingly, to quantize GR does not imply discreteness of spacetime. Making spacetime discrete is merely one possibility—and it is a possibility that comes with a cost as it actually compromises an important symmetry of physics: Lorentz-invariance.

Now, there are many different approaches to QG of this more substantial kind. According to Kiefer (2007) they may be grouped into primary and secondary theories of quantum gravity. The former employ standard procedures of quantization (canonical or covariant quantization) as it has been done in the case of quantum electrodynamics, for example. The latter comprise QG as a limit of some fundamental quantum theoretic framework, e.g. string theory. Note that this classification is based on how the approaches *proceed*. Systematically the respective approaches may nonetheless be related. For instance, Weinberg (1999) emphasizes a relation between quantum field theoretic, i.e. covariant, approaches and string theory.

But why should we seek a quantum theory of gravity at all? Usually, theoretical considerations are understood to indicate an incompleteness of present-day physics related to the issue of QG (e.g. Kiefer 2006). Frequently listed key reasons for 'quantizing gravitation' include, amongst others, cosmological considerations, black hole evolution, theoretical problems in QFT, and aiming at unification (cf. Kiefer 2006; Huggett and Callender 2001b; Wüthrich 2005). Many suspect that unification ideals are particularly crucial (e.g. Mattingly 2005), especially with respect to approaches in the context of particle physics.

This is mainly based on the following: First of all, it seems that there is no empirical need whatsoever to construct the theory. In fact, both theories (SM and GR) are in perfect agreement with all available, and—concerning quantum gravitational effects—presumably even all expectable data. The typical energy (or length) scale where quantum gravitational effects are understood to become relevant is roughly 16 orders of magnitude higher (smaller) than presently available (e.g. Arkani-Hamed 2012). So, one might argue that, pragmatically, we cannot really hope for direct experimental data—it is by no means excluded though and we particularly might hope for indirect indications.[2] Still, up to now, experiment does not suggest any need for modifications.

Second, skeptics add that also invoked theoretical arguments are in fact—or in principle—not compelling (e.g. Mattingly 2005). Finally, and probably most importantly, many share the conviction that GR and QFT are *fundamentally incompatible* for quite a simple reason: "according to GTR [general relativity], gravity simply is not a force" like the electromagnetic, the weak, and the strong interaction (Maudlin 1996). This is not to say that GR and quantum theory are incompatible in a logical sense, but to argue that they are "incommensurable (families of) theories" (Wüthrich 2005, 778).

In summary, there seems to be neither empirical ground, nor any genuine physical reason to pursue the quest for QG. That is why some suspect that internal strategies of physics alone (e.g. inductive generalization, expanding the realm of an established theory or exploiting the explanatory capacities of an established theory) cannot account for such programs. Instead, physicists are said to employ *external* arguments, for example a 'dogma of unification' (Mattingly 2005; see also Maudlin 1996; Wüthrich 2005, 2012). In this perspective, physicists would employ metaphysical principles (e.g. 'unity of nature'), metatheoretical principles (e.g. 'economy of thought') or epistemological principles (e.g. physicists pursue unification for its own sake—i.e. the mere fact of the theory dualism itself is considered a defect of theoretical physics), that is *philosophical* reasons (Mattingly 2005; Wüthrich 2006, 2012). Against this I insist that a quantum theoretic account of gravity is already part of the well-known framework of QFT and that it prompts the quest for QG.

Let me rephrase it as follows: Positions arguing that physics generally aims at unification (or a minimal theoretical system or representing an assumed 'unity of nature') can neatly explain attempts at QG. But what about positions arguing that physics aims at empirical adequacy, for example? Are such positions able to explain the quest for QG? Do physicists employ philosophical reasons, or can we understand the search for a theory of QG internally? To answer this, let us first be clear where the objections against QG typically arise: in the geometrization picture of gravity as the canonical interpretation of GR.

[2]Furthermore, actually suggestions are put forward for how theory assessment without experimental data could work (Dawid 2013)—a very interesting, but also highly controversial project (cf. Rovelli 2016).

2.3 The Canonical Picture of General Relativity

In the canonical formulation of GR already basic notions like 'metric' and 'curvature' seem to strongly suggest a reductionist view on gravitation. In fact, the interpretation of GR as a reduction of gravitation to spacetime curvature is often attributed to Einstein himself (Weinberg 1972, vii, 147)—a common misreading as Lehmkuhl (2014) insists. As a matter of fact, GR is usually presented as a geometrization of gravity in textbooks:

> General relativity (GR) is Einstein's theory of space, time, and gravitation. At heart it is a very simple subject (compared, for example, to anything involving quantum mechanics). The essential idea is perfectly straightforward: while most forces of nature are represented by fields defined on spacetime (such as the electromagnetic field, or the short-range fields characteristic of subnuclear forces), gravity is inherent in spacetime itself. In particular, what we experience as "gravity" is a manifestation of the curvature of spacetime. Our task, then, is clear. We need to understand spacetime, we need to understand curvature, and we need to understand how curvature becomes gravity. (Carroll 2004, 1)

In this interpretation, gravity reveals as a geometrized pseudo force: Gravitation is reduced to spacetime geometry and becomes a mere effect of the curvature of spacetime. As we have seen, for example Maudlin (1996) advocates this view. Undoubtedly, the textbook interpretation is very appealing. First, it remains close to the mathematical formalism that successfully unifies two apparently very different concepts: gravitation and spacetime geometry. Second, the textbook version yields quite transparent ontological commitments, most importantly that spacetime is a Riemannian manifold, M, with a metric, g,[3] and that gravitation is not an interaction (or 'force'), but simply curvature of spacetime. Altogether, this is a perfectly fine interpretation of GR.

But why adopt this interpretation? What underpins the geometrization picture except the fact that the canonical formalism of GR contains the mathematical objects mentioned above? To answer this, one needs to look at what GR is based on conceptually: At the core of GR, and at the core of the geometrization picture as well, we find Einstein's principle of equivalence. Note that the equivalence principle comes in different varieties. Essentially, there is a weak and a strong version (e.g. Carroll 2004, 48–54). The weak equivalence principle (WEP) states that the inertial mass, m_i, and the gravitational mass, m_g, of any object are equal in value. Remember the case of Newtonian mechanics: Here, the inertial mass is the constant of proportionality between some force and the acceleration of the object the force acts on. Since the value of the inertial mass of the object is the same for any force, the inertial mass is universal in character (Carroll 2004, 48). On the other hand, the gravitational mass is a specific quantity only related to the gravitational force—it is the constant of proportionality between the gradient of the gravitational potential and the gravitational force (Carroll 2004, 48). *Prima facie*, both masses are conceptually independent. Hence, m_g/m_i may differ for different objects and may

[3]Note that in light of the hole argument, the focus has shifted to the metric alone.

therefore be thought of as a 'gravitational charge' (Carroll 2004, 48). Accordingly, the behavior of different objects in a gravitational field would generally depend on the (different) gravitational charges, just as the behavior of electromagnetically charged particles in an electromagnetic field depends on the particles' charges. However, since Galilei we *empirically* know that inertial and gravitational mass are always equal in value, $m_i = m_g$. Every object in a gravitational field falls at the same rate regardless of its properties including mass. Thus, in Newtonian mechanics inertial and gravitational mass are conceptually different (or different in type), but empirically equal in value. In this sense, gravitation is *universal* in Newtonian mechanics and obeys the WEP (Carroll 2004, 48f)—without explanation. The geometrization picture of GR, on the other hand, is able to provide an explanation for the WEP by eliminating m_g from the theory altogether as we will see in a moment.

First, to better understand the geometrization rationale and to prepare for the formulation of the *strong* equivalence principle (SEP) let us rephrase the essence of the weak version in a famous thought experiment:

> Imagine [...] a physicist in a tightly sealed box, unable to observe the outside world, who is doing experiments involving the motion of test particles, for example to measure the local gravitational field. Of course she would obtain different answers if the box were sitting on the moon or on Jupiter than she would on Earth. But the answers would also be different if the box were accelerating at a constant velocity [...] (Carroll 2004, 49)

According to the WEP, it is impossible to decide whether the observed effects on freely-falling test particles stem from a gravitational field or from being situated in a uniformly accelerated frame. This is a result of the universality of gravitation. As mentioned, for the electromagnetic field such an empirical distinction is possible: we would simply have to compare the behavior of particles with different charge. Since for gravity the particle's 'gravitational charge' is universal, this does not work (Carroll 2004, 49). Note that because of possible inhomogeneities in the gravitational field this is only true for sufficiently small frames, technically speaking: it is only true *locally*. We can then formulate the WEP as follows:

> The motion of freely-falling particles are the same in a gravitational field and a uniformly accelerated frame, in small enough regions of spacetime. (Carroll 2004, 49)

Since special relativity (SR) tells us that 'mass' is a manifestation of energy and momentum, the SEP generalizes the above statement:

> In small enough regions of spacetime, the laws of physics reduce to those of special relativity; it is impossible to detect the existence of a gravitational field by means of local experiments. (Carroll 2004, 50)

This means that locally we can always 'transform away' a gravitational field and the laws reduce to the laws of SR.[4] In this sense, gravity becomes a 'pseudo force': There is no such thing as a gravitational potential in GR.

[4]Note that Carroll's definition of the SEP is not very precise. Read et al. (2017) carefully distinguish and discuss four versions of the SEP.

Now, this is not to say that gravity is fictitious. Quite the contrary, it means that gravity turns out to be *inescapable*: a 'gravitationally neutral object' with respect to which we could measure the acceleration due to gravity *does not exist* (Carroll 2004, 50). Hence, every object in the universe carrying energy and momentum is subject to gravity. In fact, every object is subject to gravity *in the same way*. Gravity does not distinguish between different types of objects. All objects, regardless of their properties including mass, are attracted *universally* (Carroll 2004, 48).

It is exactly gravity's universality that seems to strongly suggest a geometrization picture of gravity. For gravitation essentially *being* curvature of spacetime, *being* a feature of the Riemannian manifold (or the metric, respectively), *being* a geometrical background structure perfectly explains why the SEP should hold. If gravitation is curvature of spacetime, then it is obvious why we can always perform local transformations so that gravitation vanishes, why the laws of physics locally look like the laws of SR. It is then also obvious why this should affect every single object in the universe in the same way. The simple fact that gravitational effects are apparently independent of the objects' properties supports the claim that gravitation arises from spacetime itself and that the notion of gravitational mass needs to be eliminated.[5] As a result, the SEP does not only play an important role for GR, but also for the theory dualism in physics: A geometrization picture of gravity seems fairly disconnected from how we understand the other fundamental interactions (cf. Weinberg 1972, viii). While gravitation *is* spacetime, the other fundamental interactions are fields *in* spacetime.

However, this perspective on GR is not exclusive. Lehmkuhl (2008) argues that interpretations within the canonical formalism are not committed to the geometrization thesis that gravitation is reduced to spacetime geometry. Clearly, GR associates gravitation with spacetime, but the type of association is not fixed (Lehmkuhl 2008, 84). Besides the geometrical interpretation, one may as well put forward the *field interpretation* or the *egalitarian interpretation*. The former claims that—contrary to the geometrization picture—spacetime geometry is reduced to a gravitational field, i.e. the metric, which is taken as 'just another field'. Instead, in its strongest version, the latter argues for a conceptual identification of gravity and spacetime in GR (Lehmkuhl 2008, 84). Such alternative interpretations seem to reduce the conceptual differences between GR and the other field theories and may be further supported by gauge theoretic formulations of classical GR in the so-called tetrad or vielbein formalism (e.g. Carroll 2004, 483–494). Also, Brown famously argues for a *dynamical* perspective (cf. Brown 2005; Brown and Pooley 2001, 2006; Brown and Read 2016) that may be viewed as a variant of the field interpretation.

But these responses do not close the (technical) gaps between both frameworks, GR and QFT. Vague formal similarities between theories cannot be considered a substantial and physical reason for unification. Just think of the case of Newton's

[5]While Newtonian physics was unable to provide an explanation for why the equivalence principle should hold, the geometrical picture of GR provides an explanation in terms of an elimination (of gravitational potential and gravitational mass). As we will see in a moment, it is also possible to give a reductive account.

law of gravitation and Coulomb's law of electricity: The fact that both laws exhibit the exact same mathematical form does by no means imply that the phenomena of gravitation and electricity are linked in any substantial sense. Accordingly, one might still suspect that for explaining unificatory approaches like QG we need to impose additional external principles guiding physics.

However, concerning an argumentation against the geometrization picture in favor of a unified perspective another approach appears to be much more relevant: Weinberg (1964a, 1965b, 1995) and others—for example Feynman et al. (1995), and more recently Donoghue (1994, 2014)—advocated a "nongeometrical" (Weinberg 1972, viii) understanding of GR based on QFT. But let us not get ahead of ourselves and slowly approach the matter by help of Weinberg himself.

2.4 Weinberg's Conception of General Relativity

Weinberg is very clear in expressing his opposition to the geometrical understanding of GR:

> In learning general relativity, and then in teaching it to classes at Berkeley and M.I.T., I became dissatisfied with what seemed to be the usual approach to the subject. I found that in most textbooks geometric ideas were given a starring role, so that a student who asked why the gravitational field is represented by a metric tensor, or why freely falling particles move on geodesics, or why the field equations are generally covariant would come away with an impression that this had something to do with the fact that space-time is a Riemannian manifold. (Weinberg 1972, vii)

Furthermore, Weinberg considers the geometrization picture as historically contingent:

> It is certainly a historical fact that when Albert Einstein was working out general relativity, there was at hand a preexisting mathematical formalism, that of Riemannian geometry, that he could and did take over whole. However, this historical fact does not mean that the essence of general relativity necessarily consists in the application of Riemannian geometry to physical space and time. (Weinberg 1972, 3)

Weinberg argues that the geometrization picture ultimately confuses 'representation' and 'represented'. He suggests to conceive Riemannian geometry merely as a mathematical tool to account for "the peculiar empirical properties of gravitation, properties summarized by Einstein's Principle of Equivalence of Gravitation and Inertia" (Weinberg 1972, vii–viii, 3). The tool of Riemannian geometry should not be confused with the physical content of the principle of equivalence:

> In place of Riemannian geometry, I have based the discussion of general relativity on a principle derived from experiment: the Principle of the Equivalence of Gravitation and Inertia. [...] so that Riemannian geometry appears only as a mathematical tool for the exploitation of the Principle of Equivalence, and not as a fundamental basis for the theory of gravitation. (Weinberg 1972, viii)

According to Weinberg, Riemannian geometry is *one* possibility to represent the physical essence of GR, i.e. the SEP. But there are others as well. Weinberg puts forward an attitude that may be summarized as: "Don't look at the formalism, look at the physics!" However, after withdrawing the natural incorporation of the SEP via the usual foundation of GR in spacetime geometry, Weinberg then needs to come up with a proposal why gravitation should obey the SEP:

> This approach naturally leads us to ask why gravitation should obey the Principle of Equivalence. (Weinberg 1972, viii)

Interestingly, Weinberg does not expect to find an answer within the general framework of classical physics or within GR. Instead, Weinberg argues that one has to consider "the constraints imposed by the quantum theory of gravitation" (Weinberg 1972, viii). In the course of the following section, we will see what this means and how this reasoning can be spelled out.

2.5 Deriving the Principle of Equivalence

As we have seen, in the canonical interpretation of GR the SEP proves to be essential for describing the phenomenon of gravitation and for interpreting it as curvature of spacetime. Hence, it is for the SEP that gravity appears to be completely separated from the rest of fundamental physics. Accordingly, especially attempts at QG in the framework of particle physics seem to rest purely on philosophical considerations pursuing unification based on external principles. However, in the following, we will find that the SEP can be recovered in a quantum field theoretic framework. As a result, the SEP turns out to be the *link* between our theory of gravitation and particle physics. The unificatory practice of physics proves to proceed *internally*, based on genuine strategies of physics alone.

Essentially, the reconstruction of the equivalence principle according to Weinberg (1964a,b, 1965a,b, 1995, 1999) is done in three steps: First, it is argued that starting from SR and QM as 'first principles' we arrive at QFT (e.g. Weinberg 1999; Arkani-Hamed 2013). Second, while particles may in general have any spin in QFT, it turns out that in the low-energy limit (or 'long-range limit')[6] QFT provides a restrictive framework that only allows for fundamental particles with very specific spins—in the long-range limit we can only have particles with spin 0, $1/2$, 1, $3/2$, or 2. Third, analyzing this menu it can be shown that Lorentz invariant interactions with massless spin-2 particles *require* the equivalence principle to hold. So it turns out that the equivalence principle is not a fundamental principle itself, but can be *derived* from SR and QM. Actually, one can even prove that GR can be fully reconstructed in this approach as the only possible theory of gravitation at low energies (e.g. Arkani-Hamed 2010b).

[6]Here, 'low energy' means low energy with respect to the so-called Planck energy. Even the highest presently available energy scales in physics can safely be considered 'low' in that sense.

Now, we will not be able to demonstrate and appreciate all steps of the argumentation here. Instead, I will focus on a (as far as possible) non-technical presentation of step three. To set the stage, I will very briefly review some basics. To get accustomed to Weinberg's rationale, we will then—as a warm-up—consider what constraint Lorentz-invariance imposes on interactions with spin-1 particles, i.e. photons, in the low-energy regime: low-energetic photons may only participate in interactions that conserve charge. Afterwards, we will discuss the spin-2 case and show how Weinberg is able to recover the WEP and implicitly also the SEP within QFT.

In principle, there are several ways to do so. Weinberg first formulated the argument in the context of S-matrix-theory (Weinberg 1964a,b, 1965a,b). The following presentation rests on Weinberg's original work, on his textbook on QFT (Weinberg 1995, 534–539), an illuminating lecture by Arkani-Hamed (Arkani-Hamed 2010b), and lecture notes by Nicolis (2011).

To motivate what follows, recall that QFT can be understood as a theory of particles.[7] In general, quantum particles have the following essential properties: mass, charge, and spin. Still, particles may have zero mass, carry no charge, or have spin-0. In QFT, particles divide into fermions or matter particles with half-integer spin, and bosons with integer spin. Furthermore, particles can interact with each other. The interactions of the Standard Model are represented by the exchange of certain mediator particles, so-called gauge bosons with spin-1. For example, the electromagnetic interaction between two electrons is understood as an exchange of a massless spin-1 particle, the photon. The fact that it does not carry electromagnetic charge itself tells us that there is no self-interaction, its zero mass accounts for electromagnetism being a long-range interaction, and its odd spin incorporates that like charges repel. Accordingly, a hypothetical mediator particle for gravity, usually referred to as the graviton, is required to be massless as well, but to have *even* spin to account for the fact that gravity is attractive.

For the interaction processes of such particles, we can calculate so-called transition amplitudes to determine the 'cross section' or probability of the process. To calculate such amplitudes, we need to specify certain parameters. That is, most importantly, the strength of the specific interaction, the so-called coupling strength, and the masses, charges and spins of the participating particles. If two particles do not 'couple', the respective interaction is not allowed to take place.

Now, Weinberg's argument takes its departure from calculating such amplitudes of scattering processes in QFT. As Fig. 2.1a shows, in a scattering process a bunch of particles come in, interact in some way, and then a bunch of particles (the same particles or others) go out again. For any such process we can write down an amplitude, for example using the so-called Feynman rules which can be read off

[7] Of course, quantum *field* theory can be thought to be, first and foremost, a theory of fields. The corresponding particles are then derivative of the fields in the sense that they are *excitations* of the fields. Nevertheless, as the term *particle* physics stresses, we can also perceive it as a theory of particles. However, by talking about particles instead of fields I do not mean to have claimed anything substantial about the nature of QFT.

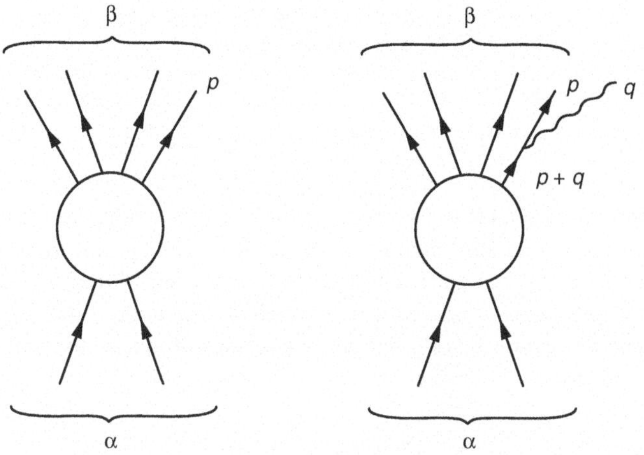

Fig. 2.1 (**a**) Arbitrary scattering process $\alpha \rightarrow \beta$ (modification of Weinberg 1995, 536). (**b**) Dominant diagram for additional emission of soft photons or gravitons from an outgoing particle in an arbitrary scattering process (Weinberg 1995, 536)

the corresponding Lagrangian. However, for our purpose it is not even necessary to write down the full scattering amplitude. We are not interested in the details of the scattering process or any specification of its interactions. Therefore, we represent the full scattering process, i.e. the sum of all possible Feynman diagrams for the process $\alpha \rightarrow \beta$, by the sphere in Fig. 2.1a. What we are actually interested in, is, without loss of generality, the analysis of a slight modification of such a generic scattering process (see Fig. 2.1b) to see if and how the corresponding interaction is constrained.

So, assume we know the amplitude, $\mathcal{M}_{\alpha\beta}(p_1, \ldots, p_\mathrm{n})$, for some arbitrary scattering process as in Fig. 2.1a. We would like to know the amplitude for the exact same process where additionally a soft massless particle with momentum q is emitted from one of the in- or outgoing particles (cf. Fig. 2.1b). Here, 'soft' means that the particle has very low energy—that is vanishing momentum, $q \rightarrow 0$. For example, the additionally emitted particle could be a photon. Note that in general the emitted particle may have any spin. We will now analyze the emission process for the photon case and learn about properties and constraints in the long-range limit. Specifically, we will explore the consequences of Lorentz-invariance for massless particles of spin-1. To mathematically account for the additional emission process, we have to multiply the original amplitude by a non-trivial factor. Because we want to model long-range phenomena, we shall only consider the most leading possible contribution that will survive 'at infinity', that is in the long-range limit. Generally, this factor will depend on the polarization vector of the photon, $\epsilon_\mu(q)$, the momentum of the photon, q, all the momenta of the other particles, p_i, and their charges, e_i.[8] Accordingly, we obtain the following amplitude for the slightly modified process of Fig. 2.1b (Arkani-Hamed 2010b):

$$\mathcal{M}_{\alpha\beta}(p_1, \ldots, p_n, q)|_{q\to 0} = \mathcal{M}_{\alpha\beta}(p_1, \ldots, p_n) \times \sum_i e_i \frac{p_i^\mu}{2p_i \cdot q} \epsilon_\mu(q) \,. \tag{2.1}$$

However, the emission factor in the amplitude is not completely arbitrary. The additional emission of a soft photon should not spoil Lorentz-invariance. Thus, we demand that Lorentz-invariance is preserved.[9] As a result, Eq. (2.1) is required to become zero for $\epsilon_\mu(q) \to q_\mu$. Since the original amplitude, $\mathcal{M}_{\alpha\beta}(p_1, \ldots, p_n)$, is assumed to be non-zero (the original process is not forbidden), the emission factor itself has to vanish:

$$\sum_i e_i \frac{q^\mu p_{i,\mu}}{2p_i \cdot q} = 0 \,. \tag{2.2}$$

Accordingly, we arrive at the fact that the sum over all charges needs to be zero, $\sum_i e_i = 0$, which means that the process is forced to obey charge conservation. So, interactions with soft massless spin-1 particles always conserve the respective charges. One could go on and derive Maxwell's equations by using perturbation theory (Weinberg 1965b), but we will stop here and turn to the next and more interesting case instead: a massless spin-2 particle, commonly referred to as the graviton.

As mentioned, such a spin-2 particle is among the quantum field theoretically allowed particles in the long-range limit. We can now essentially follow the same argumentation. Again, we want to investigate the long-range behavior, so we write down the leading contribution for our emission factor in the case of a soft graviton (Arkani-Hamed 2010b):

$$\mathcal{M}_{\alpha\beta}(p_1, \ldots, p_n, q)|_{q\to 0} = \mathcal{M}_{\alpha\beta}(p_1, \ldots, p_n) \times \sum_i \kappa_i \frac{p_i^\mu p_i^\nu}{2p_i \cdot q} \epsilon_{\mu\nu}(q) \,. \tag{2.3}$$

Here, $\epsilon_{\mu\nu}(q)$ is the polarization tensor of the graviton, and κ_i are the coupling constants for the particles with momenta p_i emitting a soft graviton (Weinberg 1965b, B989). Now, if we demand Lorentz-invariance (and again assume that the original process is allowed, i.e. $\mathcal{M}_{\alpha\beta}(p_1, \ldots, p_n) \neq 0$), we arrive at:

$$\sum_i \kappa_i p_i^\nu = 0 \,. \tag{2.4}$$

[8]Here, the charge of a particle is defined as its coupling constant for emission of soft photons (Weinberg 1965b, B989).

[9]That means that we demand the polarization vector to transform as $\epsilon_\mu(p) \to (\Lambda\epsilon)_\mu(p) + \alpha(\Lambda p)_\mu$.

So, what does this mean? According to Eq. (2.4) the sum over all momenta, p_i, weighted by the coupling constants, κ_i, is required to be conserved in all possible scattering processes. However, we know that already (unweighted) momentum conservation, $\sum_i p_i = 0$, should hold in all scattering processes. If both, momentum conservation and Eq. (2.4), are supposed to hold, there are only two options: Either the scattering between the particles of momentum p_i is trivial, that means the particles do *not* interact at all, or all coupling constants, κ_i, have to be *identical* for all particle species regardless of their properties, that is $\kappa_i = \kappa$.

So, by demanding Lorentz-invariance the coupling of a massless spin-2 particle to any other particle (including other spin-2 particles) is forced to be *universal*. This is precisely the quantum field theoretic version of the weak equivalence principle that gravitation is supposed to obey.[10] Hence, the WEP is established within QFT. Note that for obtaining the *strong* equivalence principle the coupling is usually also required to be *minimal* (cf. Read et al. 2017). This is fulfilled here because all terms violating the SEP essentially behave as high-energy corrections and are therefore absent in the low-energy limit. In this sense, the WEP effectively implies the SEP.[11]

To further appreciate this result, one can also prove that the massless spin-2 particle is *unique*. There can only be exactly *one* massless spin-2 field (Arkani-Hamed 2010b). Also, we can show that for higher spins all respective coupling constants must vanish in the long-range limit, so there are *no* Lorentz invariant theories of massless particles with spins higher than spin-2 (Weinberg 1965b, B989). In conclusion, we find that the massless spin-2 particle uniquely represents an attractive long-range interaction that universally couples to all forms of energy and momenta. Also self-interaction, another important property of gravitation, is automatically established (Weinberg 1964a). The massless spin-2 particle is therefore correctly called 'graviton'.

Before I comment on what follows regarding unification, let me first briefly summarize what follows regarding the relation between GR and QFT.

2.6 What Do We Learn from This?

In the light of Weinberg's argument, the equivalence principle, usually perceived in close connection with the geometrization thesis, turns out to be the *link* between a theory of gravitation and particle physics. The low-energy limit of our empirically best tested and theoretically most advanced framework, namely QFT, proves to

[10]Here we used a slight simplification, but for example Nicolis (2011) carefully proves that the gravitational coupling constants, κ_i, are indeed forced to be universal.

[11]Still, given that Read et al. (2017) argue that minimal coupling may violate certain versions of the SEP, there definitely remains more to be said. Ultimately, all claims involving the SEP here are in need of further clarification.

be highly constraining (Arkani-Hamed 2010a,b, 2013). In the low-energy limit it is impossible to construct a Lorentz invariant quantum theory for massless spin-2 particles that does not obey the SEP:

> In other words, by asking for a theory of a spin-2 field coupling to the energy-momentum tensor, we end up with the fully nonlinear glory of general relativity. (Carroll 2004, 299)

The equivalence principle is not merely postulated, but explained. In QFT the SEP is "not a principle, but a theorem" (Nicolis 2011, 28). The fundamental principles of locality (SR) and unitarity (QM) that ground QFT enforce the SEP to hold, they enforce a theory of gravitation (Weinberg 1999; Arkani-Hamed 2013). Hence, a reductive account of GR is obtained: GR can be *deduced* from QFT. In terms of principles, GR can be deduced from bringing together SR and QM:

> All of these things that Einstein did—Einstein thought about these falling elevators and he discovered the Principle of Equivalence and all these deep facts about classical physics that led him to think about General Relativity—all of those things could have been discovered by much more mediocre theoretical physicists who knew about Quantum Mechanics. (Arkani-Hamed 2013)

Typically, the conviction is that GR is more fundamental than SR. But according to the analysis above, it is in fact the other way around.

However, there is a well-known caveat that we have to mention. Weinberg's approach is only able to consistently account for a theory of gravitation at *low* energies. At high energies the theory is rendered non-predictive. That is the infamous problem of gravity's nonrenormalizability (Weinberg 1972, 289). Due to this QFT is not able to provide a full-fledged theory of quantum gravity. But—and this should be appreciated—the spin-2 approach is an existing quantum theory of gravitation encompassing all presently known experimental data:

> A lot of portentous drivel has been written about the quantum theory of gravity, so I'd like to begin by making a fundamental observation about it that tends to be obfuscated. There is a perfectly well-defined quantum theory of gravity that agrees accurately with all available experimental data. (Wilczek 2002)

So, we learn how to adequately understand and formulate the actual problem with QG. Usually, it is presented somehow like this: 'Combining GR and QM leads to a meaningless theory.', or 'We don't know how to combine QM as a theory of the very small and GR as a theory of the very big.', or as Wüthrich (2005, 782) states it: "In a sense, then, quantum mechanics and general relativity when combined already contain the seeds of their own destruction." In the light of Weinberg's argument, these statements prove false or at least misleading. First of all, the problem with QG is not that we have no grounds whatsoever to talk about such a theory—we actually already have one. Instead, the problem is that this theory is not valid at high energies. Accordingly, solving the problem amounts to finding the correct high energy theory (cf. Donoghue 2014). Thus, the problem with finding a (full-fledged) theory of QG is more subtle than often described.

Moreover, as discussed above, the problem is not constituted by bringing together GR and QM, but by bringing together SR and QM, since QFT is solely based on these two theories and the assumption of the cluster decomposition principle (cf. Weinberg 1999). As a consequence, we can infer that it is *SR* and QM that exhibit a very subtle conflict at high energies.[12]

2.7 Critical Remarks

There are some objections against this presentation. I will not be able to address them in detail, but they should at least be mentioned. Since I will only sketch responses, there certainly remains more to be said as the topics are fairly complex. First of all, one may be skeptical if the recovered result in Weinberg's argumentation really needs to be identified with Einstein's principle of equivalence because many general relativists would typically not state it in terms of 'universal coupling' (plus minimal coupling). This is an issue that needs to be addressed and investigated further—a task that I have to postpone.

However, it would seem dubious if we should not be allowed to translate a physical concept from one theoretical framework to another as a matter of principle. Such translations will in general involve non-trivial reinterpretations of the respective structures—business as usual in physics during theoretical progress. And, certainly, this may result in considerable deviations, but that does not necessarily undermine the new perspective's significance. In fact, I would rather suggest to read it as follows: By means of this practice, we learn something about how modern physics addresses the notion of fundamentality. In recovering and reinterpreting an established fundamental structure like the SEP within another framework, we obtain another perspective and learn more about the essential aspects of the structure. For example, the structure may reveal as nonfundamental. In this sense, physics continuously challenges alleged fundamentality of structures—again this needs to be investigated further.

Second, the issue of nonrenormalizability is certainly very important and constitutes the main objection against quantum field theoretic approaches. Still, as we have seen, this approach turns out to be quite illuminating. Also, the fact that such approaches are nonrenormalizable is not straightforwardly problematic in the light of effective theories (Donoghue 1994). Still, one has to clarify whether being an effective field theory is a serious defect in general as some do suspect (Redhead 1999). Keep in mind, however, that also classical GR is effective in that it does not contain higher powers of the curvature tensor.

---

[12]According to Maudlin (2011), there is another, very general conflict between SR and QM due to Bell's theorem. Note, however, that this is an entirely different issue closely connected to the debate on the interpretation of QM—a debate which physicists might be safe to ignore as long as the theory is empirically adequate. The high energy conflict mentioned here is not of that kind: While QFT is empirically adequate, consistent and highly predictive at low energies, it becomes non-predictive at high energies.

Finally, another frequent complaint against quantum field theoretic approaches to QG concerns the issue of background independence (e.g. Huggett and Callender 2001b; Rickles and French 2006). Especially general relativists (and many philosophers) assume that background independence constitutes the other key feature of GR besides Einstein's equivalence principle. It is typically understood to render GR fully dynamical. On the contrary, approaches in the context of QFT simply employ the Minkowski metric. Thus, QFT is argued to be background dependent.

However, first of all, background independence is a very loose concept. It already proves highly difficult to define what background independence precisely means (Belot 2011; Read 2016). Second, background independence may already be destroyed in classical GR (Belot 2011). For instance, a non-empty universe that contains matter and electromagnetic fields is not fully background independent (Belot 2011). To appreciate this result, remember that Weinberg's approach already includes matter and all interactions. On the contrary, approaches that start from GR and value background independence are typically concerned with spacetime alone. I suspect that praising background independence is closely tied to contestable philosophical premises.

2.8 Unification Revisited

Finally, what about the suspected 'dogma of unification'? I suggest to take the presented argumentation as an example for physics' *internal* capacity to promote theoretical progress by fully exploiting an existing theoretical framework and thereby uncovering substantial links to allegedly disconnected phenomena and theoretical structures.

Weinberg's quantum field theoretic derivation of the equivalence principle bridges the gulf between seemingly isolated theories. It should be emphasized that this perspective evolves *internally*. On its own, QFT provides a quantum theory of gravitation and—though this is not the final answer—resources to investigate and evaluate its flaws. As a result, we do not need to refer to any external principle to account for Weinberg's approach or to explain the quest for QG. The research program does not rely on external principles and does not commit to or execute a 'dogma of unification'. Hence, the situation is not special at all compared to previous endeavours in theoretical physics. To obtain a quantum theory of gravitation, Weinberg and others only had to take the best theoretical framework, namely QFT, seriously and analyze it thoroughly.

However, applying the same methods further and expanding the theory's realm to higher energies unfolds that QFT itself has its problems: As mentioned, Weinberg's spin-2 approach is nonrenormalizable. It does not provide meaningful, i.e. finite, results at high energies. According to particle physics, *this* is the issue of QG.

To solve this problem, one could generally proceed by either dismissing the theory as a whole (or at least one of its basic principles) and start from scratch, or try to dissolve the anomalies while keeping the basic principles (cf. Arkani-Hamed

2013). As argued in Sect. 2.6, the problem with constructing a full-fledged theory of QG arises from bringing together QM and SR at high energies. The core principles by which we were able to construct our most accurate theoretical framework prove in tension in an even more fundamental way. In this view, a full-fledged theory of QG may result as a *by-product* of better understanding QFT (which again is a completely internal endeavour of physics).

Furthermore, if approaches to QG do not rely on external principles, but turn out to be a mere by-product of physical research applying its internal methods, it seems that the objections, doubts, and worries raised against these approaches become less persuasive. The theory itself tells us that there is more to come.

The presented argumentation was concerned with Weinberg's approach only. What about other takes on QG? Do they also solely rely on internal principles or are external principles involved? One would have to investigate their practice, too. But let me add the following: By help of Weinberg's approach we saw that QFT is already able to incorporate gravity. In fact, we saw that GR can be reduced to QFT. That does not seem to be the case the other way around. The classical framework of GR does not provide links to particle physics. It seems to be a perfectly consistent and self-contained theory. Also arguments concerning singularities do neither prove GR inconsistent, nor do they hint at a quantum theory of gravity (Wüthrich 2012, 2). In the light of the presented argumentation, this should come as no surprise, since GR is a classical theory deduced from QFT. As a result, one may argue that it somehow seems odd to start approaches to QG from GR (cf. Weinberg 1999).

References

Arkani-Hamed, N. 2010a. *The Future of Fundamental Physics. Space-Time Is Doomed; What Replaces It?*, Messenger Lecture Series at Cornell University. Lecture.

Arkani-Hamed, N. 2010b. *Robustness of GR. Attempts to Modify Gravity. Part I.* Prospects in Theoretical Physics Program. Cornell University. Lecture.

Arkani-Hamed, N. 2012. The Future of Fundamental Physics. *Dædalus* 141(3): 53–66.

Arkani-Hamed, N. 2013. *Philosophy of Fundamental Physics*. Andrew D. White Professors-at-Large Program. Cornell University. Lecture.

Belot, G. 2011. Background-Independence. *General Relativity and Gravitation* 43(10): 2865–2884. arXiv:grqc/1106.0920.

Brown, H. 2005. *Physical Relativity: Spacetime Structure from a Dynamical Perspective*. Oxford: Oxford University Press.

Brown, H.R., and O. Pooley 2001. The Origins of the Spacetime Metric: Bell's 'Lorentzian pedagogy' and Its Significance in General Relativity. In *Physics Meets Philosophy at the Planck Scale*, ed. C. Callender and N. Huggett, 256–272. Cambridge: Cambridge University Press.

Brown, H.R., and O. Pooley 2006. Minkowski Space-Time: A Glorious Non-entity. In *The Ontology of Spacetime, Volume 1 of Philosophy and Foundations of Physics*, ed. D. Dieks, 67–89. Amsterdam: Elsevier.

Brown, H.R., and J. Read 2016. Clarifying Possible Misconceptions in the Foundations of General Relativity. *American Journal of Physics* 84: 327.

Carlip, S. 2008. Is Quantum Gravity Necessary? *Classical and Quantum Gravity* 25: 154010.

Carroll, S. 2004. *Spacetime and Geometry. An Introduction to General Relativity*. San Francisco: Addison Wesley.
Dawid, R. 2013. *String Theory and the Scientific Method*. Cambridge: Cambridge University Press.
Donoghue, J. 1994. General Relativity as an Effective Field Theory. The Leading Quantum Corrections. *Physical Review D* 59: 3874–3888.
Donoghue, J. 2014. General Relativity as an Effective Field Theory. PSI Summer School 'More than Higgs – Effective Theories for Particle Physics'. Zuoz. Lecture. blogs.umass.edu/donoghue/files/2009/06/Zuoz-3.pdf.
Eppley, K., and E. Hannah 1977. The Necessity of Quantizing the Gravitational Field. *Foundations of Physics* 7: 51–68.
Feynman, R., F.B. Morinigo, W.G. Wagner, and B. Hatfield 1995. *Feynman Lectures on Gravitation*. Reading, Massachusetts: Addison-Wesley.
Huggett, N., and C. Callender 2001a. Introduction. In *Physics Meets Philosophy at the Planck Scale. Contemporary Theories in Quantum Gravity*, ed. N. Huggett and C. Callender, 1–33. Cambridge: Cambridge University Press.
Huggett, N., and C. Callender 2001b. Why Quantize Gravity (Or Any Other Field for that Matter)? *Philosophy of Science* 68(Proceedings): S382–S394.
Kiefer, C. 2006. Quantum Gravity: General Introduction and Recent Developments. *Annals of Physics* 15(1–2): 129–148.
Kiefer, C. 2007. *Quantum Gravity*. Oxford: Oxford University Press.
Lehmkuhl, D. 2008. Is Spacetime a Gravitational Field? In *Philosophy and Foundations of Physics, Volume 4: The Ontology of Spacetime, Vol. II*, ed. D. Dieks and M. Redei, 83–110. Amsterdam: Elsevier.
Lehmkuhl, D. 2014. Why Einstein Did Not Believe that General Relativity Geometrizes Gravity. *Studies in History and Philosophy of Science Part B: Studies in History and Philosophy of Modern Physics* 46(Part B): 316–326.
Mattingly, J. 2005. Is Quantum Gravity Necessary? In *The Universe of General Relativity*, ed. A.J. Kox and J. Eisenstaedt, 327–338. Basel: Birkhäuser. Talk at the 5th International Conference on the History and Foundations of General Relativity in 1999.
Mattingly, J. 2006. Why Eppley and Hannah's Thought Experiment Fails. *Physical Review D* 73: 064025.
Maudlin, T. 1996. On the Unification of Physics. *The Journal of Philosophy* 93(3): 129–144.
Maudlin, T. 2011. *Quantum Non-locality and Relativity*. Chichester: Wiley-Blackwell.
Nicolis, A. 2011. General Relativity from Lorentz Invariance. Lecture notes. phys.columbia.edu/~nicolis/GR_from_LI_2.pdf.
Read, J. 2016. Background Independence in Classical and Quantum Gravity. B.Phil. Thesis, University of Oxford.
Read, J., H.R. Brown, and D. Lehmkuhl 2017. Two Miracles of General Relativity. Manuscript in Preparation.
Redhead, M. 1999. Quantum Field Theory and the Philosopher. In *Conceptual Foundations of Quantum Field Theory*, ed. T.Y. Cao, 34–40. Cambridge: Cambridge University Press.
Rickles, D., and S. French 2006. Quantum Gravity Meets Structuralism: Interweaving Relations in the Foundations of Physics. In *The Structural Foundations of Quantum Gravity*, ed. D. Rickles, S. French, and J. Saatsi, 1–39. Oxford: Oxford University Press.
Rovelli, C. 2016. The dangers of non-empirical confirmation. arXiv:1609.01966.
Wald, R.M. 1984. *General Relativity*. Chicago: The University of Chicago Press.
Weinberg, S. 1964a. Derivation of Gauge Invariance and the Equivalence Principle from Lorentz Invariance of the S-Matrix. *Physics Letters* 9(4): 357–359.
Weinberg, S. 1964b. Photons and Gravitons in S-Matrix Theory: Derivation of Charge Conservation and Equality of Gravitational and Inertial Mass. *Physics Review* 135(4B): B1049–B1056.
Weinberg, S. 1965a. Infrared Photons and Gravitons. *Physics Review* 140(2B): B516–B524.
Weinberg, S. 1965b. Photons and Gravitons in Perturbation Theory: Derivation of Maxwell's and Einstein's Equations. *Physics Review* 138(4B): B988–B1002.

Weinberg, S. 1972. *Gravitation and Cosmology: Principles and Applications of the General Theory of Relativity*. New York: Wiley.

Weinberg, S. 1995. *The Quantum Theory of Fields. Volume I: Foundations*. Cambridge: Cambridge University Press.

Weinberg, S. 1999. What Is Quantum Field Theory, and What Did We Think It Is? In *Conceptual Foundations of Quantum Field Theory*, ed. T. Y. Cao, 241–251. Cambridge: Cambridge University Press.

Wilczek, F. 2002. Scaling Mount Planck III: Is That All There Is? *Physics Today* 55: 10–11.

Wüthrich, C. 2005. To Quantize or Not to Quantize. Fact and Folklore in Quantum Gravity. *Philosophy of Science* 72: 777–788.

Wüthrich, C. 2006. *Approaching the Planck Scale from a Generally Relativistic Point of View: A Philosophical Appraisal of Loop Quantum Gravity*. Ph.D Thesis, University of Pittsburgh.

Wüthrich, C. 2012. In Search of Lost Spacetime: Philosophical Issues Arising in Quantum Gravity. In *La philosophie de la physique: d'aujourd'hui à demain*, ed. S. Le Bihan. Paris: Vuibert. arXiv:1207.1489v1.

Chapter 3
On Predictions and Explanations in Multiverse Scenarios

Keizo Matsubara

Abstract Many contemporary physicists suggest that we should take the claim that we live in a multiverse seriously. When doing this they often invoke arguments based on the controversial anthropic principle. Critics argue that this leads to untestable and unscientific theories. In this paper criteria are suggested that need to be satisfied before a multiverse theory should be considered scientifically respectable. One important point is that the multiverse is described in sufficient detail. In some cases, a multiverse theory can be testable; however, to properly test a multiverse theory it is important to distinguish new predictions from explanations based on the multiverse.

Keywords Prediction · Explanation · Multiverse · String theory · Scientific methodology

3.1 Introduction

In contemporary physics many researchers take the idea that we live in a multiverse of some sort seriously. What reasoning is behind their advocacy of such a view? After all, the idea that our universe is just one of many in a multiverse seems rather weird and outlandish to most people. Typically, ideas about multiverses are in the current discussion in physics closely connected with arguments using the rather controversial anthropic principle. Many critics find the use of reasoning based on the anthropic principle and multiverses to be unscientific.

While currently proposed multiverse scenarios do not yet live up to criteria strong enough to be counted as part of well established science, I argue that in principle we could find good scientific reasons for accepting a theory entailing that we live in a multiverse. Multiverse theories can, if sufficiently developed, give rise to testable

K. Matsubara (✉)
Department of Philosophy, Uppsala University, Uppsala, Sweden
e-mail: keizo.matsubara@filosofi.uu.se

A. Christian et al. (eds.), *Philosophy of Science*, European Studies in Philosophy of Science 9, https://doi.org/10.1007/978-3-319-72577-2_3

predictions. For this to be the case the specific multiverse needs to be properly implied by a well-articulated and precise theory.

Given the account here presented it is not sufficient to just rely on vague arguments based on considering the anthropic principle and the observation of apparent fine-tuning of physical quantities so that they allow for our existence. What I am interested in is the question how we in principle can test *specific* multiverse theories, as opposed to evaluating the generic idea that we live in a multiverse. What this amounts to in more detail will be explained below. In addition it will be argued how one needs to think differently when it comes to how a multiverse theory is used for *predictions* compared to how it can be used as part of an *explanation*.

The term "multiverse" is in this paper used in a rather broad and loose sense. This is in accordance with how the term is typically used among physicists. What is considered to be separate universes do for instance not have to be completely spatiotemporally disconnected.[1] Hence, a multiverse in this broad sense could according to some philosophers still just count as one big and complex universe.

A multiverse in itself does not have to be based on string theory and the considerations in this paper are applicable to other multiverse theories as well. However, it is fair to say that the way string theory has developed has been important for the way people now think about the possibility that we might live in a multiverse. One important aspect of the kind of multiverse theories I am here considering is that they allow for a variation among the universes so that different universes can have different values for certain physical quantities, namely the ones that are considered in discussions about fine-tuning and the anthropic principle.

String theory, with its version of a multiverse, does currently not live up to the criteria presented in this paper. On the other hand in the future this could very well change. String theory is the main contemporary example for which the analysis in this paper is relevant; for this reason a brief non-technical discussion of string theory now follows.

3.2 String Theory and the Landscape

String theory is a very influential and important research program in fundamental theoretical physics.[2] The basic assumption is that what we have previously thought of as point particles are instead strings; that is extended one-dimensional entities. Different particles are supposed to correspond to different vibrational patterns of these strings. String theorists want to provide us with a theory for quantum gravity and unify all the fundamental forces in physics within one overarching framework. In earlier days hope was widespread that string theory would be able to give us

[1]This is required in David Lewis's account on multiple worlds; see Lewis (1986).

[2]Standard textbooks are Green, Schwarz and Witten (1987), Polchinski (1998) and Zwiebach (2004). Accounts written for the layperson are Greene (1999, 2004) and Susskind (2005).

a more or less unique prediction on how our world must be, for instance why the particles we see have the properties they have. Thus, the abundance of parameters of the standard model would be explained.

While some still maintain the old hope described above, many string theorists now endorse a different view. The new view is that there are an immense number of different sufficiently stable solutions within string theory.[3] Each such solution would describe a different world with its own effective physics and particle content. The set of solutions is often called the "Landscape" of string theory. The number of sufficiently stable solutions of string theory is extremely large, a number often mentioned in the literature is 10^{500}, but this number is a rather rough estimate and should be taken with a pinch of salt.

It is important to distinguish on the one hand between the Landscape of string theory in terms of the mathematically possible solutions in string theory and the further question whether there are physical realizations of these solutions. Susskind himself distinguishes between the Landscape as the purely mathematical structure and the set of physically realized solutions. He uses the term "megaverse" instead of "multiverse" for the latter and writes,

> The two concepts – *Landscape* and *megaverse* – should not be confused. The Landscape is not a real place. Think of it as a list of all the designs of hypothetical universes. Each valley represents one such design... The megaverse, by contrast, is quite real. The pocket universes that fill it are actual existing places, not hypothetical possibilities. (Susskind 2005, 381)

What is needed for the string theory multiverse or megaverse to be realized – assuming for the time being that string theory is more or less correct – is a suitable mechanism to populate the landscape with physically realized worlds. The most popular suggested mechanism is eternal inflation. The empirical evidence for inflation is by most physicists considered good, but the exact theoretical explanation behind inflation is not yet understood. Many different versions have been proposed.

Among the proposed versions of inflation many suggest that inflation will go on forever. Parts of the inflating totality will stop inflating thus producing isolated pocket or bubble universes.[4] These would be the relevant universes that populate the multiverse. A problem is that the discussions on inflation and eternal inflation is not directly connected to string theory, they are due to independent observations and theoretical considerations. Also, while inflation is presumably compatible with string theory and attempts have been made to describe inflation within the string theoretic framework, it is not the case that string theory, as we now understand it, implies or singles out a specific detailed account of what version of inflation that would occur.

Another possible candidate mechanism for populating the Landscape could be based on an Everettian view on quantum mechanics. Of course just accepting an Everettian picture already amounts to accepting a kind of multiverse. However, in

[3]See Kachru et al. (2003), Susskind (2007) and Schellekens (2008).

[4]See for instance Linde (2007).

the discussion of this paper the Everettian picture of quantum mechanics plays a different role compared to the kind of multiverse theories I am here interested in. The reason for this is that it is not a specific quantum theory; instead it is a general view on quantum mechanics. This view could if accepted then be applied to any specific underlying quantum theory.

Whether an Everettian multiverse give rise to a variation among the physical quantities that are considered to be fine-tuned depends on the details of the specific underlying quantum theory on which the Everettian view is supposed to be applied. The Everettian view is still controversial and can at least not yet be accepted as part of well-established science, even though more and more philosophers and physicists nowadays embrace it.

Thus, an Everettian view could play a part as the relevant *mechanism* for a specific multiverse theory but should not *in itself* be seen as a multiverse theory of the kind I am here considering.

According to the string theorists, with a multitude of physically existing worlds the apparent fine-tuning of certain physical quantities and parameters of our universe can be explained.[5] Critics have said that such a theory will not be able to predict anything; they claim this is a "theory of anything" rather than a theory of "everything".[6]

3.3 On Fine-Tuning and the Anthropic Principle

One of the main reasons for why people are sympathetic to a multiverse is that it supposedly explains the apparent fine-tuning of the values of a number of physical quantities. It seems to be the case that had the values been slightly different conscious life forms like us could not exist. Other people have instead thought that the fine-tuning is a good reason to accept a theistic explanation.

Here the anthropic principle must be discussed, including which role it is supposed to play for the explanation of fine-tuning. There are many different versions of the anthropic principle.[7] The main focus here will be on the Weak Anthropic Principle (**WAP**). This is how Barrow and Tipler formulates this principle,

> Weak Anthropic Principle (WAP): The observed values of all physical and cosmological quantities are not equally probable but they take on values restricted by the requirement that there exists sites where carbon-based life can evolve and by the requirement that the Universe be old enough for it to have already done so. (Barrow and Tipler 1986, 16)

---

[5]See Susskind (2005, 2007). A good review is Schellekens (2008), especially the extended Internet version.

[6]See for instance Smolin (2006, 2007) and Woit (2006).

[7]For an extensive discussion on anthropic reasoning see Barrow and Tipler (1986).

In Carter (1974) as quoted in Weinstein (2006) the principle is presented even more succinctly,

> **WAP**: "What we can expect to observe must be restricted by the condition necessary for our presence as observers."

The principle is in a way trivial and amounts to little more than a tautology. Roughly it states that we can only observe a world, which allows for our existence. This seems obviously true and should hardly be thought of as controversial. In itself, the weak anthropic principle – being quite empty – is however neutral with respect to what further physical or metaphysical conclusions to draw. The problem that turns the principle controversial is that people tend to also add further general explanations of one sort or another.

It is important to emphasize that in this paper neither the anthropic principle nor the presence of fine-tuning are by themselves taken as evidence for a multiverse theory. They are rather just basic initial observations to be taken into account when trying to figure out how *specific* multiverse theories are to be tested.

A *generic* explanation in terms of a multiverse, or a *generic* theistic explanation are both too unclear, vague and imprecise to qualify as scientific. This does of course not show that such explanations are necessarily wrong. The point is rather that for these views to be taken seriously as part of science much more must be said in addition to the generic kind of explanation.

Given that some have seen the fine-tuning as evidence for a theistic explanation, questions of a theological nature have appeared in discussions about the anthropic principle. However my view is that theological considerations is better kept aside from the more strictly scientific discussion. The reason for this is that the theistic view is inherently too unspecific and imprecise. Assuming a deity, as an explanation is not precise enough to constitute part of a scientific theory, it is such a vague and unspecific idea that it can be adjusted to basically anything. For instance Page (2007) and Collins (2007) are both sympathetic to *combining* a multiverse with a theistic viewpoint even though multiverse theories or scenarios are typically used as an alternative to a theistic view of the world. This, I believe, shows how scientifically vacuous a generic theistic explanation is. Furthermore, I have no idea how a generic theistic explanation could be supplemented and made precise in a way analogous to how a more specific multiverse theory could in principle be formulated along the lines sketched in this paper.

Just stipulating a multiverse to explain the fine-tuning is also not by itself something scientific; it is on par with just deferring to a deity. However, a specific detailed account of a theory with clearly specified solutions and a mechanism for populating the multiverse might be acceptable if it is sufficiently testable and well supported.

Smolin has argued that the anthropic principle in itself is scientifically empty and that it cannot be used to provide evidence in favour of the existence of a multiverse of the kind suggested by string theory.[8] In Smolin (2007) it is argued as follows,

[8]See Smolin (2006, 2007).

The standard form of purported examples of using anthropic reasoning is as follows,

(i) X is necessary for life.
(ii) X is true in our world
(iii) Using the laws of physics as presently understood, together with other observed facts Y, we deduce that if X is true of our universe, then so is Z
(iv) We therefore predict that Z is true.

Do we actually use anthropic reasoning here? No we do not, since we do not need to invoke the first premise at all to derive the conclusion. If we assume that we live in a multiverse (i) can be replaced by,

(i′) We live in one member of the multiverse in which the laws of physics vary. X is necessary for life, so by a selection effect we must live in a universe in which X is true.

Still this modification does not alter the fact that (i′) is irrelevant for the conclusion. I agree with Smolin's argument to the effect that we do not really need to consider whether something we observe in our universe is necessary to allow for our existence, for the deduction to go through in this scheme. And this holds with or without throwing in a multiverse into the initial premise. This observation will be further explored when I discuss how to make predictions in a multiverse.

There are, however, some purported uses of the anthropic principle to make predictions, such as the prediction of a resonance in carbon by Hoyle (1953) and the prediction of a small cosmological constant by Weinberg (1987). The example by Hoyle fit the pattern described above and this is also the case with Weinberg's prediction but some further comments need to be made. One point is that prior to the discovery of a small positive cosmological constant, most physicists thought that it was precisely zero due to some theoretical reason.

Weinberg did not take a zero value of the cosmological for granted and instead calculated a bound for how large the cosmological constant could be given prior observations and still allow for our existence. The prediction that the cosmological constant is within this bound is just an example of the argument structure described above. In addition, Weinberg does, however, talk about a multiverse as explaining the observation and make some rough statistical estimates about what we should expect to see. This part of the argument is shakier and relies on questionable assumptions and a too loose description of the multiverse. Furthermore, the predictions and estimates would be the same if one assumed that only one universe was physically realized among the allowed possibilities.

One might think that an explanation in terms of a multiverse is more preferable and intellectually attractive compared to the alternative. This can be discussed as part of a more general philosophical discussion. However, that is not sufficient for making the multiverse picture scientific at least according to the criteria suggested in this paper. What I am interested in is to figure out how specific theories that entail the existence of a multiverse can be testable and satisfy traditional scientific standards of testability. For this to be the case the multiverse theory needs to live up to a few criteria. These criteria will now be presented.

3.4 Criteria for Scientifically Respectable Multiverse Theories

So what are the criteria I suggest need to be satisfied before a multiverse theory should be counted as scientifically respectable?

First the set of solutions or roughly "allowed universes" needs to be properly understood and well described. This means that we know what the values of physical quantities would be in the different solutions or universes.

Traditionally the existence of many different solutions to a theory has not led us to believe that all the different solutions have to be physically realized. Thus a multiverse theory requires that a mechanism such that many solutions or universes are physically realized is part of the theory. Remember that the different "universes" in the multiverse do not have to be completely physically and spatiotemporally disconnected. It is important for this mechanism to be specified in sufficient detail. It is for instance important to know if some kinds of universes are produced more often than others. Furthermore, the mechanism should be an integral part or consequence of the theory and it should be well understood how it is supposed to work.

Sometimes, the argument is given that we should not accept multiverse theories since it will be impossible for us to observe other universes than our own. However, if we have sufficiently good reasons to trust a theory then further consequences of the theory ought to be taken seriously. Hence, if the mechanism is well understood and properly embedded as a central consequence of a well supported theory and thus the theory inherently predicts the existence of multiple universes then we ought to take the multiverse theory seriously; at least if we accept some sufficiently strong version of scientific realism.

But how could we find reasons to trust the theory in the first place? One obvious answer is that the theory also succeeds in making precise predictions that can be empirically confirmed. These predictions could for instance be of such a nature that they turn out the same in all universes.

Or a multiverse theory might specify a distribution of universes with sufficient constraints so that predictions can be extracted; this will be discussed later in this text.

To make empirical predictions, which are subsequently confirmed, is the traditional way for theories to receive scientific legitimacy, but could a theory be accepted as part of established science without such empirical success? In Dawid (2013) it is argued that we can have theoretical reasons for trusting a theory prior to successful empirical tests. Dawid argues specifically that we already have good non-empirical theoretical reasons to trust string theory. One example of how non-empirical considerations are supposed to give support to a theory is to evaluate how the theory connects to and explains previously empirically well-established theories. Of course Dawid agrees that it would be even better if the theory also becomes empirically successful and that we should continue to search for ways to empirically test the theory. He also stresses that the non-empirical ways in which a theory can get support is not completely independent on empirical considerations. As mentioned above the empirical successes of previous theories are important. In addition to that Dawid also considers empirical data at a meta-level; here for instance observations about whether or not people have been able to find alternatives are considered.

While I accept that there are theoretical reasons for in some sense take a theory seriously I am more cautious than Dawid. Theoretical reasons of the kind Dawid proposes can be good to motivate further work on the theory. It can give us reasons to accept something as a respectable research program that it is worth investing more time on. However, in the absence of empirical success I do not think we ought to accept the theory as part of well-established science. For this to be the case I maintain the view that a theory must be empirically successful. Thus the view advocated here is more conservative and traditional with regards to scientific method compared to Dawid's.

Below I discuss how a multiverse theory in principle can generate testable predictions and how to make predictions in the right way. In addition I point out the difference between using a multiverse theory for *predictions* as opposed to using it for the purpose of providing an *explanation.* This is something that has often not been carefully distinguished the way it should be in the literature.

For it to be possible to make predictions from a theory that imply that there exists a multiverse consisting of many universes, we need to have a detailed understanding of that multiverse and how the different observable physical properties and quantities are distributed over the set of solutions, or possible universes according to the multiverse theory.

Ideally there will be some sharp predictions, which can be found if certain observable properties or quantities are correlated according to the multiverse theory in such a way as to give sufficient constraints. If a not yet measured physical quantity X must have a given value according to the multiverse theory based on the fact that other already measured physical quantities have the values they have, then we have a sharp prediction. The kind of predictions described here could be done even if we did not have a multiverse theory but just a theory with a set of allowed solutions where the values of different physical quantities are sufficiently constrained. Due to this it is important that the existence of a multiverse really is a consequence of the theory – with a specific mechanism for populating the multiverse with universes – so that we do not just have a well-specified set of constrained solutions.

The above was an ideal case but what is the general strategy to use when extracting predictions from a multiverse theory. An article that has dealt with this problem is Weinstein (2006). He has argued that the weak anthropic principle is ambiguous and can be understood in two different ways.

WAP$_1$: "What we can expect to observe must be restricted by the condition necessary for our presence."

WAP$_2$: "What we can expect to observe must be restricted by the condition necessary for the presence of observers."

Weinstein argues that we should prefer **WAP$_1$**. The point is that while both principles are true, **WAP$_1$** is the one to use for *predictions.* The reasoning is as follows if it is the case that the theory implies that there really exist a large number of universes and if we are to make a prediction of some *additional* not yet measured physical quantity we already know that we are in a universe allowing for our presence. To predict a further not yet observed quantity of our universe we should only consider the set of universes that allow for our existence not the ones that allow for all kinds of radically different kinds of observers.

The above is sensible but I think we should take it one step further. When the multiverse is a consequence of the theory we are testing and we furthermore want to calculate the probability of observing a value of an observable, but not yet observed, quantity in our universe, we should use *all* the information we have about our universe, whether or not it is relevant for allowing for our existence. This is just straightforward use of conditional probability. It is important to note that when doing this our existence or consciousness does not play a specific part compared to any other properties of the universe we happen to inhabit. For this to work we need to have a very clear understanding of the specific multiverse that is suggested. When this strategy is applied the apparently mysterious use of considering our existence and consciousness in traditional anthropic arguments disappear. Instead we are just conditioning on everything we can observe. We should thus replace Smolin's scheme with the following,

(i) X = (Everything we have observed) is true in our world.
(ii) Based on the multiverse theory we are testing and conditioning on X we make a prediction P.

Here, I have not spelled out the form of the prediction, for instance whether it is a sharp prediction or a prediction about a range of allowed values of a quantity with associated probabilities. What the case would be depends on the specific multiverse that is tested. Note, that with this schema no version of the anthropic principle is really used for the purposes of making the prediction.

The general problem of making predictions in a multiverse has been discussed in Aguirre (2007). He considers a large number of possible strategies and what the right one to use is. The problem is described as finding the right measure for predictions. Among the strategies discussed by Aguirre the one here suggested namely to condition on everything is considered. He however sees some problems with this strategy. Namely, that it may happen that we have to accept a theory for which the observables we see in our universe are extremely rare. I do not see this as relevant for the question of *predictions*, we might turn out to be atypical; the important question is what the theory predicts given the observations we have already observed.

If predictions fail we have reason to abandon the theory. This is definitely the case if we observe something forbidden according to the theory. If we often observe something highly unlikely according to the theory we have some evidence against the theory although it is not completely falsified. The new observations will be taken into account for when further observations are done.

The worries Aguirre present and possible considerations of other strategies might be relevant for assessing whether a theory constitutes a good *explanation*. The way a theory has explanatory power can of course also be relevant for our acceptance. Tests of predictions are however more important and the success of a sharp prediction or even more so the failure of such a prediction can overrule considerations based on explanations.

The important point is to clearly distinguish between how predictions are made compared to questions about explanation. For predictions the strategy suggested here must be the right one. Still just giving the general strategy for predictions does not completely specify what the relevant measure we should use is. To have any chance of answering the question about the measure of course a detailed description of how the mechanism that is supposed to populate the universe must be given on the basis of the assumptions of the specific theory whose predictions are tested.

3.5 Summary and Conclusions

The very failure of string theory to provide a unique prediction and thus arguably point in the direction of a multiverse can of course *not* in itself be taken as evidence for string theory. However, the rhetoric of string theorists as expressed for instance in Susskind (2005, 2007) and Schellekens (2008) is deceptively used so as to turn the non-uniqueness of string theory predictions to be seen as counted in favour of string theory. In Schellekens (2008) it is argued that the view that our universe is unique is similar to previous misconceptions that our status as humans is special or unique. He compares this with the emperor's clothes from the classic fairy tale. Given that string theory itself has not yet been successfully tested empirically, sceptics might instead wonder whether the comparison with the emperor's clothes is not more apt for describing the status of string theory research itself. It seems that string theorists have made a virtue out of a necessity.

However, if a better and deeper understanding of the Landscape of string theory is achieved, combined with a proper understanding of what the mechanism is that populates the Landscape, then this might change. For this to work out and lead to testable predictions as suggested in this paper then the Landscape needs to be sufficiently constrained. On the other hand if the Landscape allow for too many solutions without any constraints string theory might turn out to be untestable. This is the scenario Smolin (2007) find to be most plausible. We will have to see what further investigations and calculations tell us; presently I think it is an open question whether or not the Landscape will turn out to be sufficiently constrained.

So all in all, string theory and its multiverse does not yet qualify to count as well-established science. On the other hand string theorists have at least some understanding of what the dynamics are supposed to be that determines the quantities that may differ in different universes, this is not something that can be said about most alternative multiverse theories.

In the future string theory, or some other alternative theory, might manage to live up to the criteria of this paper and thus make a multiverse view of reality truly scientifically respectable.

Acknowledgments The author would like to thank Richard Dawid and Christian Wüthrich who also contributed to the symposium on *Methodological challenges in quantum gravity*, which was part of the conference *Philosophy of Science Between the Natural Sciences the Social Sciences and the Humanities* organized by GWP and held in Düsseldorf, March 8–11, 2016.

This work was performed under a collaborative agreement between the University of Illinois at Chicago and the University of Geneva and made possible by grant number 56314 from the John Templeton Foundation and its contents are solely the responsibility of the author and do not necessarily represent the official views of the John Templeton Foundation.

References

Aguirre, A. 2007. Making predictions in a multiverse. In *Universe or multiverse?* ed. B. Carr (2007). Cambridge: Cambridge University Press.

Barrow, J.D., and F. Tipler. 1986. *The anthropic cosmological principle*. Oxford: Oxford University Press.

Carr, B., ed. 2007. *Universe or multiverse?* Cambridge: Cambridge University Press.

Carter, B. 1974. Large number coincidences and the anthropic principle in cosmology. In *Confrontation of cosmological theories with observational data*, ed. Longair. M (1974). Dordrecht: D. Reidel.

Collins, R. 2007. The multiverse hypothesis: A theistic perspective. In *Universe or multiverse?* ed. B. Carr. Cambridge: Cambridge University Press.

Dawid, R. 2013. *String theory and the scientific method*. Cambridge: Cambridge University Press.

Green, M.B., J.H. Schwarz, and E. Witten. 1987. *Superstring theory: 2 volumes*. Cambridge: Cambridge University Press.

Greene, B. 1999. *The elegant universe*. London: Jonathan Cape.

———. 2004. *The fabric of the cosmos: Space, time and the texture of reality*. London: Allen Lane.

Hoyle, F. 1953. On the fragmentation of gas clouds into galaxies and stars. *The Astrophysical Journal* 118: 513.

Kachru, S., R. Kalloch, A. Linde, and S.P. Trivedi. 2003. De Sitter vacua in string theory. *Physical Review D* 68: 046005.

Lewis, D. 1986. *On the plurality of worlds*. Oxford: Basil Black well.

Linde, A. 2007. The inflationary multiverse. In *Universe or multiverse?* ed. B. Carr. Cambridge: Cambridge Universtiy Press.

Longair, M., ed. 1974. *Confrontation of cosmological theories with observational data*. Dordrecht: D. Reidel.

Page, D.N. 2007. Predictions and tests of multiverse theories. In *Universe or multiverse?* ed. B. Carr. Cambridge: Cambridge University Press.

Polchinski, J. 1998. *String theory: 2 volumes*. Cambridge: Cambridge University Press.

Schellekens, A.N. 2008. The emperors last clothes. *Reports on Progress in Physics*, *71*, 072201. Extended version: http://arxiv.org/pdf/0807.3249v3

Smolin, L. 2006. *The trouble with physics*. Boston: Houghton Mifflin Company.

———. 2007. *Scientific alternatives to the anthropic principle*. In *Universe or multiverse?* ed. Carr, B (2007). Cambridge: Cambridge University Press.

Susskind, L. 2005. *The cosmic landscape: String theory and the illusion of intelligent design*. New York: Little, Brown and Company.

———. 2007. *The anthropic landscape of string theory*. In *Universe or multiverse?* ed. Carr, B (2007). Cambridge: Cambridge University Press.
Weinberg, S. 1987. Anthropic bound on the cosmological constant. *Physical Review Letters* 59: 2067.
Weinstein, S. 2006. Anthropic reasoning in multiverse and string theory. *Classical and Quantum Gravity* 23: 231.
Woit, P. 2006. *Not even wrong*. London: Jonathan Cape.
Zwiebach, B. 2004. *A first course in string theory*. Cambridge: Cambridge University Press.

Chapter 4
The Clock Paradox: Luise Lange's Discussion

Andrea Reichenberger

Abstract In her articles on the clock paradox and the relativity of time Luise Lange (1891–1978) defends the theory of relativity against philosophical refutations, by showing that the apparent clock paradox is not a paradox, but merely conflicts with common sense and is based on a misunderstanding of the theory. The following study explores, contextualizes and analyzes Lange's clear and sophisticated contribution to the debate on the clock paradox for the first time.

Keywords Clock paradox · Twin paradox · Theory of relativity · Time dilation

4.1 The Clock Paradox: Luise Lange's Solution and Its Context

At the beginning of the 1920s great interest in Einstein's theory of relativity flared up in broad circles. Amidst this "Relativitätsrummel," as it was called by Arnold Sommerfeld (qtd. Hentschel 1990, 67), two papers on "The Clock Paradox of the Theory of Relativity" (Lange 1927a) and "On a Misconception of the Relativity of Time" (Lange 1927b) were published. The author was Luise Lange (1891–1978).[1]

[1]Luise Lange was born in Braunschweig on April 14, 1891. She studied at the Braunschweig University of Technology and at the Göttingen University and completed her doctorate in 1923. The examiners were Peter Josef William Debye (physics), Adolf Windaus (chemistry), and Constantin Carathéodory (mathematics). She also prepared David Hilbert's *Vorlesung zur Statistisches Mechanik* of the summer semester 1914 (Hilbert 1914). After her emigration to the USA, she worked at the Oxford Female College, a small women's college in Oxford, Ohio and later at the Woodrow Wilson College, Chicago. I gratefully acknowledge Katie Ehrlich, Adjunct Archivist for Access and Outreach at New York University Archives, who provided me insight

A. Reichenberger (✉)
Department of Social and Human Sciences: Philosophy: Center for the History of Women Philosophers and Scientists, Paderborn University, Paderborn, Germany
e-mail: andrea.reichenberger@upb.de

A. Christian et al. (eds.), *Philosophy of Science*, European Studies in Philosophy of Science 9, https://doi.org/10.1007/978-3-319-72577-2_4

In both works Luise Lange argued that the clock paradox does not invalidate that theory, but arises "by one or several misapplications of the theory of relativity" (Lange 1927a, 25).

Consider clocks brought together in the same inertial reference frame and synchronized. What happens if one clock moves away in a spaceship and then returns? The answer is well known today: The clock which had undergone the traveling would be found to lag behind the clock which stays put. This phenomenon is a necessary consequence of the relativity of simultaneity and time dilation. As Albert Einstein put it (Einstein 1905, 904 f.):

> If there are two synchronous clocks at A, and one of them is moved along a closed curve with constant velocity until it has returned to A, which takes, say t seconds, then this clock will lag on its arrival at A by ½ t(v/V)^2 seconds behind the clock that has not been moved.[2]

The core of Einstein's 1905 paper is the derivation of Lorentz invariance on just the two basic principles of relativity and light-speed invariance. It is often claimed that special relativity has replaced the conventional notion of an absolute universal time with the notion of time that is dependent on reference frame and spatial position. However, it is not a new concept of time, but a new conception and conceptual framework of spatiotemporal relations that emerges from special relativity, i.e. the Lorentz transformations preserve the space-time interval, an algebraic combination of space and time intervals when viewed from any inertial reference frame, not the invariant time interval between two events. Suppose an observer measures two events as being separated by a time Δt and a spatial distance Δx. Then the space-time interval Δs^2 between the two events that are separated by a distance Δx in space and a duration Δt in time is $\Delta s = \sqrt{c^2 + \Delta t^2 - \Delta d^2}$.

In a lecture to the *Naturforschende Gesellschaft* on 16 January 1911, Einstein noted that the same effect of time dilation would also apply to living organisms travelling at very high velocities (Einstein 1911, 12):

> We must add that what holds true for this clock we have introduced as a simple representative of all physical progress also applies to any closed physical system. If we placed a living organism in a box [...] one could arrange that the organism, after any arbitrary lengthy flight, could be returned to its original spot in a scarcely altered condition, while corresponding organisms which had remained in their original positions had already long

into the Richard Courant Papers MC 150, including the correspondence between Luise Lange and Richard Courant from 1939 to 1942. I would also like to express my grateful thanks to Ulrich Hunger from the Göttingen University Archive and Renate Tobies, Friedrich Schiller University of Jena, for their helpful comments and hints. A special thanks is also extended to Edgar E. Enochs from University of Kentucky Lexington, KY USA, who got to know Luise Lange during his time at the University of Chicago, from 1958 to 1960. The publication of my research results on Luise Lange is in preparation.

[2] "Befinden sich in A zwei synchron gehende Uhren und bewegt man die eine derselben auf einer geschlossenen Kurve mit konstanter Geschwindigkeit, bis sie wieder nach A zurückkommt, was t Sek. dauern möge, so geht die letztere Uhr bei ihrer Ankunft gegenüber der unbewegt gebliebenen um ½ t(v/V)^2 Sek. nach." Unless otherwise noted, all translations in the text are my own.

> since given way to new generations. For the moving organism, the lengthy time of the journey was a mere instant, provided the motion took place with approximately the speed of light.[3]

In his talk « L'evolution de l'espace et du temps » on 11 April 1911, Paul Langevin used a similar example, supposing two arbitrary events in the history of an element of matter (Langevin 1911, 48):

> Suppose that two portions of matter meet for the first time, separate and meet again. We can say that observers attached to the portions during separation, will not evaluate that duration in the same way, as some have not aged as much as the others.[4]

Langevin's considerations stimulated Einstein to consider time delays as more than just affecting clocks. In an unpublished manuscript written after Langevin's talk Einstein assumed that relativity theory described "the temporal course of no matter which process" (qtd. Canales 2015, 57). Hermann Weyl, among others, supported this assertion in his famous book *Raum. Zeit. Materie. Vorlesungen über allgemeine Relativitätstheorie*: "The life processes of mankind may well be compared to a clock" (Weyl 1919, 187). Weyl was probably the first who mention the twins in the context of the so-called paradox (ibid., 158):

> Suppose we have two twin-brothers who take leave one another at a world-point A, and suppose one remains at home (that is, permanently at rest in an allowable reference-space), whilst the other sets out on voyages, during which he moves with velocities (relative to "home") that approximate to that of light. When the traveler returns home in later years he will appear appreciably younger than the one who stayed at home.[5]

Langevin's identification of the time of clocks with human life times triggered a heated debate which reached its climax in the early 1920s, when Henri Bergson attacked Albert Einstein at a meeting of the *Société française de philosophie* in Paris. Bergson found Einstein's definition of time in terms of clocks completely inappropriate. Clocks, by themselves, could not explain either simultaneity or time.

---

[3]"Man muss hinzufügen, dass das, was für diese Uhr gilt, welche wir als einen einfachen Repräsentanten alles physikalischen Geschehens eingeführt haben, auch gilt für ein in sich abgeschlossenes physikalisches System irgendwelcher anderer Beschaffenheit. Wenn wir z.B. einen lebenden Organismus in eine Schachtel hineinbrächten [...], so könnte man es erreichen, dass dieser Organismus nach einem beliebig langen Fluge beliebig wenig geändert wieder an seinen ursprünglichen Ort zurückkehrt, während ganz entsprechend beschaffene Organismen, welche an den ursprünglichen Orten ruhend geblieben sind, bereits längst neuen Generationen Platz gemacht haben. Für den bewegten Organismus war die lange Zeit der Reise nur ein Augenblick, falls die Bewegung annähernd mit Lichtgeschwindigkeit erfolgte!"

[4]«Supposons que deux portions de matière se rencontrent une première fois, se séparent, puis se retrouvent. Nous pouvons affirmer que des observateurs liés à l'une et à l'autre pendant la séparation n'auront pas évalué de la même manière la durée de celle-ci, n'auront pas vieilli autant les uns que les autres.»

[5]"Von zwei Zwillingsbrüdern, die sich in einem Weltpunkt A trennen, bleibe der eine in der Heimat (d.h. ruhe dauernd in einem tauglichen Bezugsraum, der andere aber unternehme Reisen, bei denen er Geschwindigkeiten (relativ zur Heimat) entwickelt, die der Lichtgeschwindigkeit nahe kommen; dann wird sich der Reisende, wenn er dereinst in die Heimat zurückkehrt, als merklich jünger herausstellen als der Seßhafte."

Bergson argued that the concept of universal time arises from our own "proper" experienced time which he called "real duration" in contrast to the mechanistic time of science (see Bergson 1922). Einstein replied that the notion of simultaneity of individual perception should not be confused with the simultaneity of objective events independent of individuals as it is used in physics.

Luise Lange defended the theory of relativity against such philosophical refutations, by showing that Bergon's criticism is based on a misunderstanding of the theory. Her clear and sophisticated contribution to the debate is completely forgotten and was never discussed nor mentioned in the literature. Lange writes (1927a, 24):

> In 1911 P. Langevin formulated and solved the problem in the since well-known way: Peter stands on the earth, Paul, enclosed in a bullet, is fired away with tremendous velocity, but due to a suitable encounter with some star his path is directed back to the earth. On both journeys from and to the earth his clock is slow in comparison to those at rest to the earth; (as say the L. E.); hence in returning his time is behind earth-time in the ratio $(1 - v^2c^{-2})^{-1/2}$: 1. If he stayed away 2 years according to his time that elapsed on the earth is, say, two hundred years, if the velocity with which he performed his cosmic journey was such that $(1 - v^2c^{-2})^{-1/2}$: 0.01. While Paul has aged only slightly during his absence, the sixth generation of Peter's descendants are already populating the planet.

One might think that each twin should see the other aging more slowly, and so the paradox arises that each believes the other should be younger at their reunion. However, whereas time dilation is symmetrical, the travelling twin's trajectory involves two different inertial frames, and so there is no symmetry between the space-time paths of the two twins. The paradox centers, as Luise Lange rightly stated, on the assertion that, in relativity, either twin could regard the other as the traveler, in which case each should find the other younger – a logical contradiction. The logical fallacy lies in the incorrect assumption that the twins' situations are symmetrical and interchangeable which is indeed false. To quote Luise Lange (ibid.):

> This result is frequently spoken of as Langevin's Paradox, but unjustly; for there is nothing self-contradictory in it, contrary as it may be to customary notions. It turned paradox only in the later development.

In this context Luise Lange attacked the "fanciful stories" of Ernst Gehrcke (1920) and Karl Vogtherr (1923) who advocated an anti-relativity propaganda campaign (Lange 1927b, 501):

> We now want to show that, no matter what revolutions in our concepts of space and time the theory of relativity may have brought about it is wholly innocent as regards the above fanciful stories. That it neither demands nor even admits the dependence of sense perception on the relative state of motion of the observer.

Luise Lange pointed out that the notion "observer" is a technical one. It has nothing to do with the sense perception of human beings, but refers to measurement procedures, clocks and measuring rods, based on the invariance of the speed of light, whereby Lorentz transformation is a method of measurement of a distance in space-time. Thus, the concept of proper time does not imply the claim that just the present moment experienced by the observer is real (ibid., 506):

> Observers then in relative motion disagree on the position of the clock hands of a distant clock not because they receive different visual impressions from it, but because they account differently for the time of light transmission. The misunderstanding of this point seems to have arisen because this one fact is not kept in mind: the time of a distant event is never – in classical mechanics as little as in relativity mechanics – directly observed or experienced; but it is determined only by a combination of measuring (which involves perception) and *computation*. In the terminology of the theory of relativity, it is true, the computation is frequently disguised as a "nothing but observation," due namely to the underlying scheme if imagining observers at every place "right next to clock" synchronized with all others by means of a light signal. That way the computation is simply performed in advance and once for all by all the observers who at places x set their clocks on $t = \frac{x}{c}$ on receiving the signal sent from $x = 0$ at $t = 0$. According to our theory it is only this computation, this dating back into the past of an event observed in the present, which contains the relative element, not the sense perception and thus we have come to see that the theory of relativity in no way implies the doctrine of solipsism.

To conclude, special theory of relativity is not inconsistent as it might appear at first glance. As Luise Lange put it, "there is nothing self-contradictory in it" (Lange 1927a, 24). The paradox arises from a false premise of the motional symmetry of the twins.

4.2 Two Incorrect Solutions

In her discussion of time dilation Luise Lange presented not only the correct solution of the clock/twin paradox. She also examined two false solutions: (i) acceleration is essential in order to explain this thought experiment; (ii) a complete solution of the paradox can just be found in the frame of general relativity. Luise Lange argued that both claims are false. Neither acceleration nor general relativity are necessary ingredients for the resolution of the twin paradox. Nevertheless, they are possible and interesting cases.

In the standard formulation of the twin paradox acceleration is responsible for the asymmetric aging effect. But it is also possible to explain the scenario in another way, be switching inertial reference frames which slows the clock of the traveling twin down. There is a difference between the trajectories of the twins: the trajectory of the traveling twin is equally divided between two different inertial frames, while the Earth-based twin stays in the same inertial frame. Another way to look at it is in terms of what are known as word lines which are the path an object follows in space-time. The shift of the word line is such that the resulting world line of the travelling twin is shorter than the one of the stay at home twin.

Max von Laue, to whom Luise Lange refers, used this argumentation in order to elaborate on Langevin's explanation. He demonstrated that the world lines of the inertially moving bodies maximize the proper time elapsed between two events and concluded that the asymmetric aging is completely accounted for by the fact that the astronaut twin travels in two separate frames, while the Earth twin remains in one frame, and the time of acceleration can be made arbitrarily small compared with the time of inertial motion (Laue 1911).

Thus, the resolution of the paradox can be given purely in special relativity within a flat (Minkowski) space-time. However, Luise Lange also asked: "What about the general theory?" In his "Dialog über die Einwände gegen die Relativitätstheorie" (Einstein 1918), Einstein used gravitational time dilation to resolve the twin paradox. According to general theory of relativity, clocks low in a gravitational field run slower than clocks high in a gravitational field. Einstein invoked the equivalence principle and a homogenous gravitational field equivalent to the acceleration as the physical cause of asymmetrical time dilation.

In her discussion of Einstein's dialogue, Luise Lange made clear that the compensating gravitational time dilation is due to the difference in gravitational potential at two points in the field rather than being an effect of the accelerating motion itself. In the case of special relativity, it is the difference in the paths that results in a time dilation for the accelerated twin; analogously, in the case of general relativity, the compensating gravitational time dilation is due to the difference in gravitational potential at two points in the field rather than being an effect of the accelerating motion itself. Luise Lange showed that Max Born (1924) and August Kopff (1923) misunderstood this fact.

4.3 Conclusion

The twin paradox, or, more general, the clock paradox has been a subject of a heated discussion for over hundred years and continues to attract physicists and philosophers presenting new insight or defending old theses. In his highly-regarded book *Relativity: Special, General and Cosmological* Wolfgang Rindler (2001, 67) writes:

> Reams of literature were written on it unnecessarily for more than six decades. At its root apparently lay a deep psychological barrier to accepting time dilatation is real. From a modern point of view, it is difficult to understand the earlier fascination with this problem or even to recognize it as a problem.

Even though nowadays many physicists may consider the matter only of educational interest, the debate is ongoing. Why? Is it just because of the unfamiliarity with the historical discussion and context? I don't think so. The key to a solution of a paradox often lies in clarifying misconceptions and identifying mistaken premises. In a certain manner this is a never-ending process, beyond pure pedagogical challenges. To conclude with an example, namely the concept of "proper time." Proper time, expressed as an integral over the world line, is invariant, i.e. it takes the same value in all inertial references. Luise Lange noted with regard to proper time (Lange 1927a, 29f.):

> Thus our survey has led us to the conclusion that the question regarding the difference in duration experienced in two relatively moving systems between two encounters is as yet

unanswered. It may be, as many relativists assert, that in worlds of different state of motion the derôlement of time is different, like on motion pictures reeled through slower and faster, with the inert system being the fastest ones, but we do not see that the complete proof for it has yet been given.

The crucial point here is: Proper time is independent of coordinates, but depends on world lines. It is expressed as an integral over the world line. The time difference when the twins meet again does not come from comparing local rates of time along their world lines; it results from integrating the proper time along the world lines and concluding that their paths through space-time have different lengths. In other words, we evaluate a path integral along two different paths, and get two different results. An accelerated clock will measure a smaller elapsed time between two events than that measured by a non-accelerated (inertial) clock between the same two events. The twin paradox is an example of this effect. In this sense, the solution of the paradox is given by clarifying a common misunderstanding regarding the concept of proper time. This is exactly what Luise Lange did in her papers on the clock paradox and the relativity of time.

References

Bergson, H. 1922. *Durée et simultanéité. A propos de la théorie d'Einstein*. Paris: Alcan.

Born, M. 1924. *Einstein's theory of relativity*. London: Methuen.

Canales, J. 2015. *The physicist and the philosopher: Einstein, Bergson, and the debate that changed our understanding of time*. Princeton: Princeton University Press.

Einstein, A. 1905. Zur Elektrodynamik bewegter Körper. *Annalen der Physik* 322 (10): 891–921.

———. 1911. Die Relativitäts-Theorie. *Naturforschende Gesellschaft* 56: 1–14.

———. 1918. Dialog über die Einwände gegen die Relativitätstheorie. *Die Naturwissenschaften* 6 (48): 697–702.

Gehrcke, E. 1920. *Die Relativitätstheorie – Eine wissenschaftliche Massensuggestion*. Berlin: Schriften aus dem Verlag der Arbeitsgem. Dt. Naturforscher zur Erhaltung reiner Wissenschaft.

Hentschel, K. 1990. *Interpretationen und Fehlinterpretationen der speziellen und der allgemeinen Relativitätstheorie durch Zeitgenossen Albert Einsteins*. Basel: Birkhäuser.

Hilbert, D. 1914. *Statistische Mechanik. Vorlesung SS 1914* (1914) Ausarb. v. Luise Lange. 81 Bl.o.O. xi, 5; Maschinenschr. Bibliotheksarchiv Georg-August-Universität Göttingen.

Kopff, A. 1923. *The mathematical theory of relativity*. London: Methuen.

Lange, L. 1927a. The Clock Paradox in the theory of relativity. *The American Mathematical Monthly* 34: 22–30.

———. 1927b. On a misconception of the relativity of time. *School Science and Mathematics* 27 (5): 500–506.

Langevin, P. 1911. L'évolution de lespace et du temps. *Scientia* 19 (3): 31–54.

Laue, Max von. 1911. Zwei Einwände gegen die Relativitätstheorie und ihre Widerlegung. *Physikalische Zeitschrift* 13: 118–120.

Rindler, Wolfgang. 2001. *Relativity: Special, general and cosmological*. Oxford: Oxford University Press.

Vogtherr, Karl. 1923. *Wohin führt die Relativitätstheorie? Kritische Betrachtungen vom physikalischen und erkenntnisreichen Standpunkt aus*. Leipzig: Hillmann.

Weyl, Hermann. 1919. *Raum. Zeit. Materie. Vorlesungen über allgemeine Relativitätstheorie*. 3rd ed. Berlin: Springer.

Part II
Philosophy of Life Sciences

Chapter 5
Bio-Agency and the Possibility of Artificial Agents

Anne Sophie Meincke

Abstract Within the philosophy of biology, recently promising steps have been made towards a biologically grounded concept of agency. Agency is described as bio-agency: the intrinsically normative adaptive behaviour of human and non-human organisms, arising from their biological autonomy. My paper assesses the bio-agency approach by examining criticism recently directed by its proponents against the project of embodied robotics. Defenders of the bio-agency approach have claimed that embodied robots do not, and for fundamental reasons cannot, qualify as artificial agents because they do not fully realise biological autonomy. More particularly, it has been claimed that embodied robots fail to be agents because agency essentially requires metabolism. I shall argue that this criticism, while being valuable in bringing to the fore important differences between bio-agents and existing embodied robots, nevertheless is too strong. It relies on inferences from agency-as-we-know-it to agency-as-it-could-be which are justified neither empirically nor conceptually.

Keywords Bio-agency · Biological autonomy · Metabolism · Normativity · Embodied robotics · Artificial life

5.1 Introduction

Within the philosophy of biology, recently promising steps have been made towards a biologically grounded concept of agency. A growing number of philosophers take it that the key to an appropriate understanding of human agency is to acknowledge human agency as a sophisticated form of *bio-agency*: the agency performed by bio-agents, i.e., by organisms of different kinds (Barandiaran and Moreno 2008; Barandiaran et al. 2009; Skewes and Hooker 2009; Arnellos et al. 2010;

A. S. Meincke (✉)
Department of Sociology, Philosophy and Anthropology, University of Exeter, Exeter, UK
e-mail: a.s.meincke@exeter.ac.uk

A. Christian et al. (eds.), *Philosophy of Science*, European Studies in Philosophy of Science 9, https://doi.org/10.1007/978-3-319-72577-2_5

Campbell 2010; Arnellos and Moreno 2015; Moreno and Mossio 2015). This new trend within the philosophy of biology relies crucially on recent (non-reductionist) endeavours in systems biology to gain a comprehensive understanding of the organism as a functioning system (Alberghina and Westerhoff 2005; O'Malley and Dupré 2005; Noble 2006; Boogerd et al. 2007), in particular on the concept of biological autonomy (Rosslenbroich 2014, Moreno and Mossio 2015), which itself draws on the older influential concept of 'autopoiesis' (self-production) (Varela 1979; Maturana and Varela 1980). The claim is that bio-agency arises from the biological autonomy of organisms, that is, from the ability of organisms to actively maintain themselves through an on-going interaction with the environment so as to adjust to changes in the environment. Bio-agency is some form of adaptive behaviour, grounded in biological autonomy, and, depending on the level of complexity of autonomy, being more or less sophisticated.[1]

Acknowledging human agency as bio-agency has far-reaching implications for the philosophy of action as a subfield of metaphysics. Most importantly, it nicely addresses the prominent problem of how actions could fit into the natural world (Bishop 1989) by accounting for these in fully biological terms, avoiding the eliminativist bias of common physics-orientated approaches (Spann 2014; Meincke in preparation).[2] Metaphysicians have just started to discover the biological nature of human agency (Steward 2012), without, however, taking into account any insights from systems biology and without connecting themselves to the lively debate in the philosophy of biology on these findings and the resulting models of bio-agency. Filling this lacuna is overdue in order to lastingly overcome the tacit anthropocentrism and Cartesianism of the debate.[3]

However, there are also lessons for the philosophy of biology to be learnt from more general philosophical, or metaphysical, considerations about agency, as I shall argue in this paper. Given that the most common instance of agency in our world is bio-agency, one might feel tempted to generalise the bio-agency approach. But is *all* agency bio-agency?

In what follows, I shall discuss this question with respect to recent attempts in AI research to build artificial agents, especially in so-called embodied robotics.

[1]Barandiaran and Moreno (2008) and Moreno and Mossio (2015) distinguish between adaptive agency (manifest in the generic ability of organisms to adjust interactive processes to changes in the environment) and behavioural agency (adaptive agency involving motility). As a form of minimal (behavioural) agency they discuss the chemotaxis of bacteria, such as E. coli. What exact distinctions within the spectrum are drawn depends on the criteria deployed, which is a subject of debate and varies among authors. See also the following discussion in this chapter.

[2]It also offers a new and attractive stance on the problem of free will, by demystifying free will as a means of the survival and well-being of organisms (Skewes and Hooker 2009; Brembs 2011; Dupré 2013; Meincke in preparation b).

[3]Even though it is widely assumed that today's philosophy of mind has moved beyond classical Cartesian dualism, when it comes to agency, the debate tends to follow the old tracks by either treating human agency as a unique exception to an otherwise presumably completely mechanistic universe or rather explaining it in mechanistic terms and thereby effectively eliminating it, see Meincke (in preparation b) and Spann (2014).

The project of embodied robotics rests upon the idea that cognition is embodied in the sense that it depends on the whole of an agent's physical body rather than only on the brain.[4] As such, this project is itself importantly inspired by biology: instead of emulating the higher cognitive functions typical of humans, the aim is to artificially realise the basic motility skills found in animals, which are taken to be a prerequisite for any higher cognitive functions (Brooks 1991a)[5] and thus to pose the true and even harder challenge for AI (Brooks 1991a; Varela 1995).[6] Embodied robots are meant to be artificial agents whose actions are situated in an environment via sensorimotor loops so that action provides occasions for perception and perception guides action (Brooks 1991a, b; Steels and Brooks 1995; Clark 1997). This corresponds to an important extent with the tenets of so-called enactivism. Enactivism, which substantially contributed towards the 'embodiment turn' in AI, is the view that cognition is a biological function, arising through a dynamic interaction between an acting organism and its environment, namely, according to the original thesis, as part of an organism's autopoiesis (Varela et al. 1991; Thompson 2007; also Varela 1996, 1997).[7]

Embodied robotics as a research programme has recently become the subject of criticism, by philosophers endorsing a (systems) biological view of agency consistent with (autopoietic) enactivism. Do embodied robots really qualify as agents? Ezequiel di Paolo (2003, 2010), Alvaro Moreno and Arantza Etxeberria (2005) have denied this, claiming that embodied robots as constructed so far fail to be agents to the extent that they fail to fully realise biological autonomy. As di Paolo puts it, "biologically-inspired robots are not necessarily *organismically-inspired*, i.e., inspiration has not come from the defining organisation and conditions of the living, or some of its corollaries" (di Paolo 2003, 4f./2010, 135; italics in the original). Situated activity via sensorimotor loops as realised by embodied robots is necessary but, according to di Paolo, Moreno und Etxeberria, not sufficient for agency; something crucial is missing: the realisation of 'basic autonomy' or autopoiesis.[8]

[4]See Shapiro (2010) for a recent comprehensive discussion of the field of embodied cognition.

[5]"I believe that mobility, acute vision and the ability to carry out survivalrelated tasks in a dynamic environment provide a necessary basis for the development of true intelligence" (Brooks 1991a, 141).

[6]"This part of intelligence" – "the ability to move around in a dynamic environment, sensing the surroundings to a degree sufficient to achieve the necessary maintenance of life and reproduction" – "is where evolution has concentrated its time – and it is much harder" (Brooks 1991a, 141). Compare also Varela's twofold motto of the "disenchantment of the abstract" and the "re-enchantment of the concrete" (Varela 1995).

[7]'Autopoietic enactivism' was the historically first version of 'enactivism' but nowadays competes with 'sensorimotor enactivism' and 'radical enactivism'; see de Jesus (2016) for a recent critical discussion.

[8]The criticism of embodied robotics' failure to artificially realise agency is part of a more general criticism according to which embodied robotics fails to artificially realise cognition (Barandiaran

Against the background of this recent systems biological criticism of embodied robotics, the question raised above as to the level of generality of the bio-agency approach – 'Is *all* agency bio-agency?' – transforms into a question about the possibility of artificial agency – 'Can there be artificial agents?'. The aim of this paper is to get some idea about the answer to the former by investigating the latter with respect to the research programme of embodied robotics. For this purpose, I shall look in particular at the arguments put forward by Moreno and Etxeberria against the possibility of artificial agents within the framework of embodied robotics. Moreno and Etxeberria (2005) claim that the thesis that there cannot be artificial agency without fully realised biological autonomy comes down to the thesis that there cannot be agency without metabolism.

I shall proceed as follows: In Sect. 5.2, I shall first introduce two aspects of biological autonomy distinguished by Moreno and Etxeberria, 'basic autonomy' and adaptive interaction. As I shall show, this distinction plays an important role in their analysis of the evolution of bio-agency from metabolism (Sect. 5.2.1) and provides the basis for the thesis that embodied robots do not act because they do not metabolise (Sect. 5.2.2). In Sect. 5.3, I shall review Moreno and Etxeberria's 'argument from metabolism against artificial agency' (as I would like to call it) by examining its premises. I shall argue that Moreno and Etxeberria doubtlessly deserve credit for drawing attention to a fundamental difference between the ontological constitution of the behaviour of embodied robots on the one hand and of bio-agents on the other (Sect. 5.3.1), but that their categorical dismissal of embodied robots as agents is nevertheless too strong. It is not clear that the – overall convincing – arguments offered to elucidate the nature of agency as-we-know-it can be generalised so as to apply also to agency as-it-might-be (which reveals the case of artificial agency to be parallel to the case of artificial life).[9] In particular, Moreno and Etxeberria seem to make some questionable assumptions about the nature of matter in connection with their emphasis on metabolism as pivotal for basic autonomy (Sect. 5.3.2) and to operate with a concept of agency that, according to general philosophical considerations, is too closely tied to metabolism and what the authors call intrinsic normativity (Sect. 5.3.3). I shall conclude, in Sect. 5.4, by explicating some of the implications of this result for the project of embodied robotics, for the bio-agency approach as well as for our general understanding of agency.

---

and Moreno 2006; Ziemke 2016). In this paper I shall focus on the agency-part of the systems biological criticism of embodied robotics. A careful and instructive discussion of both aspects of this criticism is offered by Froese and Ziemke (2009).

[9]Compare Chris Langton's well-known distinction between "life-as-we-know-it" and "life-as-it-could-be", the latter of which he claimed to be the object of study of artificial life (so-called ALife) (Langton 1989, 1).

5.2 The Systems Biological Criticism of Embodied Robotics

What are "the conditions for agency in natural and artificial systems" (Moreno and Etxeberria 2005, 161)? According to Moreno and Etxeberria, we will not come to the right conclusions regarding the latter if we start from wrong assumptions about the former, and that is in particular, if we fail to make an important distinction. This is the distinction between so-called 'basic autonomy' achieved through metabolic self-constitution on the one hand and adaptive interaction with the environment on the other as two aspects of the same phenomenon: 'biological autonomy', which Moreno and Etxeberria, in accordance with the bio-agency approach, take to be the ontological source of natural agency (2005, 162).

By insisting on the necessity to distinguish two aspects of biological autonomy, Moreno and Etxeberria aim at avoiding a "confusion (2005, 162)" which they find likely to arise from Maturana and Varela's concept of autopoiesis. Autopoiesis, in Moreno and Etxeberria's formulation, denotes the process in which "an autonomous system creates and maintains its identity as a dynamic network of component production that builds a membrane or boundary between the system inside and the dynamics of the environment" (2005, 162). Insofar as the boundarys, and, hence, the autonomous system, exists only as long as the system interacts with the environment, "all interactions with the environment may be ultimately referred to the task of preserving an autonomous being" (2005, 162). However, it would be mistaken to conclude from this that cognition – which enactivism takes to arise from an organism's interactions with the environment – and life – which Maturana and Varela define in terms of autopoiesis – are "properties of the same kind of organization" (2005, 162). This would overlook the emergence of the nervous system in the course of the evolution of motility-based living systems.

5.2.1 The Emancipation of Agency from Metabolism

According to Moreno and Etxeberria, the evolution of sensorimotor systems in organisms has to be described as a stepwise decoupling of adaptive interaction from basic autonomy and that is, as a stepwise evolution, or emancipation, of agency from metabolism. Moreno and Etxeberria speculate that the first organisms on earth were "minimal agents" in the sense that in them material self-construction and adaptive interaction indeed coincided. These organisms were chemical systems which just acted *by metabolising*, i.e., by performing a selective transport of matter and energy through a semi-permeable membrane (2005, 164). "Active transport", using energy currencies (ATP, PpI) and catalysts, Moreno and Etxeberria take to be "the most basic form of agency" (2005, 164). As a consequence, "saying", with respect to those hypothetical systems, "that a given process has to do with the maintenance of the system (life) or with dealing with the environment (adaptation) is mostly a conceptual distinction" (2005, 162). Moreno and Etxeberria point out that the situation is not much different with existing and slightly more complex agents which possess some sort of sensorimotor systems. In prokaryotes, such as bacteria, sensorimotor responses to the environment are sustained by similar mechanisms as metabolic responses so that distinguishing both appears to be "arbitrary" (2005, 165). Motility is here not more than "an extension of the set of mechanisms required for self-maintenance"; "all adaptivity is based on the maintenance of a chemical organization through the control of energy flows" (ibid.).

This changes in eukaryotes where, due to the increase of size, a different organisation of motility needed to be developed so as to ensure fast sensorimotor coordination. The possibilities of sustaining motility solely through more complex sensorimotor devices (microtubules, different and more complex structure of eukaryotic cilia or flagella) together with an internal functional specialisation of cells (organelles, anterior-posterior symmetries) (2005, 165) are limited given the energetic and material cost which conflicts with accomplishing the very function of metabolic organisation, i.e., self-maintenance (ibid.). The solution to this problem is the evolution of the nervous system, made possible by the invention of a special type of cell: the neuron. Neurons are "specialized in connecting sensorimotor surfaces in a plastic, fast, and (metabolically speaking) cheap way" (2005, 166f.). Most importantly, rather than directly interacting with metabolic processes, neurons primarily interact with one another so that these "recurrent interactions within the network" give rise to the nervous system as an "independent (autonomous) domain of patterns" (2005, 167).

It is here that the distinction between basic autonomy and interaction with the environment ceases to be merely conceptual or arbitrary. "[I]ndividual lifetime interactions" of multicellular organisms "do not rely mainly on changes in body structure, but on the neural network", with behaviour – "functional body movement" – being "the most important adaptive capacity" (2005, 167). Agency in more complex organisms is supported by a system that is 'decoupled' in the sense that the rules guiding its interactions "are underdetermined by the dynamics of the remaining system"; the nervous system's dynamic is "not governed by the general metabolic organization, although it is structurally maintained by it" (ibid.). This is, according to Moreno and Etxeberria, what allowed for the evolution of ever more complex forms of agency, accompanied by a continuous complexification of body plans (e.g., evolution of the muscular system, changes of the body shape) (2005, 168). The coevolution of nervous system and body includes, at a later stage in evolution, a further decoupling within the nervous system itself, "leading to a self-sustained process of encephalization and producing the emergence of new levels of cognitive phenomena, such as emotions and awareness" (2005, 170). The mind has to be "understood as a kind of slower, second-order neurosomatic activity by means of which the animal is able to attain a basic sense of self" and which mainly relies on the "anticipatory (non-reactive) control of sensorimotor behavior", engaging in "virtual interactions with the environment" (ibid.).

As this latter characterisation of the mind indicates, it would be mistaken, according to Moreno and Etxeberria, to overemphasise the autonomy of the nervous system alongside with the cognitive functions it affords, and i.e., to overemphasise the emancipation of agency from its evolutionary origin, metabolism. The autonomy of the nervous system is "non-basic, or *derived*" (2005, 167; italics in the original), namely derived from the basic autonomy of metabolic self-constitution. This means that "[t]he nervous system cannot generate its own functionality by itself" (ibid.). Instead, its "dynamics only acquire functional meaning if embodied in a self-maintained organization, within a given environment", and that is, "when recruited for the tasks of (1) sensorimotor coordination of the animal in its environment, and (2) coordination and fine control of the organism's metabolic processes" (ibid.). Thus, even though what Moreno and Etxeberria call 'neural agency' (reappearing as 'behavioural agency' in Barandiaran and Moreno (2008, 336ff.)) relies on a system different from metabolic self-constitution, it does not follow "that the metabolic substrate of behaving organisms is irrelevant" (Barandiaran and Moreno 2008, 337). The nervous system's functions are locally decoupled from, but globally still coupled to metabolism "so that adaptive behaviour will ultimately have to satisfy the demands of metabolism" (ibid.). In other words, tigers don't go hunting for fun but in order to keep their metabolism going. And they couldn't go hunting if their metabolism did not keep going while they hunt.

5.2.2 No Agency Without Metabolism

On the basis of this analysis of "the origin and evolution of agency in living systems" (Moreno and Etxeberria 2005, 162), Moreno and Etxeberria put forward their criticism of embodied robotics. Biological autonomy, we have been told, comprises two aspects: basic autonomy through material self-constitution and adaptivity through interaction with the environment. In higher organisms, these two aspects are supported by two different subsystems: metabolism on the one hand and the nervous system on the other, the latter giving rise to cognition as opposed to mere life as being provided by metabolism. This differentiation within the organisation of living systems is what Maturana and Varela's autopoiesis model, which focussed on minimal systems, left out. However, Maturana and Varela were nevertheless right, according to Moreno and Etxeberria, in presenting the organism's interactions with the environment as serving the means of self-constitution ('autopoiesis').

Moreno and Etxeberria argue that acknowledging the latter prevents us from overlooking how embodied robots fundamentally differ from organisms despite the sensorimotor embeddedness in the environment which they have in common with organisms. Organisms are fully or "truly" (2005, 162) autonomous in that their adaptive interactions with the environment are grounded in their basic autonomy. Their agency is grounded in metabolism; cognition is continuous with life. Embodied robots, however, are autonomous only in the second sense. By adaptively interacting with the environment they imitate the nervous system, so to speak, without realising its dependence on metabolism. "In the case of robots, the continuity between life and cognition is missing" (ibid.). Therefore they fail to be agents.

Given this diagnosis, we can think of two types of therapy. We could try to artificially create 'truly' autonomous robots by grounding their behaviour in a process of material self-constitution. Alternatively, we could simply deny that basic autonomy is needed for an embodied robot's interaction with the environment to qualify as agency.

According to Moreno and Etxeberria, the viability of the first of these options turns out to crucially depend on what matter scientists use. They put the question like this: is it "possible to build an organization similar to that of organisms on a different scale and with different material building blocks" (2005, 171)? Moreno and Etxeberria deny this, explaining that no "physical system", made from "inert aggregates", could ever achieve "(full) constructive closure" (ibid.). Only "biomolecular [. . .] components" (2005, 172) possess the "plasticity" (2005, 173) and "interactive capacities" (2005, 171) required for building "a recurrent network of component production", i.e., metabolism (ibid.). If this is true, it seems that in order to build agents we would have to build organisms in the first place. We would have to artificially create life.

However, maybe such high ambitions are not needed after all if we could show that "other organizing principles, different from the ones we have described for living systems, can produce systems that are similarly autonomous" (2005, 172). More specifically, Moreno and Etxeberria ask, can the decoupling of the nervous system and its dynamics from metabolism "be interpreted in constructive terms" and "autonomous behaviour [. . .] be successfully emulated at this sensorimotor level only" (ibid.)? Can there be agency without metabolism?

Again, Moreno and Etxeberria's answer is No. The fact that agency in living systems is grounded in metabolism is essential to its being agency. In organisms, "agency[] is continuous with [. . .] autopoiesis" in the sense that actions are both "a consequence *and a cause* of the structure of the system" (2005, 171; italics added). This "reciprocal causality" between "what it *is* and what it *does*" (2005, 173; see also di Paolo 2003, 4) is missing in the embodied robot. The robot does not need to act in order to survive; its material existence is independent of its actions. As a result, its actions lack *meaning* (Moreno and Etxeberria 2005, 162 and 173). Despite being situated in an environment by means of "real (or simulated) perception-action cycles", "it is not clear that the system has its own purposes or goals, as it is not the need to maintain its structure that constrains situatedness in its environment" (2005, 172).[10] It thus seems to follow that as long as embodied robots are not able to metabolise, and that is, as long as they are not exact copies of organisms, we cannot praise them for reproducing agency – at least not for "reproducing *true* agency" (2005, 173; italics added).

[10]Similarly di Paolo argues that the robot does not "*care*[] about what it is doing" (di Paolo 2003, 10; italics in the original) and, hence, fails to be an "intentional agent" (di Paolo 2003, 11). Its behaviour, despite being "embedded in a sensorimotor loop of situated interaction[,] [. . .] can be fully described simply as movement as opposed to action" (di Paolo 2003, 9).

5.3 No Agency Without Metabolism?

It is time to review Moreno and Etxeberria's 'argument from metabolism against artificial agency' which I take to have the following logical structure:

P1 Unlike the adaptive interaction with the environment performed by organisms, the adaptive interaction with the environment performed by embodied robots is not grounded in metabolic self-constitution and, i.e., not grounded in 'basic autonomy'.

P2 Embodied robots are not made from materials suitable for realising basic autonomy, i.e., they are not made from bio-molecular components that are able to actively organise themselves into a metabolic network.

P3 True agency must be grounded in basic autonomy realised via metabolism.

C Embodied robots are no agents and will never be agents.

5.3.1 A Fundamental Difference

Starting with P1, I think that Moreno and Etxeberria tightly direct our attention to a fundamental difference between the ontological constitution of the behaviour of organisms on the one hand and of embodied robots on the other. In the case of organisms, adaptive interaction with the environment, even though in higher organisms relying on a decoupled subsystem, is part of the process of organismic self-maintenance, which has its basis in metabolism. If the organism fails to interact with the environment in the right way, and that is, if it fails to perform the right actions, it will die and cease to exist; while the standard embodied robot happily exists whether or not it behaviourally interacts with the environment.[11]

This difference might perfectly well strike one as so fundamental as to appear to be likely also to impose a difference on the quality of the behaviour of organisms on the one hand and of embodied robots on the other. One therefore might reasonably think that embodied robotics, as it stands, cannot claim to build "physical artificial systems endowed with agential capacities similar to the ones that (natural) organisms exhibit" (Moreno and Etxeberria 2005, 170; see also 173), at least not if we assume that 'similar' means something like 'very much alike' or 'in principle the same'.[12] Embodied robotics, by emulating situated sensorimotor interactions with the environment rather than higher cognitive functions, is (to repeat di Paolo's

[11]The embodied robot's dependence for functioning on energy supply is no valid objection to this, given that functioning and existing in the case of the embodied robot, unlike in the case of the organism, do not coincide and given, furthermore, that dependence on energy supply does not qualify as adaptive interaction in the (active) sense of behaviour.

[12]I shall come back to some problematic aspects of this assumption in the conclusions of this chapter.

formulation) 'biologically-inspired', but it is not 'organismically-inspired' to the extent that it does note realise the metabolic basis of agency as to be found in the bio-agency of organisms.

Moreno and Etxeberria's emphasis on the ontological conditions of bio-agency brings to the fore an important insight in the general ontological nature of organisms. Organisms are processes, not things (Dupré 2012; Nicholson and Dupré forthcoming; Meincke forthcoming; Meincke in preparation a). The correct observation that it is impossible to distinguish between what the organism is and what it does ('reciprocal causality') reflects the fact that an organism just is the process of its own constitution via interacting with environmental processes in specific ways. More precisely, an organism – i.e., a living body – is a stabilised higher-level process relying on a complex network of lower-level processes, as being investigated by systems biology.

In this vein, Tom Ziemke has recently criticised embodied robotics for still "operat[ing] with a view of the body that is largely compatible with mechanistic theories in psychology and a view of control mechanisms that is still largely compatible with computationalism" (Ziemke 2016, 7). According to Ziemke, rather than granting "the living body [. . .] some fundamental role in constituting embodied cognition" (2016, 6), "the robot body is typically viewed as some kind of input and output device that provides physical grounding to the internal computational mechanisms" (2016, 7).[13] On the basis of this diagnosis, Ziemke calls for a methodological renewal of embodied robotics that (*seriously*) takes seriously the biological roots of cognition as in particular revealed by enactivism. This renewed embodied robotics would accordingly shift the focus from emulating the "sensorimotor interaction between agents and their physical and social environments" to modeling the "layered/nested networks of bodily self-regulation mechanisms", in which this sensorimotor interaction is rooted, including developing "computational cognitive architectures for robotic systems taking into account some of these mechanisms" (2016, 9). Ziemke expects an important contribution towards this programme to be made by synthetic biology through "modeling/understanding/synthesizing the nature of organisms as such layered networks" (2016, 9).

5.3.2 The Wrong Matter?

Does this mean embodied robotics ultimately has to hand over to synthetic biology and that is, to so-called wet ALife?[14] Moreno and Etxeberria seem to insinuate this by arguing for P2, the claim that only biomolecular components are suitable for

[13]This actually resonates with the continuing popularity of mechanist approaches to biology, including the common understanding of organisms as (very complicated) machines (on the latter, see Nicholson 2013).

[14]'Wet ALife' is different from both software-based 'soft ALife' and hardware-based 'hard ALife' (robotics).

realising the constructive basis of agency as to be found in organisms: "autopoiesis (self-production) realized via metabolism" (Moreno and Etxeberria 2005, 171).

The question of what type of matter is needed is a topic familiar from debates about what life is and whether and how it might be possible to artificially create life. Robert Rosen has famously distinguished two opposing main approaches to the explanation of life, one of which follows the maxim "throw away the organization and keep the underlying matter" and the other of which follows the maxim "throw away the matter and keep the underlying organization" (Rosen 1991, 119). Maturana and Varela's autopoiesis theory of life, just as Rosen's so-called relational biology, belong to the latter formalist or functionalist approach: any system realising the defining organisational structure is taken to qualify as a living system, no matter what its matter is.[15]

Against this background, Moreno and Etxeberria's emphasis on the "special materiality" on which "living organization" is supposed to depend (Moreno and Etxeberria 2005, 163) might look surprising. As Pim Haselager puts it, "[g]iven that" – according to the concept of autopoiesis, on which Moreno and Etxeberria rely – it is the organization of the components and not their material constitution that matters", one should assume that "the question is open whether autopoiesis could be realized in artificial matter" (Haselager 2005, 528); roboticists "can hardly be expected to work with organic material and living creatures" (ibid.). However, even though Haselager does "not think that autopoiesis provides a *principled* obstacle for robotics", he admits that spelling autonomy out in terms of autopoiesis "does indicate a further constraint on robotics", raising the question of whether the materials used so far – metals and plastics – "allow for a genuinely autopoietic organization" (ibid.; italics in the original). Referring to Aristotle's idea that form is constrained by the potentialities of matter which it actualises, Haselager envisages "a *mild* functionalism that pays attention to material aspects of the body to a higher degree than currently is customary" (ibid.; italics in the original).

Much of what Moreno and Etxeberria say indicates that they, too, wish to put forward not more than a slight qualification of the formalist approach. Thus they conclude their article by clarifying that they do not intend to invoke "something extra in living matter that we do not know (an idea that brings old vitalisms back to life)" (Moreno and Etxeberria 2005, 173) or "a mysterious capacity of living matter" (2005, 174) as "[t]he reason for the (in spite of all progress) still evanescent expectation of reproducing true agency and cognition in robots" (2005, 173). Instead, they want to draw attention to "the deep and radical interrelation between forms of organization and materiality", i.e., to the fact that "[t]he material structure of robots is very different from that of living systems, and [that] this imposes qualitative differences on the form of organization" (2005, 174).

[15]"[T[he phenomena they [autopoietic unities] generate in functioning as autopoietic unities depend on their organization and the way this organization comes about, and not on the physical nature of their components" (Maturana and Varela 1987, 51). "Life is the manifestation of a certain kind of (relational) model. A particular material system is *living* if it realizes this model" (Rosen 1991, 254; italics in the original).

All this granted, the question still remains whether in principle there *could* be a robot, rather than some artificially created wet biochemical system, realising basic autonomy in the sense defined by Moreno and Etxeberria. And here the actual arguments given by Moreno and Etxeberria point towards the assumption of an indeed 'radical' interrelation between forms of organization and materiality, one that in fact rules out this possibility, provided that we take a 'robot' to be an artificial system made, at least for the most part, from synthetic, anorganic materials.

Moreno and Etxeberria's main complaint is that any such system would fail to realise the "(full) constructive closure" achieved by autopoietic systems through metabolising, insofar as it would be a 'physical system' that uses 'inert' building blocks, as opposed to a 'chemical system' that works with "intrinsically active elements" (2005, 171). As an illustration, they invite us to imagine a "fully automatized factory, which could reproduce itself starting from simple components" (ibid.). "This system's organization would not be really self-constructive", they say, "because it would depend on a set of external devices whose design required more complex processes than those produced by themselves, and so on, in an infinite regression" (ibid.).

To start with, it seems to me that if it really were possible to build a thing "which could reproduce itself starting from simple components", it should not matter where this ability comes from, i.e., whether it possibly depends on a 'device' that was "externally designed" (ibid.), such as a microchip whose "structure [. . .] does not emerge from its intrinsic composition, but requires external shaping" (ibid.). To be sure, it is not clear that this description would not, in a certain sense, even apply to organisms. Thus someone could point out, if the idea is that the information stored on the microchip governs the self-construction of the hypothetical fully automatised factory, then the microchip's function is actually entirely analogous to the function of the DNA, which no organism creates itself but inherits from its parents.

Thanks to the rise of epigenetics, we now know that the DNA does in fact not act like a computer program. However, we still might be in doubt as to whether "(full) constructive closure" is something ever brought about by any organism, namely exactly because of the importance of the interaction with the environment in which the organism is embedded. An organism is no *causa sui*. It is not true that it "as a whole, is cause and effect of itself" (Barandiaran and Moreno 2006, 174) – not only because an organism needs to be created by its parents in the first place, but also because its persistence crucially depends on environmental conditions, including the right kind of interactions with the (the right kind of) environment to be performed by the organism.[16] I take it that in particular recent research in symbiosis shows that the distinction between 'intrinsic' and 'extrinsic' with respect to organisms is

[16]Barandiaran and Moreno acknowledge this via more recent paper by stressing that "interactive dynamics are constitutive of the [autonomous] system and not something to be added, a posteriori, in the form of structural coupling" as which it appears, according to the authors, in Varela's concept of autopoiesis (Barandiaran and Moreno 2008, 327(f.)). See also Moreno and Mossio (2015, 5), footnote 7. For a discussion of the concept of (organisational) closure with respect to the role of interactive dynamics and to its ontological implications, see Meincke (in preparation a).

anything but clear-cut and that the boundaries of living systems might be wider and more flexible than we used to think (Gilbert et al. 2012; Gilbert and Tauber 2016). It thus seems we cannot deny that the fully automatised factory performs 'real' self-construction just because its self-construction (which we have presupposed as part of its imagined character of being a fully automatised factory) depends for its persistence on 'external devices'. If we did, we would need to do the same with respect to organisms.

What about complexity? Why should it be the case that the "organizational logic" of the imagined fully automatised factory "requires that the external degree of complexity be always greater than the internal one" (Moreno and Etxeberria 2005, 171)? Again, one could wonder about whether the case of organisms is really so different. A foetus is hardly more complex than the maternal organism in which it resides and on which it depends. The same goes certainly for a microbe in my gut. And do not forget about the complex balance of the ecosystem on which we all depend.

The reason why Moreno and Etxeberria think that *there is* a difference between the hypothetical fully automatised factory and an organism with respect to complexity has to do with the assumed pivotal role of metabolism for basic autonomy, together with certain ideas about matter. The claim is that the building blocks of a truly basic autonomous system must be such that they do not "need additional design (i.e., an external level of organization) to become functional constraints harnessing the flows of matter and energy so as to achieve self-construction" (ibid.). In other words, the building blocks must be suitable for realising metabolism in a spontaneous, self-guided manner, and Moreno and Etxeberria are convinced that only "intrinsically active elements" can do the job, while "inert aggregates" fail (ibid.). From this they conclude that basic autonomous agents must be made from "biomolecular" components (2005, 172).

Again, I am not persuaded by this argumentation. First of all, it is not clear what the distinction between "inert aggregates" and "intrinsically active elements" is about. Obviously it is not just about the macroscopic qualities of the materials used, i.e., about the difference between rigid and soft materials, such as plastic and metals, as traditionally used in robotics, on the one hand, and elastomers and hydrogel, as used by the new 'soft robotics', on the other.[17] It seems what is at stake are rather microscopic qualities of inorganic versus organic matter. However, 'inertness', as a technical term in chemistry, denotes the low tendency of a chemical substance to react with other chemical substances under standard environmental conditions. From a chemical point of view, there is no sense in which organic matter is less 'inert' than inorganic matter.[18] What, then, do Moreno and Etxeberria have in mind when attributing to the matter of organisms a special activity, lacked by the matter of embodied robots?

[17]'Soft robotics' explicitly names biology as the key source of inspiration (Trivedi et al. 2008; Kim et al. 2013). It is therefore also called 'organic robotics', see https://orl.mae.cornell.edu/index.html

[18]In fact the only chemical substances that actually are regarded as 'inert' are noble gases, also called 'inert gases'.

The answer points towards the "power of self-assembly", which Moreno elsewhere, in its characteristic combination with "dissipative self-organization", locates in "the (bio-)chemical domain" (Barandiaran and Moreno 2008, 340, endnote 4). However, the theory of self-assembly is not without problems, since the alleged self-assembly phenomena occur only under special circumstances, i.e., depending on interactions of the 'self-assembling' substances with surrounding substances. What I said earlier[19] about the fuzziness of the 'intrinsic'–'extrinsic' distinction with respect to the organism and its environment applies equally on the molecular level.[20]

Against this background, I also hesitate to welcome the further suggestion, made by Moreno, Etxeberria and Umerez in an earlier paper, that information might be contained in the matter of organisms which is not contained in the organisation of the matter of organisms. The "materiality of biological systems", they speculate, might "impl[y] certain characteristics that cannot be formalised" in the sense that "a great part of the information necessary for the process of self-reproduction is implicit in the specific materiality that compose [sic] it" (Moreno et al. 1994, 407f).[21] If the idea is that one of these implicit, non-formalisable properties of living matter, as opposed to non-living matter, is its power to self-assemble spontaneously and without guidance from 'the outside', then we are in fact not far from vitalism.[22]

Be this as it may, what type of matter is needed for the realisation of metabolic self-constitution will of course also depend on how we define metabolism. Moreno and Etxeberria are not all too explicit about their understanding of the term. When reconstructing the emergence of agency from metabolism in the course of evolution of life on earth, metabolism is broadly referred to as "the maintenance of a chemical organization through the control of energy flows" (Moreno and Etxeberria 2005, 165). Later we read: "the functional autonomy of living systems emerges from a

---

[19]See this section on page 20.

[20]This is evident, for instance, from recent insights in the formation of cell membranes. Lipid bilayers build up only in an aqueous environment, which has led chemists "to regard the 'active volume' of molecules such as proteins as extending beyond their formal boundary (the van der Waals surface, say), by virtue of the way they shape and manipulate the shell of water that surrounds them. Moreover, the structure and dynamics of this hydration shell seem to feed back onto those aspects of the proteins themselves so that biological function depends on a delicate interplay between what we have previously regarded as distinct entities: the molecule and its environment" (Ball 2008, 75). Bagatolli et al. (2010, 381) likewise stress that "[c]ooperative phenomena are the reason in the first place for the formation of an aqueous supra-molecular aggregate like a bimolecular lipid-bilayer membrane." I am grateful to my colleague Stephan Güttinger for bringing this literature and the related questions to my attention.

[21]This is meant to be a critique of a purely functionalist or formalist approach to life: "If it were the case that the relations among components arise from their material properties, the complex organisation of living systems could not be fully understood except by recourse to the properties of living matter" (Moreno et al. 1994, 407). "If we suppose that some of the essential relations of the logic of the living are necessarily implicit, then only those material components that can bear such relations can constitute life" (Moreno et al. 1994, 408f.).

[22]For an affirmative adoption of the idea of 'molecular vitalism' see Kirschner et al. (2000).

structure-process relation that is intrinsically dependent on rates, energy flows, and mass" (2005, 173). These rather vague characterisations could arguably also be met by a mobile petrol lamp endowed with a sensor that regulates the usage of petrol in response to the availability of oxygen and petrol and triggers the lamp to refill its petrol tank at a petrol station when needed (especially if we assume that the lamp can only move via burning petrol).

However, it has already become obvious that Moreno and Etxeberria actually (without explicating it) presuppose a very specific meaning of metabolism: metabolism is supposed to be the mechanism which brings about basic autonomy in living beings. As such, it is thought to entail 'reciprocal causality' – the interdependency of the material structure and the activity of the system –[23] and to require certain 'intrinsically active', i.e., presumably biochemical, at least chemical materials if we assume that arranging the latter in the right way could "trigge[r] a process of spontaneous organization" leading to metabolic self-constitution in virtue of these materials' alleged intrinsic activity (2005, 173).[24]

This understanding of metabolism is not without alternative. Margaret Boden (1999) has distinguished three concepts of metabolism, two of which she finds weak enough to be actually compatible either with *both* strong ALife, i.e., "virtual creatures having physical existence in computer electronics, but not bodies", and ALife in the sense of "some non-biochemical A-Life robots" or *at least* with the latter (Boden 1999, 231). These are the notion of metabolism as energy dependence (fulfilled by any computer) and the notion of metabolism as energy dependence plus a mechanism of controlled usage of individual energy packets for performing energy-dependent behaviour (as instantiated, e.g., by robots that store and budget their energy including recharging themselves) (1999, 234ff.). Only the third and strongest notion of metabolism, according to Boden, excludes both forms of ALife. This notion is supposed to "capture what biologists normally mean by the term", referring "to the use, and budgeting, of energy for bodily construction

---

[23]This is clearly stated in an earlier paper co-authored by Moreno and Kepa Ruiz-Mirazo: basic autonomy is defined as "the capacity of a system to *manage* the flow of matter and energy through it so that it can, at the same time, regulate, modify, and control (i) internal self-constructive processes and (ii) processes of exchange with the environment", and the "interdependence [...] between the constructive and interactive aspects of the phenomenon is precisely the cornerstone for the constitution of any real metabolic organization" (Ruiz-Mirazo and Moreno 2004, 240; italics in the original).

[24]The passage reads in full: "So far, the research program in robotics and artificial intelligence has focused on autonomy at the sensorimotor level, without basic autonomy. Instead of trying to reconstruct the conditions for triggering a process of spontaneous organization starting from chemical components, computers and other complex technical tools are the building blocks" (Moreno and Etxeberria 2005, 173). In the aforementioned co-authored paper we are similarly warned not to "forget[] about the material mechanisms that are crucial to trigger off a biological type of phenomenon/behavior", followed by the demand that "research should be directed to implement those – or very similar – interactive processes with molecular components and tools that are alternative to (and, perhaps, less complex than) the biochemical machinery present in known living beings" (Ruiz-Mirazo and Moreno 2004, 238).

and maintenance, as well as for behaviour" (1999, 236) rather than meaning "mere energy dependency (with or without individual energy packets)" (1999, 245).

As it happens, despite her instructive discussion of possible alternative definitions of metabolism, Boden finally makes a case for regarding metabolism in the strongest sense as a criterion for life. Hereby she like Moreno and Etxeberria, assumes that strong metabolism "involves the autonomous use of matter and energy in building, growing, developing, and maintaining the bodily fabric of a living thing" (1999, 237) and "requir[es] a complex biochemistry to effect the[] vital functions" of "anabolism and catabolism" (1999, 238). We thus end up with the same close link between metabolism and organic materials ('living matter') as well as with the same corresponding hermeneutic circle that defines basic autonomy in terms of metabolism while defining metabolism in terms of basic autonomy, followed by the same strong dismissal of embodied robotics.[25]

Clearly, if it is true that metabolism requires organic matter, and if it is further true that basic autonomy can only be realised by metabolism, then it follows that in order to build basic autonomous systems, we have to use organic matter and, i.e., we have to create wet ALife. But only *if*.

As the foregoing considerations indicate, there are reasons to be sceptical about whether the first half of this double conditional truly holds[26] – even if one were happy to admit that metabolism, indeed, "is not an abstract functionalist concept, divorced from the specific material realities" (Boden 1999, 231). Acknowledging the contingency of the material basis of life on earth, and refraining from assigning dubious intrinsic active powers to organic ('living') matter as opposed to 'inert' inorganic matter, I feel strongly driven towards a 'mild functionalism' as suggested by Haselager. Such a moderate functionalist position sits well with the plausible assumption that, ontologically speaking, any properties we ascribe to matter, including those exploited by life, result from certain modes of organisation (or at least supervene on these).[27] This is not to deny that matter is active in some sense, but rather to stress that this activity, which is not restricted to organic matter only, to a large extent is itself a result of organisation. If any, it is activity all the way down because it is organisation all the way down.

[25]Boden goes so far as to deny that so-called 'embodied' robots are in fact embodied: "[B]eing embedded does not necessitate being (truly) embodied. [...] a body is not a mere lump of matter, but the physical aspect of a living system, created and maintained as a functional unity by an autonomous metabolism. If this is right, then these robots do not have bodies" (Boden 1999, 239). The very same claim "that artificially created adaptive agents are not fully embodied" we find in Barandiaran and Moreno (2008, 339), based on the diagnosis that embodied robotics focuses on 'sensorimotor embodiment' only at the expense of 'biological embodiment' (Barandiaran and Moreno 2008, 338ff.; see also Ziemke 2016) or 'organismic embodiment', including 'emotional embodiment' (Barandiaran and Moreno 2006, 180).

[26]On the second half see the next section of this chapter.

[27]I thus agree with Langton's claim on which he founded the research agenda of Artificial Life, that life is "a property of the organization of matter, rather than a property of the matter which is so organized" (Langton 1989, 2).

Saying that embodied robots are made from the 'wrong matter' to realise the kind of self-recursive processuality which we know from organisms can only mean then that we have not yet sufficiently understood how to organise matter accordingly. This, I dare to speculate, might change, though, without this implying that the creatures fabricated according to these hypothetical new procedures necessarily would become exact copies of organisms. Current research on life *in silico* impressively demonstrates this.

5.3.3 *Metabolism, Intrinsic Normativity and Agency*

But would those future robots need to realise the familiar organic kind of processuality at all? Could there not be other kinds of organisation which we equally could accept as giving rise to agents? As we have seen, Moreno and Etxeberria deny this, endorsing P3, according to which true agency must be grounded in basic autonomy realised via metabolism.

Agency, as a manifestation of autonomy, implies that an agent "has its own purposes or goals" (Moreno and Etxeberria 2005, 172), and this is what we, according to Moreno and Etxeberria, cannot say about embodied robots exactly because they do not metabolise. Metabolism, as has been first pointed out by Hans Jonas (1968, 2001), is the mechanism in nature that gives rise to needs, most importantly to the need for the supply of matter and energy so as to keep the metabolising system stable in a state far from thermodynamic equilibrium (Barandiaran and Moreno 2008, 327ff.). These needs transform, at higher levels of organisation, into goals which the system seeks to achieve by means of suitable actions – an endeavour in which it might succeed or fail. The crucial point already noted above is that whether it succeeds or fails decides nothing less than the system's existence. In virtue of this 'reciprocal causality' between a metabolic system's actions and its material structure, metabolism is taken to be the source of the 'intrinsic teleology' or 'intrinsic normativity' of agency by a number of authors (e.g., Barandiaran and Moreno 2008; Barandiaran et al. 2009; di Paolo 2003, 2010; Froese and Ziemke 2009).

According to Moreno and Etxeberria and these other authors, the actions of embodied robots fall short of intrinsic normativity and, hence, do not qualify as cases of 'true' agency. They lack "meaning", which "is relative to a point of view or perspective created by the very task of self-construction while interacting with a given environment" (Moreno and Etxeberria 2005, 162). This remains true even in the case of "new functional forms of behaviour emerging from [the robot's] interactions with the environment", as these "do not affect its material structure (which is inert), but only its behavioural organization" (2005, 173). Conversely, possible variations in the robot's material organisation "do not substantially affect the organization of agency" (ibid.). Not being "intrinsically dependent on rates, energy flows, and mass" (ibid.) as organisms are, and that is, not being "subject to a permanent precariousness" which needs to be "compensated by [...] active organization" (Barandiaran et al. 2009, 375),[28] embodied robots do not "generat[e] their own norms", which is, however, what it would mean to be 'autonomous' (Barandiaran and Moreno 2006, 174). Instead, rather than pursuing "autonomous goals or purposes" (Moreno and Etxeberria 2005, 173), the robots serve the goals of their designer.

I take it that acknowledging the critical role metabolism plays in the constitution of living systems indeed paves the way for a naturalised account of the normativity characteristic of bio-agency. A large number of the norms that guide our actions do not fall from the sky but have a robust biological foundation, and it is after all overdue to grant the same normativity, in principle, to the agency of our non-human fellows in the animal kingdom. However, this, I think, does not suffice to categorically dismiss embodied robots as possible agents.

The contingent fact that the intrinsic normativity of the agency performed by living systems is rooted in these systems being metabolising systems does not rule out the possibility that intrinsic normativity could have sources different from metabolic organisation. Recognising this possibility weakens Moreno and Etxeberria's thesis while possibly preserving the main idea. True agency, according to the modified thesis, does *not* require an agent to metabolise; but it still needs to be grounded in basic autonomy, i.e., in some sort of material self-constitution, however realised.

This is *in nuce* the research programme Tom Froese and Tom Ziemke envisage for what they call 'enactive AI', an AI that, broadly following Maturana and Varela's autopoiesis theory, grounds cognition and agency in "constitutive autonomy" (Froese and Ziemke 2009, 479). The new AI is meant to radicalise the insight already present in embodied AI that the so-called problem of meaning for AI "is not a problem of *knowledge* but rather of *being*" (2009, 474; italics in the original), insofar as the "intrinsic teleology" (2009, 479) found in living systems derives from their particular "mode of being", namely the mode of "*being by doing*" (2009, 473; italics in the original).

Froese and Ziemke do not discuss the question "whether metabolism is the only way to realize this particular mode of existence" (2009, 474), but they admit that "it is still rather impractical – if not impossible – to design artificial agents that are fully metabolizing" (ibid.). Subsequently they formulate two 'design principles' for enactive AI, neither of which requires enactive robots to metabolise. Yet, the two principles correspond roughly with the two aspects of biological autonomy distinguished by Moreno and Etxeberria. In order to qualify as an agent, "the system must be capable of generating its own systemic identity at some level of description" (2009, 485) and "have the capacity to actively regulate its ongoing sensorimotor interaction in relation to a viability constraint" (2009, 486). Apparently, there are artificial systems which at least fulfil the first requirement (2009, 492ff.). The hope

[28] The idea of the 'precariousness' of life was first developed, and worked out in detail, by Jonas (2001). It was introduced to the recent debate on agency among philosophers of biology by di Paolo (2009).

is by designing more elaborate versions of such self-constituting systems to create "the appropriate conditions of emergence for an agent to self-constitute" (as opposed to "directly engineering an agent") (2009, 485). So far these approaches "focus on constitutive autonomy in the molecular domain"; however, the authors stress that "the possibility of constitutive autonomy in domains other than a chemical substrate deserves much closer study" (2009, 493).

However, one still could think that the guiding assumption of this research programme is too strong in the sense that it unnecessarily restricts the possible instantiations of artificial agency to those which resemble the kind of agency we know from living systems on earth. Surely, basic autonomy could be realised differently than through metabolism; but does agency need to be grounded in basic autonomy at all?

It seems to me, in purely conceptual terms, there is no difficulty with this. Setting aside the obvious problem case of sophisticated human agency which many consider to be 'purposeless', such as arts or philosophy, there has been a powerful tradition of seeing God (or, outside monotheism, gods) as actor(s). God is assumed in acting to pursue goals which are eminently His own and derive from His particular mode of being, even though He is neither believed to metabolise nor taken to be in some other kind of precarious state of being. Of course, if we had any means to prove the matter, it might turn out that our assumptions are wrong (God might not act, He might actually metabolise and/or be a precarious system; or He might not even exist). However, the point is not whether the idea of divine agency is plausible, but rather whether it is conceptually coherent, and I cannot see why the latter should not be the case. The concept of intrinsic normativity does not in itself entail any constraints as to the ontological mechanism by which, or on the basis of which, an agent generates its own norms. Material self-constitution, realised by metabolism, is the mechanism we know; but there might be others as well.[29]

We could even go a step further and ask whether normativity really needs to be intrinsic in order for actions to qualify as being meaningful. Is it really the case that to act in a meaningful way and, i.e., according to Moreno, Etxeberria and others, to truly act, means to pursue one's own goals? What about acting according to someone else's wishes? Human agents do this occasionally; some human agents even do it, pathologically, on a regular basis. Consider, for instance, a man who, as a rule, would do what his mother wants him to do. This poor guy might even jump out of a window if his mum asks him to do so or if he feels this is what she wishes to happen. In that case, he would clearly act against what one might regard as the most important natural goal of any living system: the goal to survive. He

[29] I am grateful to the anonymous reviewer of this chapter for drawing my attention to another example of intrinsically normative agency that is not grounded in metabolism, namely that of collective and institutional agents like companies and nations or states. However, apart from the fact that it is contentious whether there really are such collective and institutional agents in a non-metaphorical sense, there is also room for the speculation that those agents' actions actually serve the purpose of self-maintenance, i.e., are ontologically grounded in some sort of precariousness. I therefore take the concept of divine agency to be the more instructive one for my argument.

would act against what would have been his intrinsic natural goals in accordance with that main natural goal if he had determined his goals himself. Yet his action is not meaningless. We know why the man who died after jumping out of a window in the fifth floor did what he did: he wanted to please his mum.

If we think that this poor guy truly acted even though, in acting, he did not pursue his own goals,[30] how then can we deny an embodied robot to act *just because* the goals of the robot's behaviour are not generated by the robot himself but rather by its designer? What we usually take to decide about whether or not a bodily movement counts as an action is not whether the moving entity, by moving, pursues its own or some other entity's goal, but rather whether the moving entity, by moving, pursues *any goals at all*. The crucial point is whether the moving entity actually *pursues* goals (of whatever origin) as opposed to merely *serving* goals.[31] Imagine a robot that is programmed such that it always wants what its designer wants. This robot would be no different, with respect to its actions, from the mummy's boy considered above.

Alternatively, imagine a robot that always wants what its designer wants without even knowing that it wants what it wants just because its designer wants it to want it. According to Pim Haselager, such a robot would have to be regarded as being 'brainwashed' just in the same sense in which I would have to be regarded as being brainwashed if I decided to sneak into a beer-factory at night and drink a whole lot of beer because "external agents make me think that I want to drink beer because it is good for me (or to stop the aliens, or whatever)" (Haselager 2005, 520). However, being brainwashed does not prevent me from being an agent, even though it doubtlessly is in tension with the idea of autonomy (which is why we could argue about whether I am to be held responsible for my actions). As long as there is some sense in which I *want* what I do for some purpose, and that is, as long as my body movements are *intentional*, I act; and the same arguably holds for the hypothetical brainwashed robot.[32]

[30]The guy's generic wish to please his mum allows for attributing his observable behavior to him as his actions, but it does not suffice to explain why he performs specific actions at specific times in specific situations; instead, these specific actions are fully explained only by taking into account the specific wishes of the guy's mum. The imagined case, hence, is *not* a case of intrinsic normativity, despite the fact that, insofar as the guy in question qualifies as an agent (and, i.e., insofar as his body movements qualify as intentional (see the following considerations)), his mum's goals could be seen as the guy's 'own' goals in a *derivative* sense.

[31]Exactly this is contentious in the case of embodied robots because it is contentious whether embodied (or any other robots) actually possess intentionality, alongside with other mental states. The following considerations, rather than presupposing that embodied robots are subjects of knowledge and wishes, have to be understood as thought experiments, designed to bring out a deficiency in the concept of intrinsic normativity as tied by Moreno and Etxeberria to the concept of agency.

[32]'Hypothetical', as it an open question whether there are brainwashed robots; to claim that there are would be to concede that these robots are agents. This in contrast to an objection against the idea that robots are autonomous as imagined by Haselager: "[A] philosopher might argue that robots are in a situation comparable to that of people who are brainwashed and that therefore robots are not even close to being candidates for any serious degree of autonomy. Robots are not autonomous

From a philosophical point of view, any movement that is intentional under some description counts as action (Davidson 1980). Moreno and Etxeberria do not mention intentionality throughout their paper and as far as I can see, intentionality does not figure prominently in Moreno's other writings either. The reason for this seems to be that the concept of intentionality is believed to be linked to the concept of representation, as evident from a footnote in a paper co-authored with Xabier Barandiaran where the authors claim their concept of teleology to be "devoid of any intentional or representational character", stressing that it rather "simply denotes the participation or contribution of a process into a self-maintaining organization" (Barandiaran and Moreno 2008, 341, footnote 9).[33] Not being able to address the issue of representationalism here, I have to confine myself to emphasising the key role of intentionality for a philosophically robust, non-deflationary notion of agency, even if we, as I would like to suggest, invoke a fairly liberal understanding of what it means for a given behaviour to be intentional.[34]

Against this background, the concept of 'intrinsic normativity', as used in the debate about artificial agency as a condition for agency, appears to be at least insufficiently precisely defined. In one sense it is clearly too narrow as it excludes all cases where agents act according to extrinsic goals. This applies also to cases of acting under coercion. If I show up at an administrative meeting because my boss forces me to do so, this action is neither meaningless nor no action at all nor in fact my boss's action. To be sure, it is not even clear that what one would regard as my 'natural' goals is properly 'intrinsic' to me as I have not myself chosen them; they were rather imposed upon me by my biological nature. I might occasionally feel forced to eat in a situation in which I would rather prefer to do something else. Yet, going to the kitchen, opening the fridge and starting to cook are actions of mine.

In another sense the concept of intrinsic normativity as a condition for agency is clearly too wide as it includes cases which from a philosophical point of view do actually not qualify as actions as they lack intentionality. Is intrinsic normativity just functionality? And is agency just any sort of survival-securing process? According to Moreno and Etxeberria, "[a]n agent has the capacity to interact with the environment so as to contribute to its own maintenance" (Moreno

because they themselves don't choose their own goals and they do not even know that it is us that set their goals for them" (Haselager 2005, 520). Haselager subsequently rightly points out that the issue of choosing one's own goals has to be distinguished from the "issue of intrinsic *ownership*" (2005, 522; italics in the original); however, like Moreno, Etxeberria and colleagues he thinks that intrinsic ownership is strictly tied to the maintenance of homoeostasis: "Fundamentally, what makes my goals mine, is that I myself am at stake in relation to my success or failure in achieving them" (2005, 523).

[33]See also Moreno and Mossio (2015, 92) for an explicit exclusion of "rational and conscious intentionality" from their definition of agency.

[34]See Schlosser (2015) for a summary of the state-of-the-art of the philosophical debate on the matters of representationalism and intentionality with respect to agency.

and Etxeberria 2005, 163).[35] This definition of agency applies to a tiger's hunt for an antelope just as well as for the tiger's digesting of the antelope.

As we have seen in their analysis of the evolution of agency as an emancipation of agency from metabolism, in a sense this is part of the point Moreno and Etxeberria want to make: that full-blown agency arises from primitive forms, most fundamentally from so-called metabolic agency, i.e., active transport of chemical substances through a semipermeable membrane. There is supposed to be a spectrum from primitive to sophisticated forms of agency. From a philosophical perspective, however, we have to make sure that this story, which is persuasive indeed, does not seduce us to overlook crucial shifts in quality in the course of the evolution of agency. Moreno and Etxeberria themselves draw attention to the appearance of the nervous system as what supports motility in higher animals. Should we not expect to find a distinctively new quality in those adaptive interactions with the environment which rely on the decoupled dynamics of a neuronal network? A quality that marks the beginning of agency in a *proper* sense, as opposed to agency in a mere metaphorical sense?[36]

I take it that one of the obstacles for a fruitful dialogue between traditional philosophy of action and the research on agency in recent philosophy of biology is the mismatch between the ways the term 'agency' is used in both fields. The traditional philosophical term, which essentially relies on intentionality, is in an important respect more demanding than the term used by philosophers of biology, which is associated with intrinsic normativity (somewhat vaguely defined). For philosophers of action, not even something like the corneal reflex can count as agency, despite it being a kind of bodily behaviour as opposed to mere organic processes such as digestion (it is 'mere' behaviour as it is often said). What matters is whether or not a given behaviour is intentional under some – suitable – description (and there is no such suitable description for the corneal reflex).

This has implications for the view on the possibility of artificial agents. Moreno and Etxeberria argue that because embodied robots realise only the higher-order part of biological autonomy, their movements do not qualify as agency. However, from the traditional point of view of philosophy of action, the lack of lower-level, 'basic' autonomy could seem entirely irrelevant because it seems doubtful that the supposed forms of 'agency' other than what Moreno and colleagues call 'neural' or 'behavioural agency' are actually proper cases of agency at all. For the philosopher of action, if any, then only the nervous system-based type of behaviour is agency. But if so, why then, she asks, should we put so much emphasis on the lower-level biological conditions upon which the nervous system rests?

---

[35]See also Barandiaran and Moreno (2008, 330): "We use the term 'agents' for those systems that interact with their environments, so that the changes produced between the system and its environment contribute to its self-maintenance."

[36]In Spann (2014) I make a similar point with respect to Skewes and Hooker's (2009) account of bio-agency.

Agency, as we usually use the term, need not be grounded in intrinsic normativity (not even according to the best possible definition of that latter concept), and the normativity which we indeed take to be entailed does not have to derive from material self-constitution, unless we can convincingly argue that intentionality can only be realised by biological systems. This case has not yet been made. Surely, Braitenberg's famous artificial vehicles have demonstrated that we are easily misled by certain observable behaviours to postulate that these are intentional or even emotional (Braitenberg 1984). However, this itself should not mislead us to rule out any possibility of artificially creating intentionality and emotion. What this possibility could look like is an open question, and I concede that there are strong arguments that at least emotions are genuinely tied to some precarious mode of existence – the precarious mode of animals that need to search for food – in which they serve the function of bridging the gap between need, i.e., desire, and satisfaction (Jonas 2001, 99ff.). However, there are attempts being made to design artificial systems which, though materially non-precarious, develop a desire to conserve their way of life in the sense of habit formation (di Paolo 2003 and 2010).

Only if we from the start equate agency with bio-agency, and i.e., with agency-as-we-know-it, 'true' agency appears to be necessarily grounded in basic autonomy realised via metabolism. But why should we think so? The classical philosophical notion of agency, even though in some respect more demanding than its biological competitor, does not justify such a restriction. *Any* intentional behaviour is a case of agency. It thus as if by equating agency with bio-agency we seems just unhelpfully pre-decide the answer to the question of whether there could be artificial agents.

5.4 Conclusions

As we have seen, Moreno and Etxeberria use the key distinction between basic autonomy and adaptive interaction with the environment for arguing in two different directions. For one thing, they say, the fact that we, with respect to living systems, can legitimately make this distinction explains why it appears as if embodied robots were agents; but for the other, the fact which we have to bear in mind, namely that basic autonomy and adaptive interaction with the environment are not literally different phenomena but rather aspects of the very same phenomenon – the biological autonomy of living systems – makes it clear that this impression is illusory.

Looking at the details of this argument, we have found that Moreno and Etxeberria have indeed a convincing story to tell about the evolution of bio-agency on earth, which they present as a stepwise emancipation of agency from the constraints of metabolic self-constitution. However, it is not clear that the history of the evolution of bio-agency on earth enters into the concept of agency *as such*,[37] a concept that we actually have and that appears to work independently of assumptions about matter, metabolism and material self-constitution, as evidenced by the fact that we can coherently conceive of the possibility of divine agency. According to the general concept of agency, which is the one that dominates debates in philosophy of action, agents do not even have to pursue their own goals in acting as long as they pursue some goals at all in the sense of acting intentionally.

If there is no justification for requiring agency necessarily to be grounded in material self-constitution via metabolism (which was the third premise of Moreno and Etxeberria's argument), then we do not need to worry too much any more about what kind of matter to use for creating artificial agents (which, in the second premise of the argument, was restricted to biomolecular components), even though a mild functionalism that takes into account certain constraints imposed on higher-level organisations by lower-level organisations of matter, assuming that any properties of matter are organisation-based, still seems reasonable. If so, while the observation of a fundamental difference with respect to the ontological constitution of the behaviour of organisms on the one hand and of existing embodied robots on the other (i.e., the first premise) remains true, the conclusion of Moreno and Etxeberria's argument from metabolism against the possibility of artificial agency nevertheless does not follow. On the basis of this argument we can neither categorically deny all present embodied robots to be agents nor rule out in principle the possible occurrence of future artificial agents who are not simply instances of wet ALife.

This result of our analysis of the systems biological criticism of embodied robotics has interesting implications for the project of embodied robotics, for the bio-agency approach and for our understanding of agency in general.

To start with the first, it might lead us to reconsider not only the principles and methods, but also the very target of embodied robotics as a research programme. Moreno and Etxeberria assume that "what we want is to build physical artificial systems endowed with agential capacities *similar* to the ones that (natural) organisms exhibit" (Moreno and Etxeberria 2005, 170; italics added). On the basis of this assumption they put forward their criticism that the behaviour of embodied robots "is not yet self-maintaining or truly autonomous *in the same sense* as for living beings" (2005, 170; italics added) and for fundamental reasons never will be, which is why "we should not expect that work in artificial systems will generate forms of agency *similar* to that in living ones" (173; italics added). As we have seen, for Moreno and Etxeberria this is sufficient for actually dismissing embodied robots as agents altogether, or at least as 'true' agents.

However, 'similar' is an elastic concept, open to a variety of interpretations. *How* similar need embodied robots be to organisms in order to match with the aim of embodied robotics? To what extent, and in what sense, must embodied robotics turn itself from being biologically-inspired to being organismically-inspired in order to

[37] It is not even clear that it enters into the concept of *bio*-agency as such, given that the bio-agency on earth possibly could also have developed differently and that there might be extraterrestrial forms of bio-agency with entirely different evolutionary histories.

be successful? What exactly is it that allows us to differentiate between success and failure? Is emulating organismic agency a desirable aim at all, given that there is a concept of agency available which is wider than the concept of agency used by philosophers of biology? Taking seriously this latter point, we might come to think that concepts, such as embodiment, material self-constitution, autonomy or (intrinsic?) normativity, are useful in the first place as heuristic means without obliging us to exact reproduction, this in accordance with the reasonable maxim that in order to explore what we don't know it makes sense to start with what we (at least to some degree) do know.

If the target of embodied robotics is to build agents *exactly alike* (terrestrial) bio-agents, i.e., to emulate agency-as-we-know-it, then we have good reasons to regard this project as unsuccessful so far. If the target of embodied robotics, however, is to build *some sort* of agents, and i.e., to emulate agency-as-it-might-be, then the question of whether roboticists have succeeded, or may succeed in the future, is open. How we are to answer this question depends, as indicated, crucially on general philosophical considerations about the meaning of the concept of agency, just as the answer to the question of whether and how artificial life is possible depends on general philosophical considerations about the meaning of the concept of life. The case of artificial agency turns out to be parallel to the case of artificial life in this respect: we cannot from agency or life as we know it extrapolate how it might be.

As explained in the introduction, the aim of this paper was, by assessing the project of embodied robotics, to determine the scope of the theory of bio-agency. This theory is becoming increasingly popular amongst philosophers of biology and possesses the potential to revolutionise the way philosophers of action used to think about agency. So far, philosophers of action have equated 'agency' with *human* agency, thereby tacitly following the Cartesian conviction that human mental performances stand out as singular exceptions within a homogeneous mass of purely mechanistic happenings in nature. The theory of bio-agency tells us that this picture is severely wrong. The power to act is not unique to humans; as a biological capacity it is rather manifested by non-human organisms as well. This development in philosophy is echoed by a parallel development in AI research: the traditional focus on higher cognitive functions as to be found in humans has been replaced by attempts to artificially reproduce basic motility-related skills as to be found in non-human animals, acknowledging the continuity between life and mind.

Yet, exactly the fact that, as I have argued in this paper, neither on empirical nor on conceptual grounds can we rule out these attempts – the attempts of biology-inspired embodied robotics – to be feasible ways to artificially generate agency, demonstrates the limits of the bio-agency approach. Even though this approach is promising and overdue with respect to the task of elaborating an empirically informed and appropriate metaphysical understanding of agency, we should not hastily equate agency with bio-agency. We have to be aware of the possibility of unforeseen, and unforeseeable, instantiations of agency.

At this point someone might feel tempted to make one last effort to defend Moreno and Etxeberria's argument from metabolism against artificial agency. Is this argument not merely meant to show that embodied robots fail to be *true* agents, while still allowing for them to be agents *in some weaker sense*? Whatever the answer to this question is, I hope to have shown that there is no justification for reserving the predicate 'true' for (terrestrial) bio-agency and dismissing all other possible forms of agency as 'not true' without giving any further reasons apart from those already offered. One might even wonder what such a distinction between 'true' and 'not true' agency would amount to. What is the supposed antonym of a 'true' agent? A 'fake' agent? Something that only pretends to be an agent? In that case, saying that embodied robots do 'not truly' act is actually and amount to saying that they don't act at all. If, however, the idea is that embodied robots are somehow deficient agents, then we are in need of an argument that explains why we should take bio-agency to be the paradigmatic and 'perfect' case of agency, compared to which all other (actual or hypothetical) forms of agency look deficient.

There is nothing in principle wrong with assuming that the concept of agency is gradable. Agency as we find it in organisms on earth clearly comes in degrees. Complex organisms are agents in a stronger sense than primitive ones. However, we might ask whether artificial and natural agents are actually located in the same spectrum or whether they rather belong to two different kinds of agency, both subsumed under the generic term 'agency'? This touches on a fundamental question concerning the conceptual profile of the concept of agency: is 'agency' an umbrella term, comprising various kinds of agency, such as bio-agency, robot agency and divine agency, which each are of distinct quality and follow their own rules? Or is the only valid distinction actually between full or strong versus reduced or weak agency, implying that there is only one spectrum in which all forms of agency are to be located? Could there be a way to combine these two views?

I take it that these questions have to be answered in a joint effort by the philosophy of action, general metaphysics, the philosophy of biology and embodied robotics. On such interdisciplinary footing, research in Artificial Agency will substantially contribute towards a better understanding of agency in general, just as Langton predicted that "Artificial Life can contribute to theoretical biology by locating life-as-we-know-it within the larger picture of life-as-it-could-be" (Langton 1989, 1).

Acknowledgments This paper was funded by the European Research Council, grant agreement number 324186 ("A Process Ontology for Contemporary Biology"). Previous versions of this paper were presented in September 2015 at the Internal "Physics and Philosophy" workshop of the Research Project "Agency and (Quantum) Mechanics" (funded by the Templeton World Charity Foundation) in Reichenau at Lake Constance in Germany, in March 2016 at the 2016 Conference of the German Society of the Philosophy of Science (GWP) in Düsseldorf in Germany and in June 2016 at the 2016 meeting of Philosophy of Biology in the UK (PBUK) in Bristol in the United Kingdom. I am grateful to the organisers and audiences, especially Hans F. Briegel, Thomas Müller and Josef Quitterer, for helpful discussions and suggestions. I am also grateful to John Dupré and an anonymous referee for helpful comments on an earlier draft of this paper.

References

Alberghina, L., and H.V. Westerhoff, eds. 2005. *Systems biology: Definitions and perspectives*. Berlin/Heidelberg: Springer.

Arnellos, A., and A. Moreno. 2015. Multicellular agency: An organizational view. *Biology and Philosophy* 30: 333–357.

Arnellos, A., T. Spyrou, and J. Darzentas. 2010. Towards the naturalization of agency based on an interactivist account of autonomy. *New Ideas in Psychology* 28: 296–311.

Bagatolli, L., J.H. Ipsen, A.H. Simonsen, and O.G. Mouritsen. 2010. An outlook on organization of lipids in membranes: Searching for a realistic connection with the organization of biological membranes. *Progress in Lipid Research* 49: 378–389.

Ball, P. 2008. Water as an active constituent in cell biology. *Chemical Reviews* 108: 74–108.

Barandiaran, X., and A. Moreno. 2006. On what makes certain dynamical systems cognitive: A minimally cognitive organization program. *Adaptive Behavior* 14: 171–185.

———. 2008. Adaptivity: From metabolism to behavior. *Adaptive Behavior* 16: 325–344.

Barandiaran, X., E. di Paolo, and M. Rohde. 2009. Defining agency: Individuality, normativity, asymmetry, and spatio-temporality in action. *Adaptive Behavior* 17: 367–386.

Bishop, J. 1989. *Natural agency. An essay on the causal theory of action*. Cambridge: Cambridge University Press.

Boden, M. 1999. Is metabolism necessary? *British Journal for the Philosophy of Science* 50: 231–248.

Boogerd, F.C., F.J. Bruggeman, J.-H.S. Hofmeyr, and H.V. Westerhoff, eds. 2007. *Systems biology: Philosophical foundations*. Amsterdam et al.: Elsevier.

Braitenberg, V. 1984. *Vehicles: Experiments in synthetic psychology*. Cambridge, MA: MIT Press.

Brembs, B. 2011. Towards a scientific concept of free will as a biological trait: Spontaneous actions and decision-making in invertebrates. *Proceedings of the Royal Society* 278: 930–939.

Brooks, R. 1991a. Intelligence without representation. *Artificial Intelligence* 47: 139–159.

———. 1991b. Intelligence without reason. In *Proceedings of 12th international joint conference on artificial intelligence*, 569–595. Sydney, Australia: Morgan Kaufman publishers.

Campbell, R. 2010. The emergence of action. *New Ideas in Psychology* 28: 283–295.

Clark, A. 1997. *Being there: Putting mind, world, and body back together*. Cambridge, MA: MIT Press.

Davidson, D. 1980. *Essays on actions and events*. Oxford: Oxford University Press.

De Jesus, P. 2016. Autopoietic enactivism, phenomenology and the deep continuity between life and mind. *Phenomenology and the Cognitive Sciences* 15: 265–289.

Di Paolo, E. 2003. Organismically-inspired robotics: Homeostatic adaptation and teleology beyond the closed sensorimotor loop. In *Dynamical systems approach to embodiment and sociality*, ed. K. Murase and T. Asakura, 19–42. Adelaide: Advanced Knowledge International.

———. 2009. Extended life. *Topoi* 28: 9–21.

———. 2010. Robotics inspired in the organism. *Intellectica* 53 (54): 129–162.

Dupré, J. 2012. *Processes of life. Essays in the philosophy of biology*. Oxford: Oxford University Press.

———. 2013. How much of the free will problem does (the right kind of) indeterminism solve? *Philosophical Inquiries* 1: 79–92.

Froese, T., and T. Ziemke. 2009. Enactive artificial intelligence: Investigating the systemic organization of life and mind. *Artificial Intelligence* 173: 466–500.

Gilbert, S.F., and A.I. Tauber. 2016. Rethinking individuality: The dialectics of the holobiont. *Biology and Philosophy* 31: 839–853.

Gilbert, S.F., J. Sapp, and A.I. Tauber. 2012. A symbiotic view of life: We have never been individuals. *Quarterly Review of Biology* 87: 325–341.

Haselager, W.F.G. 2005. Robotics, philosophy and the problems of autonomy. *Pragmatics & Cognition* 13: 515–532.

Jonas, H. 1968. Biological foundations of individuality. *International Philosophical Quarterly* 8: 231–251.

———. 2001 (1966). *The phenomenon of life. Toward a philosophical biology*. Evanston: Northwestern University Press.

Kim, S., C. Laschi, and B. Trimmer. 2013. Soft robotics: A bioinspired evolution in robotics. *Trends in Biotechnology* 31: 287–294.

Kirschner, M., J. Gerhart, and T. Mitchison. 2000. Molecular 'Vitalism'. *Cell* 100: 79–88.

Langton, C.G. 1989. Artificial life. In *Artificial life*, ed. C.G. Langton, 1–47. Redwood City: Addison-Wesley.

Maturana, H.R., and F.J. Varela. 1980. *Autopoiesis and cognition: The realization of the living (Boston Studies in the Philosophy and History of Science)*. Dordrecht et al.: Reidel.

———. 1987. *The tree of knowledge: The biological roots of human understanding*. Boston: Shambhala.

Meincke, A.S. forthcoming. Persons as biological processes. A bio-processual solution to the personal identity dilemma. In *Everything flows. Towards a processual philosophy of biology*, ed. D. Nicholson and J. Dupré. Oxford: Oxford University Press.

Meincke, A. S. in preparation a. *Autopoiesis, biological autonomy and the process view of life*.

Meincke, A. S. in preparation b. Bio-powers and free will.

Moreno, A., and M. Mossio. 2015. *Biological autonomy. A philosophical and theoretical enquiry*. Dordrecht: Springer. (*History and theory of the life sciences* 12).

Moreno, A., and A. Etxeberria. 2005. Agency in Natural and Artificial Systems. *Artificial Life* 11: 161–175.

Moreno, A., A. Etxeberria, and J. Umerez. 1994. Universality without matter? In *Artificial life IV*, ed. R.A. Brooks and P. Maes, 406–410. Cambridge, MA/London: MIT Press.

Nicholson, D. 2013. Organisms ≠ machines. *Studies in History and Philosophy of Science. Part C: Studies in History and Philosophy of Biological and Biomedical Sciences* 44: 669–678.

Nicholson, D., and J. Dupré, eds. forthcoming. *Everything flows. Towards a processual philosophy of biology*. Oxford: Oxford University Press.

Noble, D. 2006. *The music of life*. Oxford: Oxford University Press.

O'Malley, M.A., and J. Dupré. 2005. Fundamental issues in systems biology. *BioEssays* 27: 1270–1276.

Rosen, R. 1991. *Life itself. A comprehensive inquiry into the nature, origin and fabrication of life*. New York: Columbia University Press.

Rosslenbroich, B. 2014. *On the origin of autonomy. A new look at the major transitions in evolution*. Cham et al.: Springer.

Ruiz-Mirazo, K., and A. Moreno. 2004. Basic autonomy as a fundamental step in the synthesis of life. *Artificial Life* 10: 235–259.

Schlosser, A. 2015. Agency. In *The Stanford encyclopedia of philosophy*, ed. E.N. Zalta, Fall 2015 edn. https://plato.stanford.edu/archives/fall2015/entries/agency/

Shapiro, L. 2010. *Embodied cognition*. London/New York: Routledge.

Skewes, J.C., and C.A. Hooker. 2009. Bio-agency and the problem of action. *Biology and Philosophy* 24: 283–300.

Spann, A. S. (née Meincke). 2014. Bio-agency: Können Organismen handeln? In *Vermögen und Handlung. Der dispositionale Realismus und unser Selbstverständnis als Handelnde*, ed. A.S. Spann and D. Wehinger, 191–224. Münster: Mentis.

Steels, L., and R. Brooks. 1995. *The artificial life route to artificial intelligence: Building embodied, situated agents*. New Haven: Lawrence Erlbaum Assoc..

Steward, H. 2012. *A metaphysics for freedom*. Oxford: Oxford University Press.

Thompson, E. 2007. *Mind in life: Biology, phenomenology, and the sciences of mind*. Cambridge: Harvard University Press.

Trivedi, D., C.D. Rahn, W.M. Kier, and I.D. Walker. 2008. Soft robotics: Biological inspiration, state of the art, and future research. *Applied Bionics and Biomechanics* 5: 99–117.

Varela, F.J. 1979. *Principles of Biological Autonomy (The North-Holland Series in General Systems Research 2)*. New York: Elsevier.

———. 1995. The re-enchantment of the concrete. In *The artificial life route to artificial intelligence: Building embodied, situated agents*, ed. L. Steels and R. Brooks, 11–20. New Haven: Lawrence Erlbaum Assoc.
———. 1996. Neurophenomenology: A methodological remedy for the hard problem. *Journal of Consciousness Studies* 3: 330–349.
———. 1997. Patterns of life: Intertwining identity and cognition. *Brain and Cognition* 34: 72–87.
Varela, F.J., E. Thomson, and E. Rosch. 1991. *The embodied mind*. Cambridge, MA: MIT Press.
Ziemke, T. 2016. The body of knowledge: On the role of the living body in grounding cognition. *Biosystems* 148: 4–11.

Chapter 6
When Mechanisms Are Not Enough: The Origin of Eukaryotes and Scientific Explanation

Roger Deulofeu and Javier Suárez

Abstract The appeal to mechanisms in scientific explanation is commonplace in contemporary philosophy of science. In short, mechanists argue that an explanation of a phenomenon consists of citing the mechanism that brings the phenomenon about. In this paper, we present an argument that challenges the universality of mechanistic explanation: in explanations of the contemporary features of the eukaryotic cell, biologists appeal to its symbiogenetic origin and therefore the notion of symbiogenesis plays the main explanatory role. We defend the notion that symbiogenesis is non-mechanistic in nature and that any attempt to explain some of the contemporary features of the eukaryotic cell mechanistically turns out to be at least insufficient and sometimes fails to address the question that is asked. Finally, we suggest that symbiogenesis is better understood as a pragmatic scientific law and present an alternative non-mechanistic model of scientific explanation. In the model we present, the use of scientific laws is supposed to be a minimal requirement of all scientific explanations, since the purpose of a scientific explanation is to make phenomena expectable. Therefore, this model would help to understand biologists' appeal to the notion of symbiosis and thus is shown to be better, for the case under examination, than the mechanistic alternative.

Keywords Scientific explanation · Mechanistic explanation · Scientific laws · Eukaryotic cell · Symbiogenesis · Symbiosis

---

This work is fully collaborative; the authors are listed alphabetically.

R. Deulofeu (✉)
Department of Philosophy, University of Barcelona, Barcelona, Spain
e-mail: rogerdeulofeu@ub.edu

J. Suárez
Department of Sociology, Philosophy and Anthropology, University of Exeter, Exeter, UK
e-mail: javier.suarez@exeter.ac.edu

A. Christian et al. (eds.), *Philosophy of Science*, European Studies in Philosophy of Science 9, https://doi.org/10.1007/978-3-319-72577-2_6

6.1 Introduction

In recent years, mechanistic talk has become very popular among philosophers of science. Particularly, mechanistic talk has displaced the traditional approach to scientific explanation in terms of scientific laws (Nicholson 2012). Mechanists claim that scientific explanation consists of looking for a *causal process* –in this sense, the mechanistic movement is just the other side of the coin of traditional causal models of explanation– such that, through connecting the different entities and activities that participate in the process, the phenomenon that we aim to explain simply emerges. This claim is in contrast with the claim made by defenders of nomological expectability models of scientific explanation who generally claim that "to explain a phenomenon is to make it expectable on the basis of non-accidental regularities" (Díez 2014, 1414). Mechanists usually put forward biology as their main counterexample against defenders of nomological models: when biologists claim to have explained a phenomenon, they do so on the basis of having found a mechanism that brings that phenomenon about (Machamer et al. 2000). Biologists do not appeal to laws of nature, logical arguments, or any other kind of logic: they simply appeal to mechanisms. Thus, scientific explanation is, on this view, mechanistic explanation. In this paper, we contend this claim on its own terms, by presenting an example from biological practice. Specifically, we present the case of the origin of the eukaryotic cell and argue that the explanation of the salient features of this peculiar case is more suited to be understood in terms of a nomological expectability model of scientific explanation than in terms of mechanisms. For this purpose, we make explicit a kind of general regularity that biologists seem to be assuming when they provide explanations of the origin of the eukaryotic cell, and which forms the basis of the kind of proposals that they take as explanatory of certain facts that they consider particularly salient and in need of explanation (see Alleva et al. 2017, for a similar line of reasoning applied to the case of allosterism).

The paper is organised as follows: In Sect. 6.2, we introduce the symbiosis theory (ST, hereafter) of the origin of the eukaryotic cell, nowadays considered the canonical model for explaining the origin of eukaryotic cells, and we introduce a classification of the questions that ST provides answer to. In Sect. 6.3, we introduce the mechanistic account of scientific explanation complemented with Woodward's account of causality and provide evidence that suggests that the appeal to mechanisms is not the most appropriate way to justify the explanatory character of ST of the origin of the eukaryotic cell and why this is so. In Sect. 6.4, we present a nomological expectability model of scientific explanation that we then use to provide an understanding of the explanatory character of the ST of the origin of eukaryotic cells by considering that ST appeals to scientific laws. Finally, in Sect. 6.5 we conclude by defending the superiority of the nomological approach over the mechanistic approach in providing an understanding of the explanatory practices of biologists in the context of the theories of the origin of the eukaryotic cell and we propose future lines of research.

6.2 Symbiosis Theories of the Origin of Eukaryotic Cells

The biological world is populated by different kinds of entities, ranging from cells, to all kinds of multicellular forms of life. Cells are normally taken to be the basic and most fundamental unit of life, of which all the other entities are made up (Archibald 2014; Audesirk et al. 2008; Stearns and Hoekstra 2000). There are two main types of cells, classified according to the location of their DNA: prokaryotic cells (subdivided into the domains of Archaea and Bacteria) and eukaryotic cells. The main structural difference between prokaryotic cells and eukaryotic cells is that in the former, the genetic material is dispersed throughout the cytoplasm; whereas in the latter it is encapsulated within a membranoid-structure called the "nucleus". Apart from this, there are many other structural differences between the two types of cells, concerning aspects such as their size (eukaryotic cells generally being bigger), the types of membranes and the presence or absence of organelles. This last different constitutes a salient feature of eukaryotic cells, since only they host organelles within their bodies. Organelles are structural subunits, analogous to organs in humans, which perform certain functions within the body of the cell they belong to. Two of the organelles within eukaryotic cells are mitochondria (present in all eukaryotic cells) and chloroplasts (present only in plant eukaryotic cells); these two organelles bear their own DNA. Mitochondria are the site of cell respiration. Photosynthesis, in contrast, takes places within chloroplasts. Eukaryotic and prokaryotic cells are quite distinct from each other, and there does not seem to be any record of an intermediate form between the two types of cells, which is why certain biologists have referred to the origin of the eukaryotic cells as "the greatest single evolutionary discontinuity to be found in the present-day living world" (Stainer et al. 1963, quoted in Sagan 1967, 226). This immediately triggers a serious question for biologists: how did the first eukaryotic cell appear, given that all organisms share a common ancestor, and therefore eukaryotes and prokaryotes must have originated from the same ancestor?

Answering this question about the origin of the eukaryotic cell consists, among other things, of explaining the origin of cellular organelles, as the most salient subunits that allow for the distinction between eukaryotes and prokaryotes, and particularly of answering questions about the origin of mitochondria and chloroplasts. Mitochondria and chloroplasts are, then, one of the hallmarks of "eukaryocity" and, as Martin and his collaborators have put it, "the invention of eukaryotic specific traits required more metabolic energy per gene than prokaryotes have at their disposal, and (...) mitochondria afforded eukaryotic cells an order of magnitude increase in the amount of energy per gene, which (finally) explains *why* the origin of eukaryotes corresponds to the origin of mitochondria" (Martin et al. 2015, 2; also Williams and Embley 2015, Sect. 6.1).[1] Furthermore, it consists of justifying

[1]This point is however controversial, as some people have also defended the idea that other processes such as a phagocytosis might also be considered as the starting point of eukaryocity (e.g. Cavalier-Smith 1989). However, that would not remove the need to explain the origin of

the lack of continuity in the fossil record between eukaryotes and prokaryotes, the biochemical differences between the two types of cells, the different capabilities of one type of cells with respect to the other, etc. Explaining the origin of eukaryotic cells consists, therefore, of providing satisfactory answers to a series of why-questions (facts) about the particular features of the two kinds of cells and especially answering certain questions about the particular nature of each type. The family of surprising facts that a theory of the origin of the eukaryotic cell has to provide explanations of can be roughly classified as:

- *Physiological and biochemical questions*. The model of the origin of the eukaryotic cells has to explain, for instance, why the membrane of mitochondria is biochemically quite distinct from the membrane of the eukaryotic cell, but biochemically closely related to the nature of the membranes of certain prokaryotes; it also has to explain why the genetic material of eukaryotes has a mosaic nature, i.e. it is composed of phylogenetically distinct classes of DNA.
- *Phylogenetic questions*. Mitochondria and chloroplasts are not phylogenetically close to eukaryotes, but they are phylogenetically close to certain prokaryotes. This fact is surprising, since mitochondria are organelle in the eukaryotic cell, so one important question to answer would be why their genetic material is distinct in nature from the one present in the eukaryotic nucleus.
- *Historical questions*. The most important question to be answered is why there is a gap in the fossil record between prokaryotes and eukaryotes, if we take evolution to be continuous with no sudden evolutionary jumps.

So, a theory of the origin of the eukaryotes (i.e. a theory that answers the question: "How did eukaryotic cells originate?") should provide satisfactory answers to a list of why-questions of different natures, and evaluating its success at doing so is fundamental for the acceptance of one theory over another.

To answer the set of question outlined above, two families of theories have been proposed: on the one hand, *self-genetic or autogenous theories*, according to which the organelles within eukaryotes appeared as a consequence of invaginations within the original pre-eukaryotic cell (Raff and Mahler 1972; Uzzel and Spolsky 1974; all reviewed in Sapp 2010, 130–131; O'Malley 2010; Archibald 2015, R912); and on the other, *symbiosis or exogenous theories*, whose main claim is that eukaryotic cells originated through the symbiotic merger of two previously extant prokaryotic cells (Margulis 1970; Martin et al. 2012; Cavalier-Smith 2013; Dolan 2013). In short, the proponents of ST argue that the eukaryotic cell evolved as a consequence of a phagocytic process in which prokaryotes "were swallowed but not digested" (Margulis 1970, 60). The difference between the two families of theories is radical, and so are the conclusions that one can derive from them. For instance, if one

mitochondria and chloroplasts in a satisfactory manner, which would lead to the same kind of questions that we mention later. For the purposes of this paper and for simplicity, we will follow Martin's proposal that equates the origin of eukaryotes with the origin of mitochondria. Thanks to Thomas Bonnin for pointing this out to us.

defends an autogenous theory, one has difficulties explaining the genetic affinities between mitochondrial DNA and the DNA of free-living prokaryotes, since one has to explain how this foreign DNA arrived in the mitochondria of present-day eukaryotes. However, if one defends a ST, this fact becomes easily explainable: the fact that in the origin of eukaryotes two different prokaryotic lineages merged makes it more likely that the primitive lineage associated with mitochondria still preserves part of its original DNA. The same logic can be applied to all kinds of questions that might be raised about the difference between prokaryotes and eukaryotes. So, the capacity to play a more effective role in scientific explanation proves to be a good feature for preferring one theory to another.

Nowadays, the ST family predominates among biologists, although the versions of it come in many different forms, with at least 20 different models that explain the origin of the eukaryotic cells appealing to symbiosis (Archibald 2015). What matters for the purposes of this paper is the general structure of the arguments that appeal to endosymbiosis to explain the origin of eukaryotes and to explain the set of why-questions that we have selected as relevant, more than the peculiarities of the different models.

In general, ST appeals to the notion of *symbiogenesis* as the process by which the eukaryotic cell originally appeared.[2] This symbiogenetic process is supposed to have given rise to an endosymbiotic relationship between the different interacting organisms. The initial organisms involved in the origin of the first eukaryote are hypothesized to have been an archaeon (although there is no definite consensus on this question), with the capacity to phagocytize other microorganisms, and an alpha-proteobacteria, which would have given rise to mitochondria as we know it today (Spang et al. 2015). The peculiar nature of *symbiogenesis* qualifies it as the reason that biologists offer to explain the surprising features that are observed in eukaryotic cells. For instance:

- Why is the membrane of mitochondria biochemically more similar to free-living proteobacteria than to its host, i.e. the eukaryotic cell itself?
  Because it originated through symbiogenesis, which means that a free-living microorganism was engulfed but not digested and therefore it is very likely that the lineage this previously free-living microorganism gave rise to still preserves some of its original biochemical properties, such as the composition of the membrane.

[2]Symbiogenesis is the process of generation of a new biological structure (organ, metabolic pathway, etc.) as a consequence of a long-term symbiotic association. In the case of the eukaryotic cell, symbiogenesis refers to the origin of a complete new biological domain as a consequence of symbiosis. Symbiotic organisms can interact in two different ways: endosymbiotically, if one organism lives within the cell(s) of the other, and ectosymbiotically, when one organism lives on the surface of the cell(s) of the other, but not within them (Archibald 2014). Symbiogenesis is thus a process, whereas endosymbiosis is a state. This distinction has to be kept in mind for the rest of the paper. Thanks to an anonymous reviewer for encouraging us to clarify this point.

- Why does the eukaryotic genome have a mosaic nature?
  Because it originated by symbiogenesis. This entails two free-living organisms suddenly stopping their free-living mode to live together as a unit. As a consequence of a long-term relationship after symbiogenesis, it is very likely that there will be genetic exchange between the partners, thereby creating the mosaic structure of the eukaryotic genome.
- Why are mitochondria phylogenetically closer to free-living alpha-proteobacteria than to their host?
  Because if mitochondria were once free-living microorganisms that, via a process of symbiogenesis, became organelles within the eukaryotic cell, it seems natural that their DNA would be phylogenetically closer to the DNA of the free-living forms from which they originated than to eukaryotic DNA.
- Why is there a gap in the fossil record between prokaryotes and eukaryotes?
  Because if eukaryotic cells appeared through symbiogenesis, it is very unlikely that intermediate forms would be found in the fossil record. Symbiogenesis is a discontinuous process.

The appeal to symbiogenesis is therefore used as a general strategy to answer a different set of why-questions concerning particular features of the eukaryotic cell, providing answers that trace these features back to their evolutionary origin. In the following sections, we analyse whether this general strategy used by biologists is more in accordance with a mechanistic theory of explanation or with a nomological expectability approach, and we argue that what biologists actually do seems closer to the latter.

6.3 Mechanistic Explanation

Mechanistic explanation is the most influential approach to explanation in biology. The view was originally presented in direct opposition to the (previously) dominant nomological models of scientific explanation. Mechanists argue that in order to explain a biological phenomenon it is necessary to describe the mechanism that brings the phenomenon about (Glennan 1996, 2002; Bechtel 2011; Bechtel and Richardson 1993; Bechtel and Abrahamsen 2005; Machamer et al. 2000; Craver 2006, 2007; Darden and Craver 2002). Describing a mechanism, they claim, is not the same as presenting a scientific law that underlies a phenomenon. In fact, they deny the possibility of explaining a phenomenon by subsuming it under laws. In other words, the explanatory character of a mechanism does not lie on its supposedly underlying regularities, but in the identification of causal relations: "while it is sometimes the case that description of the inner parts of the mechanism will entail a description of the mechanism's outward behaviour, the explanation lies not in the logical relation between these descriptions but in the causal relations between the parts of the mechanism that produce the behaviour described" (Glennan 2002, S348; see also Machamer et al. 2000 for a similar argument).[3]

There are several ways of describing what a mechanism is. For instance, Machamer et al. (2000, 3, our emphasis) claim that a mechanism is a set of "*entities* and *activities organized* such that they are productive of regular changes from starting or set-up conditions to finish or termination conditions"; Glennan (2002, S344, our emphasis) defines a mechanism by saying that it is a "complex system that produces the behavior by the *interaction* of a number of *parts*"; Bechtel (2006, 26, our emphasis) says that it is "a structure performing a function in virtue of its component *parts*, component *operations*, and their *organizations*".

It seems clear from the above definitions that all of them presuppose that a mechanism consists of a set of entities and activities (or parts and operations/interactions) plus their corresponding organization.[4] To identify a mechanism, therefore, one has to disentangle its parts (the entities), individuated by their properties, and the activities it is involved in, "the producers of change". Allegedly, the properties of the entities plus their organization are responsible for the way in which the activities come about. In the words of Machamer et al.: "Mechanisms are identified and individuated by the activities and entities that constitute them, by their start and finish conditions and by their functional roles" (2000, 6). This dualist reading of mechanisms in terms of entities and activities generates a new framework that should, in principle, be fruitful when it comes to clarifying notions such as causation, lawhood, function and explanation. In particular, the notion of activity is supposed to play the role of causes, laws and functions. For instance, if a law is supposed to be a regularity of something that acts in the same way under the same conditions, philosophers of a mechanistic bent can provide a similar reading of a mechanism: "a mechanism is the series of activities of entities that bring about the finish or termination conditions in a regular way" (Machamer et al. 2000, 7). According to such authors, these regular mechanisms are not accidental and can give support to counterfactual reasoning. Therefore, there is no need to talk of laws in biology, for their role is already played by the identification of activities within mechanisms. In the same vein, Glennan refers to the interactions within a mechanism as "*invariant change-relating generalizations*" which can support counterfactual claims (Glennan 2002, S344).

[3]Leuridan (2010) argues that for every mechanism we can find an underlying regularity. His conclusion is that the explanatory character of mechanisms lies precisely in these hidden regularities, which actually is conceding too much to the nomological expectability models mechanists were criticizing in the first place.

[4]In fact, the notions of entities and activities come from a modification of the previous description of a mechanism in terms of parts and operations/interactions. Bechtel and Glennan still define mechanisms by appealing to the notions of parts and operations/interactions. The motives for their choice can be found in Bechtel and Abrahamsen (2005, fn. 5). Machamer et al. (2000, §3) introduced the new notions of entities and activities, mainly for ontological reasons. We take this not to be a substantive distinction for the purposes of this paper.

Given this characterization of a mechanism, we can now say that to give a mechanistic explanation of a given phenomenon consists of giving a description of the mechanism that brings the phenomenon about, such that the explanans includes the set-up conditions (arbitrarily taken as the beginning of the mechanism) plus the intermediate entities and activities together with their organization.

Nonetheless, there still remains the problem of providing criteria for identifying the different parts that compose a mechanism and that should be taken as relevant for the purposes of explanation. One possible way out of this problem, adopted among others by Craver (2007, 144), is to make use of Woodward's manipulability criteria for identifying causes (Woodward 1997, 2000, 2003). Woodward's strategy is to look for a "difference-making" clause in the explanans that, if we were to change it in various possible ways, would result in the final phenomenon being different. This strategy is mainly interventionist: if we want to identify the relevant factors for the production of a particular phenomenon, we must block certain parts allegedly involved in the causal path that terminates in the phenomenon to see whether this intervention has any consequence on the final output. Following this line of reasoning, one can say that "a part is causally relevant to the phenomenon produced by a causal mechanism if one can modify the production of this phenomenon by manipulating the behavior of the part, and one can modify the behavior of the part by manipulating the production of the phenomenon by the causal mechanism" (Nicholson 2012, 160).

Woodward is conscious that the interventions he requires to uncover the causes of phenomena might not always be available (think, for example, of historical phenomena). In order to resolve this difficulty, he argues that in those contexts where such manipulation is not feasible, the manipulability strategy takes the form of a counterfactual claim: "The notion of information that is relevant to manipulation thus needs to be understood modally or counterfactually: the information that is relevant to causally explaining an outcome involves the identification of factors and relationships such that if (perhaps contrary to fact) manipulation of these factors were possible, this would be a way of manipulating or altering the phenomenon in question" (Woodward 2003, 10). In other words, even in contexts where manipulation is not possible, it is "heuristically useful" to pursue or think of causes in the same way as we do when the relevant manipulation is available.

The task now is to try to apply the mechanistic schema plus Woodward's account of causes to the explanation of the origin of the eukaryotic cell in order to test its usefulness. We will advance question by question, following the schema presented in Sect. 6.1:

- Why is the membrane of mitochondria biochemically more similar to free-living proteobacteria than to its host, i.e. the eukaryotic cell itself?
- This question is about similarity, i.e. it is about why certain biomarkers are similar in an organelle and an organism that is phylogenetically distant from the eukaryotic cell that bears the organelle, whereas those biomarkers are different between the eukaryotic cell and its organelle. The mechanist would want to look for the different entities and activities, and their organization that would

allow the phenomenon under investigation (the nature of the membrane) to occur. If we were to do that, the entities would be the membranes and their biochemical nature; the activities would be those of membrane synthesis and membrane destruction; and the organization would depend on the way in which the aforementioned parts are spatiotemporally located in standard mitochondria. Let us suppose we follow this strategy. It is highly likely that we will discover many details about membrane synthesis, the biochemical products that are produced, the way in which they relate to each other, how they become arranged within mitochondria to give rise to a new membrane, etc. However, valuable as this information might be, it does not provide us with the answer we are looking for. This line of research would isolate the causes, allow interventions and provide a better understanding of membrane composition and membrane synthesis. But this is not what we were looking for in the first place. Our question concerned the similarities between mitochondria and a free-living microorganism, and the best answer to the question lies in symbiogenesis, as we mentioned in Sect. 6.1, and nothing in the strategy that the mechanist might elaborate mentions symbiogenesis.

Nevertheless, the mechanicist might still try to argue that the explanation lies in symbiogenesis because symbiogenesis is, in this particular circumstance, a mechanism. The problem is that we are looking for a historical explanation and thus we can only apply Woodward's counterfactual strategy. But this does not seem to do the trick either. First, the notion of symbiogenesis does not look like a mechanism at all: it is a very formal and general notion which does not make any reference to entities (it is supposed to cover a wide range of them, from the eukaryotic cell to most insect microbiota), activities (also very wide and diverse, from oxidation of glucose to synthesis of essential amino acids) or organization (which can be very variable). Second, because of the complexity of symbiogenesis, one cannot even imagine a set of factors whose alteration would block the phenomenon from appearing. If the factor we blocked was the symbiotic merger itself, then the result is not that we do not have a biochemical similarity between mitochondria and certain free-living bacteria: the result is that we do not even have either mitochondria or eukaryotic cells in the first place.

- Why does the eukaryotic genome have a mosaic nature?
  The argument in this case is very similar to the previous one. The mechanist philosopher might try to isolate certain biochemical elements of the eukaryotic genome whose presence is responsible for the mosaicism. However, these different elements are merely biochemical and do not respond to the question asked in the first place; at most, the mechanist might provide us with a very good mechanistic explanation (in terms of parts, activities and arrangements) of why it is that the compounds of a eukaryotic genome admit mosaicism. But this does not explain why in fact all eukaryotes present this kind of genome and, more specifically, why they present the kind of additional genes they do with the particular functions they have. Again, we need a historical explanation to satisfy our queries and to ask "why" the genome in fact has a mosaic nature. As we said in Sect. 6.1, symbiogenesis can provide a reply to this: the fact that

distinct organisms came together to form the eukaryotic cell and have been living together for 1.5 billion years (with all the "arms races" that exist when distinct organisms live together) would explain this feature and would even explain the specific functions of the genes involved in such mosaicism (namely, these related to avoiding cheating on the part of mitochondria).

Once again, the mechanist philosopher might claim that, if, as biologists assume, the appeal to symbiogenesis provides the right answer, this is because symbiogenesis is a mechanism. But then the mechanist philosopher would have to acknowledge that the concept of a symbiogenetic process is so formal that no entities, activities or organization can be properly recognized and isolated, so as to identify a mechanism. Then the mechanist philosopher would have two options: either to relax the notion of mechanism, which would mean that the concept is made either empty or equivalent to the notion of regularity, or to accept that this fact is not explainable in mechanistic terms.

- Why are mitochondria phylogenetically closer to free-living alpha-proteobacteria than to their host?

  Here, the argument against mechanists precisely mimics that presented for question one, merely changing all the details concerning membranes for details concerning phylogenetic relations; so to save space, we will not repeat it.

- Why is there a gap in the fossil record between prokaryotes and eukaryotes?
  In this case, a defender of the mechanistic model of explanation might claim that we can always imagine a particular set-up with certain initial conditions and reason counterfactually (as Woodward proposes for historical explanations in general). Let us fix those set-up conditions. It would be a set-up where archaea and bacteria merge symbiotically. If we want to provide a reason why there is a gap, we have to isolate a factor such that, if we block it, the result would be different. Suppose for the sake of argument that symbiosis is such a factor and imagine that archaea evolve progressively until they give rise to eukaryotes.[5] Would this entail that there is no gap in the fossil record? Not necessarily. We have cases of gaps in the fossil record that are not due to symbiosis. For instance, nobody believes that the famous missing link between dinosaurs and birds is a consequence of symbiosis, despite this missing link creating a gap in the fossil record. Furthermore, there are examples of symbiotic mergers where no gap is present. Paracatenula is known to be the result of ancient endosymbiosis, but its existence does not entail that there is a gap in the fossil record between Paracatenula and other catenulid flatworms (Gruber-Vodicka et al. 2011). Therefore, reasoning counterfactually in the strict manner Woodward suggests does not help to explain this particular phenomenon. It seems that what is required is the assumption of a very particular pattern that follows a unique event (namely, a symbiotic merger). This pattern, due to

[5]Earlier in this section we argued, as part of the response to question 1, that it was quite hard to conceive of symbiosis as a factor. We still believe this, for the reasons discussed there, but we are going to assume here that it might serve as one, just for the sake of the argument.

the complexity in determining its parts, activities and organization, cannot be interpreted mechanistically. In addition, it is difficult to see what a mechanistic reading in terms of parts, activities and organization can offer to explain the actual gap.

From this reasoning, three consequences follow. First, taking for granted that the appeal to symbiogenesis explains many of the issues about the origin of the eukaryotic cell, symbiogenesis is not, and it is very far from being, a mechanism.[6] Second, symbiogenesis seems to be more a general pattern which biologists appeal to in order to explain the features they find in the eukaryotic cell. Finally, even if the reference to mechanisms might complement explanations of some of the questions asked, and it might add some precision, the real explanatory role, as biologists accept, is played by the appeal to symbiogenesis.[7] Therefore, if symbiogenesis is not a mechanism but a general pattern, then it seems that the appeal to regularities might be explanatory after all.[8] In the next section, we further explore the possibility of considering symbiogenesis as a regularity.

6.4 Symbiogenesis as a Nomological-Expectable Explanation of the Origin of the Eukaryotic Cell

Biologists' appeal to the notion of symbiogenesis, as we have argued, has the form of a general pattern: the biologists look for a general principle, which may be quite vague (in the sense that it might be applicable to a large number of entities, irrespective of their particular biological properties), that allows them to say not only how the first eukaryotic cell came about, but also why it has the properties it has (which are the answers to the four why-questions we have presented, plus other

[6]In fact it would not even be a mechanism in the sense of "expanded mechanism" as defended by Roe and Baumgaertner (2016), since the problem here is not related to incorporating "pieces" of the environment, as they suggest: the problem is related to the fact that what plays the explanatory role is a regularity.

[7]One might still wonder about the exact relationship between mechanisms and regularities in certain explanatory contexts. It is not the aim of this paper to elucidate the nature of that relationship. Nonetheless, some mechanist philosophers have already recognized the use of non-accidental regularities in mechanistic explanations (e.g. Craver and Kaiser 2013; Glennan 1996) and we believe that, in most cases of mechanistic explanation, what does the real explanatory work is the presence of background non-accidental regularities. We plan to develop this line of thought further in a future paper.

[8]Of course, the defender of mechanistic explanation might still argue that the appeal to symbiogenesis is not, after all, explanatory. A similar strategy has been pursued by Craver (2008) concerning the explanatory character of the Hodgin-Huxley model of action potential in neurons. However, we believe that pursuing that strategy would violate some basic commitments common to biologists concerning explanation.

relevant questions that might be asked). It is convenient to specify at this point why we consider symbiogenesis to work as a regularity that might be used to account for certain facts (Archibald 2014; Douglas 2010).

First of all, symbiogenesis mere implies that the process by which an actual living organism has come about is a consequence of a symbiotic merger. Furthermore, in the case of the eukaryotic cell, it is always specified that this symbiogenesis gave rise to a case of endosymbiosis, whereby one organism lives inside the other. However, nothing about the particular nature of the organisms that interact endosymbiotically is specified, nor does it require to be specified in a general definition of symbiogenesis. Symbiogenesis just says something about how the mode of life of the organisms came about. Second, and related to the vagueness of the term, symbiogenesis is supposed to cover all the different cases of structures (and species) that emerge as a consequence of symbiosis between two different organisms. This entails that the entities that can interact symbiotically and give rise to a case of symbiogenesis are very different with respect to each other: bacteria, fungi, arthropods, mammals, etc.; they can all bear endosymbionts and/or enter endosymbiotic relationships with others. Third, by its very nature and its connection with the appearance of new biological structures, when it occurs through the acquisition of endosymbionts, symbiogenesis tends to trigger certain changes in the organisms involved: genomic decay, genetic assimilation, free exchange of genes between partners, vertical transmission, the appearance of particular bodily structures to bear the symbionts, etc. The evolution of these particular traits will differ depending on the particular relationship between the organisms and their necessities, and is normally what causes endosymbiotic states to be irreversible. Fourth and finally, symbiogenesis normally leaves some traces of the previously independent life of the partners. However, these traces vary quite a lot if we consider them on a case-by-case basis. Sometimes the traces will be biochemical pathways; others, molecular properties or chromosome structure, etc.

We believe that these four characteristics of symbiogenesis justify consideration of the phenomenon as a general pattern that biologists use in order to guide their research and to explain certain features that would not be explained otherwise. Indeed, the key aspect of symbiogenesis, in relation to accounting for the features of the eukaryotic cell as mentioned above, is that it makes these "expectable on the basis of [a] non-accidental regularit[y]" (Díez 2014, 1414). Nonetheless, this pattern, though general, is not empirically empty: it says something about the past and the future of the organisms which interact, and this can be studied further (and proved to be true or false). We believe that symbiogenesis, understood as we have specified above, is a kind of scientific law in Mitchell's sense (1997, 2000, 2003). In Mitchell's account, laws are understood pragmatically, according to the role they play in scientific practice. In other words, laws are not interpreted in terms of necessary and sufficient conditions, as traditional normative approaches suppose, but in terms of what they allow scientists to do. In this vein, Mitchell argues that a scientific statement must be understood as a scientific law if it allows good predictions to be made, good explanations to be provided and feasible interventions to be designed. This flexible conception of scientific laws allows her to provide a new multidimensional framework to represent a whole set of scientific generalizations (Mitchell 2000, 259). Furthermore, scientific laws in this sense provide a certain non-zero degree of nomic necessity,[9] which is established in terms of the stability of the conditions upon which the regularity is contingently dependent.[10] Therefore, the degree of nomic necessity of regularities in physics is higher than that of regularities in biology, because the stability of the conditions upon which a regularity is contingent in physics and in biology are significantly different. However, both regularities in physic and in biology involve a certain degree of nomic necessity; which is what matters here and is relevant for considering these generalizations as legitimate scientific laws.

In the context of the symbiosis models of the origin of eukaryotes, the appeal to the concept of symbiogenesis seems to play the role of a scientific law in this sense. First, it is often supposed that endosymbiotic association between two different organisms will give rise to a tendency for a series of adaptations to evolve that will increase the tightness of the fit between the partners. These adaptations will tend to evolve due to the possible presence of "cheaters", i.e. organisms that benefit from the association without providing any benefit to its partner. This is a consequence of the fact that endosymbiotic associations that are capable of evolving adaptations that prevent the possible presence of cheaters outrun those that are not. Second, it is also assumed that the partners in an endosymbiotic association will still preserve some traces of their previous free-living state, as a consequence of the special features of the symbiogenetic process. Indeed, symbiogenesis sometimes entails (and it definitely does so in the eukaryote case) a transition in biological individuality. But, as is well known, the framework of transitions in individuality assumes the existence of individuals whose evolutionary fates align and form a higher-level entity. It is precisely the existence of independent individuals whose individualities become combined into a higher-level unit what makes it reasonable to expect that certain features of their previously independent existence will be preserved. In addition, the features that are preserved could be studied in a lab, making certain predictions possible. It is in at least these senses that we believe symbiogenesis plays the role of a nomic pattern (a pragmatic law): it allows for certain predictions, makes a set of phenomena that can be empirically tested expectable and supports counterfactuals. This nomic character seems to be the aspect of the notion of symbiogenesis that biologists have in mind when they use it for explanatory purposes.

Of course, defenders of mechanistic explanation might still question the alternative that we offer to mechanistic models of scientific explanation. As is well-known, the models that have traditionally appealed to scientific laws as the main explanatory "weapon" are conceptually flawed –they have to face numerous problems: flag-pole

[9]See also Brandon (1997) for more about biological generalizations having a limited range of nomic necessity and explanatory power.

[10]Mitchel also includes other parameters: ontological ones (strength, plus the aforementioned stability) and representational ones (degree of abstraction, simplicity and cognitive manageability), which we take not to be relevant for our purposes in this paper. See Mitchell (2003, chapter 5) for more details.

cases, contraception pills and male pregnancy, syphilis-paresis cases, vitamin C and flu recovery, etc.– and are not very popular among contemporary philosophers of science (Woodward 2017). Maybe, after all, we have to admit that, although not perfect, as our case illustrates, mechanistic explanation is the best theory of scientific explanation that we have for the moment. Nonetheless, Díez very recently proposed a new neo-Hempelian account that solves most of the conceptual problems that have been raised against nomological expectability models and –allegedly– would include mechanistic explanations as a specific subcase satisfying additional conditions (Díez 2002, 2014). As this is the only nomological alternative we know of that has these features, we now proceed to evaluate whether Díez's model can accommodate the case of the origin of the eukaryotic cell.[11]

Díez's model takes as a point of departure Hempel's thesis that "to explain a phenomenon is to *make it expectable* on the basis of non-accidental regularities" (Díez 2014, 1414). This expectability, however, is not as strict as it has traditionally been in deductive/inductive nomological models (one of the possible forms that nomological expectability models can take), where the cases in which the explanation is based on a low-probability relationship between the explanandum and the explanans were excluded. The reason for this exclusion was that explanations were taken as logical inferences; thus, in the case of inductive inferences, they demanded high probability (Hempel and Oppenheim 1948; Hempel 1965). In contrast, Díez substitutes the notion of logical inference for the less demanding notion of "embedding": according to Díez, to explain a phenomenon is to embed it "into some branch of a net of theoretical constraints" (the explanans) such that they make the phenomenon expectable (Díez 2014: 1419). The idea of embedding is the structuralist counterpart to the positivist notion of *implication* and it presupposes a distinction in scientific models/structures between *data models* and *theoretical models* (Balzer et al. 2012). A data model is a structure that describes the phenomenon to be explained; whereas theoretical models are the structures defined by their satisfying a theory's laws. A data model is embeddable in a theoretical model when the former "fits" the latter, i.e. the relevant values of the phenomenon square with those of the theoretical model. Embedding is thus a relation between models, not a relation between sentences, which allows for a weakening of the positivist demand for logical inference (for instance, making room for embedding in increasing yet law probability cases) but still preserves the core intuition behind Hempelian *expectability*. To put it in Díez's words "[e]xplanations are (at least) certain kinds of predictions" (Díez 2014, 1420).

We will now provide an example of embedding. Suppose we want to explain the movement of the Moon using Newtonian mechanics. Our data model would include the Earth, the Moon, and the space and time functions that describe the kinematic trajectory of the Moon around the Earth, [$DM_{E,M}$ = <{Earth, Moon}, space, time>]. The theoretical model would include, apart from the aforementioned components,

[11]What follows is mainly based on the analysis we already presented in Deulofeu and Suárez (2015).

the functions of *mass* and *force*, [$TM_{E,M}$ = <{Earth, Moon}, space, time, mass, force>] defined by their satisfying Newtonian laws. The idea of the embedding of the data model within the theoretical model would be the following: by using the "machinery" of classical mechanics (laws of motion) plus the relative positions of the Moon and the Earth at a particular time, the theoretical model includes the relevant positions at other times; if such values fit the measured values of the data model, the former successfully embeds the latter, otherwise the embedding fails (and the theory has a Kuhnian anomaly). In this sense, model-theoretical embedding expresses the core intuition of nomological expectability.

However, as Díez explains and the case of the Moon's trajectory exemplifies, nomological embedding, though necessary, is not sufficient for explanation, since we may still fail to have explanatory embedding in two kinds of cases. First, one may have embedding by merely descriptive/phenomenological theories that systematize data with laws describing general phenomena without explaining them (e.g. Galilean kinematics or Kepler's laws). Second, in theories with what Kuhn calls "general schematic principles" such as Newton's Second Law (Kuhn 1970), one can always construct *ad hoc* trivial "successful" embedding that cannot count as explanatory. To exclude these cases, Díez adds two further conditions: the embedding has to be *ampliative* and *specialized.* Its *ampliative* character is based on the notion of T-theoreticity (Balzer et al. 2012; related to Hempel's distinction between "characteristic" and "antecedently understood", and Lewis's distinction between old and new vocabulary). T-theoretical concepts are those introduced by a theory such that, in order to determine their extension, one has to use/accept some T-law (e.g. *mass* and *force* in classical mechanics); whereas T-non-theoretical concepts are those which are already available and that can be determined (at least on some occasions) without the help of T-laws (e.g. *space* and *time* in classical mechanics). Explanatory embedding is ampliative, as in the case of classical mechanics: classical mechanics explains why the Moon is in location X at time t through embedding the phenomenon and introducing new T-theoretical concepts/entities (masses and forces) that do not appear in the data model $DM_{E,M}$. Thus, for embedding to be explanatory, it must make use of laws that (as in classical mechanics and not in Galilean kinematics or Keplerian astronomy) appeal to new concepts/entities. *Specialization*, on the other hand, requires that we introduce non-*ad hoc* "special laws" in order to account for the phenomena.[12] As Díez points out, we always require that our explanations include something more than merely schematic, very general principles such as $\sum f = ma$. In the case of the Moon–Earth system, for example, we need to introduce the law of universal gravitation, $f = G^{*}mm'/r^2$, if we aim to explain the positions of the Moon over time.

In short, we might now say that a theory explains a phenomenon if: (1) we can embed the phenomenon in the theory, in such a way that the theory makes the phenomenon expectable; (2) the theory includes and makes use of at least one T-theoretical term; and (3) the theory incorporates and makes use of at least one special

[12]As we said before, the notion of law that we use is Mitchell's idea of pragmatic law.

law in order to account for the phenomenon (Díez 2014, 1425). We will show that the appeal to symbiogenesis that biological theory makes to explain the origin of eukaryotes and the different phenomena laid out in Sect. 6.1, which does not fit the mechanistic account, is nevertheless perfectly legitimate and can be explicated by applying Díez's re-elaborated model of explanation as nomological expectability.

First, the appeal to symbiogenesis provides a theoretical model that allows the embedding of the phenomena that constitute our data model. In the case of the origin of the eukaryotic cell, the data model would include objects such as membranes –of both cells and mitochondria– or genomes –again, both cell and mitochondrial genomes– and their respective biochemical properties –those of the lipid components of the mitochondrial membrane versus those of the lipid components of the cell membrane; circular, single-strand DNA versus linear, complex DNA, etc.– ($DM_{G,M}$ = <{genome, membrane}, biochemical properties of both>). The theoretical model would include these objects plus entities/functions that correspond to the notions of *fitness* and *symbiogenesis*, which are purely theoretical and associated with particular features of symbiosis relationships and the theory of natural selection ($TM_{G,M}$ = <{genome, membrane}, biochemical properties of both, fitness, symbiogenesis>).[13] The embedding is possible in this case because $DM_{C,M}$ happens to actually be a submodel that squares with $TM_{G,M}$, and $TM_{G,M}$ makes the phenomena we aim to explain expectable (as reviewed in Sect. 6.1 in response to questions 1–4).

Furthermore, $TM_{G,M}$ includes a couple of T-theoretical entities/functions, fitness and symbiogenesis, that play an *ampliative* role. Biologists do not explain the features of the mitochondrial genome by appealing to features of free-living bacteria. They explain them by appealing to the idea of symbiogenesis (and its specific endosymbiotic form): certain formerly free-living bacteria (that we can indicate through phylogenetic analysis) were at some point endosymbiotically acquired by an archaeon and, *symbiogenetically*, gave rise to the organelles that nowadays we call mitochondria. The preservation of the genetic and biochemical features of the mitochondrial membrane is explained by appealing to its symbiogenetic origin plus the fact that they confer fitness advantages. In this sense, it seems clear that the embedding is ampliative in the sense Díez's account requires.

Finally, the explanation in terms of symbiogenesis includes an element of specialization in relation to ST (or the concept of symbiosis): an appeal to a special law which plays a non-trivial role in the explanation of the particular features of mitochondria. Symbiogenesis is a particular form of integration that two symbiotically associated organisms could enact, if the circumstances were favourable. It is well established that there are different types of symbiotic relationship (mutualism, commensalism and parasitism); some might be long-term evolutionarily relationships that are not conducive to integration, whereas others

[13]This reconstruction is merely informal and, due to particular complexities of biological theory, it cannot be made as precise as it could be in the case of classical mechanics. In any case, it has all the elements that are supposed to provide a general idea concerning embedding.

are. If they are conducive to integration and they have the desired fitness effects (i.e. they do not lead to the extinction of the integrated lineages), then they would trigger certain changes in the lineages that evolve symbiogenetically (mosaicism, genomic decay, loss of independent modes of life, etc.), giving rise to the appearance of new biological structures (they would fall down an evolutionary "rabbit hole", as some biologists describe it, e.g. Moran and Sloan 2015). In contrast, if the symbiosis relationship does not lead to integration, even if it is a long-term relationship, it would lead to a different kind of changes that would affect to both organisms independently, such as certain phenotype changes, changes in behaviour, etc. In this sense, symbiogenesis plays the role of a special law concerning a more general principle of the expected outcomes of long-term symbiotic associations.

We believe this appeal to a special law is the crucial step in ST, it is what provides the main explanatory power and as we argued, it does not have the form of a mechanism. The special symbiosis law certainly is such in Mitchell's pragmatic sense: it provides a certain degree of nomic necessity, therefore providing biologists with a guide to what they might find. For instance, appealing to a symbiogenetic origin makes it expectable that organelles, i.e. mitochondria, still preserve a certain degree of biological individuality that might be manifested, for example, by the possibility of in vivo replication. It is important to bear in mind that this would not be expected if the origin was self-genetic: in this latter scenario, we would never expect mitochondria to have any degree of biological individuality. Furthermore, if the origin of mitochondria is symbiotic, we will not expect to find intermediate forms in the fossil record, since symbiosis gives rise to saltational evolutionary events, which would not be the case if the origin was self-genetic. This same line of reasoning might be applied to all the features that ST makes nomically expectable and, in this sense, we have something similar to a pragmatic law that provides the research field with some order.

We should still note something about the special character of the law. As we said before, the condition is introduced in order to avoid counting as explanatory cases in which we merely apply general principles to *trivially* justify why certain phenomena occur (using *ad hoc* mathematical functions in $\sum f = ma$ to explain intentional movement, for instance). One might argue that the appeal to symbiogenesis is still trivial in this last sense: it is just one general principle we could use to justify every feature we find in an organism. Nonetheless, this is not the case: the appeal to symbiogenesis rules out certain possibilities and it makes a difference (as does the appeal to $f = G^*mm'/r^2$, in the case of planetary movement). It specifies the manner in which evolutionary innovation can arise, and this is in contrast to other possibilities, such as mutation, recombination, methylation, changes in the developmental matrix, or even other types of long-term non-integrative symbiotic relationships. It specifies a very particular pattern followed by the organisms that experience this mode of generation of evolutionary novelties and precludes triviality by ruling out the appearance of certain features that other evolutionary pathways would make expectable.

In conclusion, we have provided a (partially weakened, partially strengthened) nomological expectability framework as a possible alternative to a mechanistic framework of scientific explanation that explicates why biologists consider ST a legitimate explanatory theory of the origin of the eukaryotic cell by appealing to the notion of (pragmatic) scientific laws. In this sense, we have provided reasons to justify why an account of scientific explanation in terms of laws (in the restricted sense we have given) might be appealing to gain an understanding of the explanatory practices of biologists in certain contexts; an understanding that –we have claimed (Sect. 6.3)– mechanist philosophers cannot provide.

6.5 Concluding Remarks

In this paper we have presented the symbiosis model of the origin of the eukaryotic cell together with a set of questions (phylogenetic, biochemical, etc.) that any theory of the origin of the eukaryotic cell must provide answers to. We argue that the notion of symbiogenesis, understood as the process by which a new biological structure (organ, metabolic pathway, etc.) originates as a consequence of a long-term symbiotic relationship, plays the entire explanatory role when biologists aim to provide an answer to the different questions we mention (Sect. 6.2). This said, we defend the idea that the mechanistic account of scientific explanation is not well-suited to understanding why the notion of symbiogenesis plays the entire explanatory role in these cases. First, we argue that every attempt to offer a mechanistic explanation to the questions previously mentioned turns out to be unsatisfactory, since they move to a level of detail which turns out to be unnecessary for the matters discussed; moreover, many of the causes that should be mentioned in a mechanistic account seem orthogonal to the type of phenomena that demands an explanation. Second, we show that the notion of symbiogenesis is far from being a mechanism as they are conventionally understood in the literature (in terms of parts, activities and organization): symbiogenesis is a regularity or general pattern that cannot be suitably captured in mechanistic terms (Sect. 6.3). Finally, we present Díez's nomological expectability model of scientific explanation as an alternative to mechanistic models of explanation and defend the notion that Díez's model helps in understanding the explanatory character of symbiogenesis, despite its not being a mechanism but a general pattern (Sect. 6.4). If our argument is sound, it shows how and why the appeal to general patterns –that might well be considered scientific laws in Mitchell's sense, as we argue– might be explanatory in some contexts, thus challenging the universality of mechanistic explanations. It remains to be explored, however, whether the nomological expectability approach to scientific explanation we have defended here could also be applied to other biological contexts, either as a complement to (e.g. Alleva et al. 2017) or as a substitute for mechanistic accounts.

Acknowledgments Different versions of this paper were presented at the VIII Meeting of the Spanish Society for Logic, Methodology and Philosophy of Science (University of Barcelona, 2015) and the III Conference of the German Society for the Philosophy of Science (University of Düsseldorf, 2016). We would like to thank all the participants for their helpful comments and suggestions. We would also like to thank Thomas Bonnin, Mark Canciani, José Díez, John Dupré, Çağlar Karaca, Adrian Stencel and an anonymous referee, who read previous versions of this paper and made helpful comments and suggestions. Finally, the Spanish Ministry of Economy and Competitiveness (FFI2016-767999-P) and the Fundación Bancaria la Caixa are formally acknowledged for their economic support.

References

Alleva, K., J. Díez, and L. Federico. 2017. Models, theory structure and mechanisms in biochemistry: The case of allosterism. *Studies in History and Philosophy of Science Part C: Studies in History and Philosophy of Biological and Biomedical Sciences* 63: 1–14.

Archibald, J. 2014. *One plus one equals one: Symbiosis and the evolution of complex life*. Oxford: Oxford University Press.

———. 2015. Endosymbiosis and eukaryotic cell evolution. *Current Biology* 25: R911–R921.

Audesirk, T., G. Audesirk, and B.E. Byers. 2008. *Biology: Life on earth*. Hoboken: Pearson Education.

Balzer, W., C.U. Moulines, and J.D. Sneed. 2012. *An architectonic for science: The structuralist program*. Vol. 186. Dordrecht: Springer.

Bechtel, W. 2006. *Discovering cell mechanisms: The creation of modern cell biology*. Cambridge University Press.

Bechtel, W. 2011. Mechanism and biological explanation. *Philosophy of Science* 78 (4): 533–557.

Bechtel, W., and A. Abrahamsen. 2005. Explanation: A mechanist alternative. *Studies in History and Philosophy of Science Part C: Studies in History and Philosophy of Biological and Biomedical Sciences* 36 (2): 421–441.

Bechtel, W., and R.C. Richardson. 1993. *Discovering complexity: Decomposition and localization as scientific research strategies*. Cambridge, MA: The MIT Press.

Brandon, R.N. 1997. Does biology have laws? The experimental evidence. *Philosophy of Science* 64: S444–S457.

Cavalier-Smith, T. 1989. Archaebacteria and Archezoa. *Nature* 339: 100–101.

———. 2013. Symbiogenesis: Mechanisms, evolutionary consequences, and systematic implications. *Annual Review of Ecology, Evolution, and Systematics* 44: 145–172.

Craver, C.F. 2006. When mechanistic models explain. *Synthese* 153 (3): 355–376.

———. 2007. *Explaining the brain*. New York: Clarendon Press.

———. 2008. Physical law and mechanistic explanation in the Hodgkin and Huxley model of the action potential. *Philosophy of Science* 75: 1022–1033.

Craver, C.F., and M.I. Kaiser. 2013. Mechanisms and laws: Clarifying the debate. In *Mechanism and causality in biology and economics*, ed. H.K. Chao, S.T. Chen, and R.L. Millstein, vol. 3, 125–145. New York: Springer.

Darden, L., and C.F. Craver. 2002. Strategies in the interfield discovery of the mechanism of protein synthesis. *Studies in History and Philosophy of Science Part C: Studies in History and Philosophy of Biological and Biomedical Sciences* 33 (1): 1–28.

Deulofeu, R., and J. Suárez. 2015. On the explanatory character of Margulis' theory of the origin of the eukaryotic cell. In *Proceedings of the VIII Meeting of the Spanish society for logic, methodology and philosophy of science*, ed. J.A. Díez, M. García-Carpintero, J. Martínez, and S. Oms, 276–280. Barcelona: Universitat de Barcelona.

Díez, J.A. 2002. Explicación, unificación y subsunción. In *Diversidad de la explicación científica*, ed. W.J. González, 73–96. Barcelona: Ariel.

———. 2014. Scientific w-explanation as ampliative, specialised embedding: a neo-hempelian account. *Erkenntnis* 79: 1413–1443.

Dolan, M.F. 2013. Endosymbiosis and the origin of eukaryotes. In *Evolution from the Galapagos: Two centuries after Darwin*, ed. G. Trueba and C. Montúfar, 125–128. New York: Springer.

Douglas, A.E. 2010. *The symbiotic habit*. Oxford: Princeton University Press.
Glennan, S. 1996. Mechanisms and the nature of causation. *Erkenntnis* 44 (1): 49–71.
———. 2002. Rethinking mechanistic explanation. *Philosophy of Science* 69 (S3): S342–S353.
Gruber-Vodicka, H.R., U. Dirks, N. Leisch, C. Baranyi, K. Stoecker, S. Bulgheresi, N.R. Heindl, M. Horn, C. Lott, A. Loy, M. Wagner, and J. Ott. 2011. Paracatenula, an ancient symbiosis between thiotrophic Alphaproteobacteria and catenulid flatworms. *PNAS of the United States of America* 108 (29): 12078–12083.
Hempel, C.G. 1965. *Aspects of scientific explanation and other essays in the philosophy of science*. New York: Free Press.
Hempel, C., and P. Oppenheim. 1948. Studies in the logic of explanation. *Philosophy of Science* 15 (2): 135–175.
Kuhn, T.S. 1970. Second thoughts on paradigms. In *The structure of scientific theories*, ed. F. Suppe, 459–482. Urbana: University of Illinois Press.
Leuridan, B. 2010. Can mechanisms really replace laws of nature? *Philosophy of Science* 77 (3): 317–340.
Machamer, P., L. Darden, and C.F. Craver. 2000. Thinking about mechanisms. *Philosophy of Science* 67 (1): 1–25.
Margulis, L. 1970. *Origin of eukaryotic cells: Evidence and research implications*. New Haven: Yale University Press.
Martin, W.F., M. Roettger, T. Kloesges, T. Thiergart, C. Woehle, S. Gould, and T. Dagan. 2012. Modern endosymbiotic theory: Getting lateral gene transfer into the equation. *Journal of Ensocytobiosis and Cell Research* 23: 1–5.
Martin, W.F., S. Garg, and V. Zimorski. 2015. Endosymbiotic theories for eukaryote origin. *Philosophical Transactions of the Royal Society B* 370: 20140330.
Mitchell, S.D. 1997. Pragmatic laws. *Philosophy of Science* 64: S468–S479.
———. 2000. Dimensions of scientific law. *Philosophy of Science* 67 (2): 242–265.
———. 2003. *Biological complexity and integrative pluralism*. Cambridge: Cambridge University Press.
Moran, N., and D.B. Sloan. 2015. The hologenome concept: Helpful or hollow? *PLoS Biology* 13 (12): e1002311.
Nicholson, D.J. 2012. The concept of mechanism in biology. *Studies in History and Philosophy of Science Part C: Studies in History and Philosophy of Biological and Biomedical Sciences* 43 (1): 152–163.
O'Malley, M. 2010. The first eukaryote cell: An unfinished history of contestation. *Studies in History and Philosophy of Science Part C: Studies in History and Philosophy of Biological and Biomedical Sciences* 41 (3): 212–224.
Raff, R.A., and H.R. Mahler. 1972. The non symbiotic origin of mitochondria. *Science* 177: 575–582.
Roe, S.M., and B. Baumgaertner 2016. Extended mechanistic explanations: Expanding the current mechanistic conception to include more complex biological systems. *Journal for General Philosophy of Science* 48: 517–534
Sagan, L. 1967. On the origin of mitosing cells. *Journal of Theoretical Biology* 14: 225–274.
Sapp, J. 2010. Saltational symbiosis. *Theory Biosciences* 129: 125–133.
Spang, A., J.H. Saw, S.L. Jørgensen, K. Zaremba-Niedzwiedzka, J. Martijn, A.E. Lind, R. van Eijk, C. Schleper, L. Guy, and T.J.G. Ettema. 2015. Complex Archaea that bridge the gap between prokaryotes and eukaryotes. *Nature* 521: 173–179.
Stearns, S.C., and R.F. Hoekstra. 2000. *Evolution. An introduction*. Oxford: Oxford University Press.
Uzzel, T., and C. Spolsky. 1974. Mitochondria and plastids as endosymbionts: A revival of special creation? *American Scientist* 62: 334–343.
Williams, T.A., and T.M. Embley. 2015. Changing ideas about eukaryotic origins. *Philosophical transactions of the Royal Society of London. Series B* 370: 20140318.
Woodward, J. 1997. Explanation, invariance, and intervention. *Philosophy of Science* 64: S26–S41.

———. 2000. Explanation and invariance in the special sciences. *The British Journal for the Philosophy of Science* 51 (2): 197–254.
———. 2003. *Making things happen*. New York: Oxford University Press.
———. 2017. Scientific explanation. In The *Stanford encyclopedia of philosophy*, ed. E.N. Zalta. https://plato.stanford.edu/archives/spr2017/entries/scientific-explanation/

Chapter 7
Functions, Malfunctioning, and Negative Causation

Ludger Jansen

Abstract Functional explanations apply not only in cases of normal functioning, but also in the case of malfunctioning. According to a straightforward analysis, a bearer of the function to F is malfunctioning if and only if it does not F although it should do so. This makes malfunctions and malfunctionings analogous to negative causation and thus peculiarly problematic, because they seem to involve absent dispositions and absent processes. This analysis seems also to require that the function to F cannot be identical with the disposition to F. Then we seem to be trapped in a dilemma: Either the realm of functions is separated from the realm of dispositions; then it seems that functions cannot be causally efficacious. Alternatively, functions are considered to be identical with dispositions; but then malfunctioning seems to be conceptually impossible. The paper defends and further develops the thesis of Röhl and Jansen (J Biomed Semant 5:27, 2014) that functions are not a special type of dispositions. For this purpose, it first reviews different varieties of malfunction and malfunctioning and suggests definitions of both malfunction and malfunctioning. It reviews how causal, etiological and intentional theories of functions can deal with these problems. In particular, the paper discusses the special-disposition account of the Basic Formal Ontology (BFO). Spear, Ceusters and Smith (Applied Ontology 11(2):103–128, 2016) have defended the special-disposition account of the Basic Formal Ontology (BFO) by suggesting various strategies how a special-disposition account can deal with malfunctions. On the one side, the paper evaluates these strategies and indicates several problems arising from them. On the other hand, it describes how to account for the non-optionality and the causal efficacy of functions, if functions are not dispositions. While function types are not identical to disposition types, there are important interrelations between functions and dispositions, namely

L. Jansen (✉)
Department of Philosophy, University of Rostock, Rostock, Germany

Faculty of Catholic Theology, Ruhr University Bochum, Bochum, Germany
e-mail: ludger.jansen@uni-rostock.de

A. Christian et al. (eds.), *Philosophy of Science*, European Studies
in Philosophy of Science 9, https://doi.org/10.1007/978-3-319-72577-2_7

(1) heuristically, (2) from a design perspective for artefact functions, and (3) from an evolutionary perspective for types of biological functions.

Keywords Function · Disposition · Malfunction · Negative causation · Contrastive explanation

7.1 Introduction

In science, as well as in ordinary discourse, we often seem to refer to negative causes. If the gardener does not water the flowers, he can cause them to die (Hart and Honoré 1959, 35). Non-smoking can be a cause of health (Merrill 2010, 88). Not eating anything can cause slimming, the absence of light can cause fear, and so on. There is a hot debate how an absence, i.e., something that is not there, can be causally efficacious at all – or, in short, how nothing can cause something (see, e.g., Birnbacher and Hommen 2012). At least some of the marks of causality are also displayed by these examples. Absences make a difference; they may give rise to regular connections; and they may allow for counterfactual reasoning. The shortest possible demonstration of the causal power of absences has been published by Tyron Goldschmidt in the journal *Dialectica* (Goldschmidt 2016). The paper consists of the title, the author details and – nothing. Where one would expect a long text detailing with much sophistication the argument for the causal power, there is a huge white spot. There is, that is, an absence of any body of text in this publication, causing confusion and maybe amusement on the side of the reader. This way Goldschmidt can score his goal: The absence of text causes confusion and amusement. Of course, in Goldschmidt's 'paper' the causal efficiency of absences is rather shown than stated. It is shown, indeed, by invoking the contrast to what readers normally expect when they consult a journal paper. In the present paper, I will argue that one can account for malfunctioning in an analogous way: Something does not happen although it should. Malfunctioning happens if an object does not display a certain activity although it has the function to display this activity. The object in question is not displaying this activity, or so I will argue, because it lacks the disposition to do so, although it possesses the respective function. I want to argue, that is, that the phenomenon of malfunctioning shows that something can have the function to F but not the matching dispositions to do so – for otherwise it would do so.

For this purpose, I first introduce a number of widely acknowledged desiderata for theories of functions (Sect. 7.2). The one desideratum I will extensively discuss in this paper is the possibility of malfunctioning, and thus I will continue with a review of various modes of malfunctioning (Sect. 7.3). I distinguish between having a malfunction and actually malfunctioning and propose definitions for both (Sect. 7.4). If it is true that a bearer of the function to F is malfunctioning if and only if it does not F although it should do so, then this seems to presuppose the possibility that something has the function to F but not the matching dispositions to do so – for otherwise it would do so. Some accounts of function suggest, however, that functions are dispositions – either any disposition (Sect. 7.5.1) or special dispositions (Sect. 7.5.2). Spear and colleagues have argued that their special-disposition account of function cannot only answer the malfunctioning challenge but is also superior to the non-dispositional account of function suggested by Röhl and Jansen (2014) in that it can explain the non-optionality of functions as well as their causal efficacy. I review the strategies which Spear and colleagues suggest as possible accounts of malfunctioning and conclude that there is no sustainable account of malfunctioning that is both general and dispositional (Sect. 7.6). In addition, I argue that the non-dispositional account can as well explain the non-optionality, or essentiality, of functions (Sect. 7.7), and that it is able to bridge the gap between non-dispositional functions and causal efficacy (Sect. 7.8). One of the strategies suggested by Spear and colleagues is to refer to kind membership as a source of the variety of normativity usually connected with a function. I conclude by indicating that such a reference to kind membership might as well be considered an explanation for having a non-dispositional function (Sect. 7.9).

7.2 Desiderata for a Theory of Functions

Functional explanations are common currency in biology, medicine, the social sciences, and engineering. Roughly, they work in two ways, which could be labelled "top-down" and "bottom-up". In a top-down way, features of a system's element can be explained by reference to their function and contribution to the whole system. Similarly, processes in which these system parts participate can be explained as realisations of the functions of these parts. In a bottom-up way, system processes can be explained by reference to the functions of the parts of the system. The power plant is secure because it has a cooling system that keeps temperatures low; the patient is stable because his heart functions normally. At least some of these explanatory patterns can be applied not only in cases of normal functioning, but, by way of negation, also in the case of malfunctioning. The power plant can explode because of a malfunction in the cooling system; the patient can die because of a malfunction of her heart. Both kinds of explanations are closely linked to each other:

- The patient needs his heart to function in order to be in a stable condition.
  The heart functions well.
  Hence, *ceteris paribus*, the patient is stable.
- The patient needs his heart to function in order to be in a stable condition.
  The heart does not function well.
  Hence, *ceteris paribus*, the patient is not stable.

Mind that these two explanations do not employ deductive inferences but only *ceteris paribus* reasoning. Other health issues could cause the instability of a patient; and medical devices like a heart-lung machine could stabilise a patient even if her heart is malfunctioning.

Cummins once equated functions with causal effects (Cummins 1975). By now, most authors on functions disagree with him. First, because the effects are rather conceived of as the realisations of the functions than the functions themselves. Second, because not all causal effects are considered to be realisations of functions. In biology and medicine, functions are typically ascribed to parts of organisms. It is a function of the heart to pump blood, and this is what hearts regularly do. Though hearts also regularly produce rhythmic noises, this is normally not regarded as a function of hearts. Similar observations apply to engineering, where functions are typically ascribed to artefacts and their parts. The function of a hammer is, e.g., to nail – and not to accelerate towards the centre of the earth when falling, although this is also a behaviour that hammers regularly display.

This shows that the effects of functions (or of, as they are sometimes called, "proper functions") are to be distinguished from other non-functional effects a thing may have. Hence, we have (at least) a twofold distinction: On the one hand, we have to distinguish a thing's activity or "functioning" from the function of the thing that continues to inhere in a thing when it is not functioning. The hammer continues to have the function to nail even when nobody is using it for nailing. On the other hand, we have to distinguish the function from non-functional properties. For sure, the hammer has a disposition to accelerate towards the centre of the earth even when it is not falling, but this is merely a disposition and not a function of the hammer.

The ability to draw this distinction is one of the desiderata for a theory of functions that are accepted by several authors (cf. Artiga 2011; Houkes and Vermaas 2010; Burek 2006; Röhl and Jansen 2014). Another such desideratum is that reference to functions should be able to play a role in the explanation of the existence of the function bearer; they should allow, that is, what is often considered a certain kind of teleology. Third, there should be space for innovation; i.e., it should be possible that new functions emerge in the history of life or technology. Fourth, the causally relevant properties of the function bearer should support the existence of the function or at least the function ascription. A fifth desideratum is that an account of functions needs to be able to account for the normative character of functions and thus be able to account for malfunctioning. Finally, it can be considered an advantage of a theory of functions if it is able to account for both biological and artefact functions (Röhl and Jansen 2014).

The main motivation behind the fourth desideratum is that functions should somehow be tied to the intrinsic properties of their bearers in order to be causally efficacious. This seems to exclude all accounts on which functions are external properties of some kind. There are several 'externalist' accounts to be found in the literature: Most prominent for biological functions is the evolutionary account, which seems to turn functions into historic, past-related properties, which have more to do with the fate of the function bearer's ancestors than with its actual physical make-up (Millikan 1984). Similarly, for artefacts, explaining functions in terms of the intentions of their designer (Houkes et al. 2002; Houkes and Vermaas 2010) would make them not only historic but also mind-dependent properties. Another view makes functions observer-relative properties. John Searle, for example, does not even view biological functions as properties intrinsic to their bearers. Rather, for Searle, functions are ascribed by observers (e.g., scientists) in view of certain underlying values – and the values responsible for function ascription in biology and medicine are survival and reproduction. In this perspective, the heart's function is to pump blood, contributing thus to these values. Alternative views are possible, as Searle remarks: "If we thought the most important value in the world was to glorify God by making thumping noises, then the function of the heart would be to make a thumping noise, and the noisier heart would be the better heart." (Searle 1995, 15).

The fourth desideratum seems to be of special importance for the issue of negative causality: What if there is no such support from the side of the qualities and dispositions of the purported function bearer? As the fifth desideratum seems to postulate that there are cases where this is the case, the present paper examines in particular the explanation of malfunctioning, and it is the variety of types of malfunctioning to which I will now turn.

7.3 Varieties of Malfunctioning

Some authors have misgivings with using the word "malfunction" at all. Discussing strategies for the construction of artefacts, Pahl and colleagues argue that functions are always intended purposes, and failures are never intended. For this reason, "malfunction" seems to be a misnomer, just like the German equivalents "*Fehlfunktion*" and "*Störfunktion*". Pahl and colleagues suggest using the word "*Störwirkung*" ("mal-effect") instead (Pahl et al. 2005, 226–227). Obviously, this observation is justified only on the background of the intentional framework for artefact functions. From the point of view of Cummins' causal theory, there can also be non-intended functions. Nevertheless, it should be clear that this is a mere terminological issue, as I know of no one who claims that malfunctions are special functions or special intended purposes. Even in artefacts, that is, a malfunction is not a special function and hence needs not be intended. Rather, in a case of a malfunction the function bearer fails to be able to realise the function properly.

A function can fail to be realised in various ways. Ulrich Krohs distinguishes, e.g., between malfunction and dysfunction (Krohs 2010). Malfunctions are all cases of not functioning properly. As many functions can be realised in degrees, with the proper functioning in the middle of two extremes, it is possible to distinguish between hyperfunctioning and hypofunctioning. A heart that pumps too much blood is hyperfunctioning; a heart that does not pump enough is hypofunctioning. While such hyper- or hypofunctioning of the heart is detrimental to health, an organism can live quite long with a moderately malfunctioning heart. In contrast, a heart that totally stops pumping is an immediate threat to life. It would not only be a case of malfunction, but also a case of dysfunction, i.e., a malfunction that threatens the life or health of an organism or, more generally speaking, the existence or functioning of the whole system.

In engineering, too, different varieties of malfunction can be distinguished. To start with, engineers often distinguish between failures and faults. For example, the International Electrotechnical Vocabulary (IEC 60050(191), IEV ch. 191) defines "failure" as "the termination of the ability of an item to perform a required function", and states, "After failure the item has a fault". "Fault", in turn, is defined as "the state of an item characterized by inability to perform a required function, excluding the inability during preventive maintenance or other planned actions, or due to lack of external sources". Hence, a failure is an event resulting in a fault. The IEC characterisation encloses what Luca del Frate (2012, 119) calls the "traditional definition" of function-based failure and function-based fault, respectively. According to del Frate, a *function-based fault state* obtains if an item is no longer able to fulfil its function. In addition, del Frate distinguishes two further notions of failure and fault. A *specification-based fault state* is "the state of an item characterized by inability to perform as specified under the specified operational conditions for which it is designed, excluding (i) the inability during preventive maintenance or other planned actions, or (ii) the inability due to lack of external resources, or (iii) the inability due to previous violations of specified operational conditions" (del Frate 2012, 123). A *material-based fault state* is "the state of an item resulting from any permanent change in the values of geometrical or physicochemical properties of the materials of an item which (i) renders the item unable to perform as specified or (ii) increases substantially the likelihood that the item will become unable to perform as specified" (del Frate 2012, 126). All three notions are, as del Frate argues, logically independent (del Frate 2012, 126–127). There is, however, a common core to all three notions: All of them come along with an (actual or expected) inability, i.e., with the lack of the disposition to perform in the desired way.

Just as functioning has to be distinguished from having a function, malfunctioning has to be distinguished from having a malfunction. Functioning is an activity or process, while a function is a property. Similarly, malfunctioning is a process or the non-occurrence of a process *during* time, while having a malfunction is the instantiation of certain property types *at* a time. That is, functions and malfunctions are continuant entities; they have, that is, only spatial parts and thus exist with all their parts at any time of their existence. In contrast, functionings and malfunctionings are occurrent entities; they may have temporal parts and stretch out in time (Smith 2005; Jansen 2008). Similarly, faults are continuants while failures are occurrents. This does not necessarily mean that malfunction types are irreducible universals; they rather seem to be defined classes, for (at least on the account defended in this paper) they are conjunctive complexes consisting in having a function and not having the matching proposition. In a way, malfunctionings are caused by malfunctions, but we should not say that malfunctionings are the realisations of malfunctions, at least not in the same way as we consider functionings to be the realisations of functions. This is, because a function is related to a single precise type of processes such that the realisation of the function is an instance of the process type. In contrast, malfunctioning can consist in any process instance that is not an instance of the realisation type of the function.

The most extreme case of malfunctioning is probably the case where no instance of a kind is able to realise its function. Hammers are produced, sold, bought, and used for certain typical activities – e.g., for nailing. Normally, hammers are able to display these activities; they have, e.g., the disposition to be successfully used in nailing processes. Other artefacts are, however, produced, sold, bought and even used for purposes they are not able to fulfil. For principled reasons, no perpetuum mobile function can ever be realised; and it is at least questionable whether homeopathic globuli can fulfil the functions for which they are produced, sold, bought and used. We can add more examples to this list: the anti-limescale Magnoball, the earth-radiation rectifier, the dream catcher, and amulets (all of which could be ordered via eBay in 2016). Sceptics have good reason for the assumption that none of these artefacts can live up to what they are advertised for.

If the sceptics are right, all of these examples are examples for what Ulrich Krohs has dubbed "type-malfunction" (Krohs 2010), and that Beth Preston dubs a "phantom function" – "which occurs when a type of item is regularly reproduced to serve a specific function, but no exemplar of it has ever been structurally capable of performing that function (Preston 2013, 138). No purported perpetuum mobile can ever fulfil the function ascribed to it, nor can any Magnoball or dreamcatcher do what it is advertised for. Still, it makes sense to ascribe to these artefacts the functions in question. A cultural anthropologist might want to say that certain artefacts are used as amulets, that they have the function to increase economic success, combat demons, or protect against diseases. The cultural anthropologist could corroborate his claim by reference to the intentions of the designer, maker, buyer, or user of the amulets. Thus, an intentional theory of functions would be able to account for type-malfunctions in the realm of artefacts. The truthmaker for the respective function ascriptions would be the respective intentions of people dealing with these artefacts.

In contrast, no such reference to intentions seems to be possible in biology. Note that this observation is independent from the background theory we may adopt for biology, i.e., independent from the choice between evolution and creation. Evolution works without intentions at all, and creation is standardly assumed to be done by an intelligent designer who will take care not to create things for purposes that they cannot fulfil. Thus, either there are no intentions to refer to at all, or the functional intentions fit to the actual dispositions of things created.

7.4 Malfunctions and Malfunctionings

As this review of different varieties of malfunctioning shows, a straightforward analysis of malfunctioning would be that something should perform a certain task, but it does not perform it. The cause for this is normally a malfunction, which consists in not being able to function properly. Hearts should pump blood, but some hearts, for different reasons, do not have the disposition to do so.

In order to spell this out more formally, we have to distinguish functions that are always to be performed from those that are only occasionally or maybe never realised. A function of the first kind is, e.g., the function of the human heart: It should pump blood at every time of its existence. Once a heart has started to pump blood, any time of not pumping would be a case of malfunctioning. In contrast, the human eye has at every time the function to see, but it does not have the function to see-at-every-time. When we close our eyes and see nothing, this does not imply that our eyes are malfunctioning. More extremely, the red button may have the function to be pressed in order to launch a missile with atomic weapons, but this function may never be realised.

For those functions that need always to be realised, every case of not functioning is a case of malfunctioning. In these cases, a malfunction (i.e., not being able to realise the function) then leads immediately to a malfunctioning. The situation is more complicated when a function needs not always to be realised. For the red button, the disposition to launch the missile may get lost without ever making a difference, because the button is never pressed. The broken red button would malfunction if pressed. In general:

- A material object x has a *malfunction* at t with respect to the function to F-in-situation-S, if and only if x has the function to F-in-situation-S but *would* not F in S, because, at t, it does not have the disposition to F-in-situation-S.
- A material object x is *malfunctioning* at t with respect to the function to F-in-a-situation-S, if and only if, at t, x is in situation S but does not F.

The characterisations of malfunction and malfunctioning are obviously not equivalent, but they are logically connected. The occurrence of malfunctioning implies the existence of a malfunction. Hence, it is possible to have a malfunction without actually malfunctioning, but not the other way round. Note that in accordance with the IEC definition of "fault" quoted in Sect. 7.2, a lack of external resources need not count as a malfunction. A car does not have the function to drive-without-petrol; hence, no car is expected to drive with an empty tank. It is also important how to describe functions and to distinguish them from intended effects of the realisations of the function. The red button may have the function to be pressed or to close a circuit by being pressed, or to give a signal to the missile launcher. That the missile launches and brings bombs to the enemy country is no longer part of the function of the button, but an intended causal effect of the realisation of its function.

In a case of perpetual realisation of the function, like in the case of the pumping function of the heart, changes in disposition will quickly lead to changes in the realisation. In the case of the red button, where the function may never be realised, changes in disposition may even pass unnoticed. Taken together, however, we have to distinguish three distinct aspects: activities, dispositions, and functions. An activity is what a thing *actually is doing*; disposition ascriptions tell us, what a thing *can do*; whereas function ascriptions tell us, what a thing *should do*. Precisely the phenomenon of malfunction urges us to separate the functional level from the dispositional level. Once we separated functions from dispositions, we are thrown back to the desiderata for any theory of functions: How, then, can we guarantee a causal relevance of functions? How can functions still be supported by the non-functional properties of the function bearer? It is to these questions that I will turn now.

7.5 Are Functions Dispositions?

7.5.1 Are All Dispositions Functions?

The causal relevance of functions would be obvious, if functions just were those dispositions whose realisations are the functional activities. A clear statement of this view is the so-called causal theory of functions (Cummins 1975), which could, for that reason, also be called a dispositional theory of functions (Artiga 2011). Like the causal theory in general, its identification of functions with dispositions comes along with a number of problems, which become clear once we check the desiderata for function theories outlined above. For reference to dispositions in general does not support the kind of teleological explanations we expect from reference to functions: Sure, both hearts and hammers have the disposition to accelerate towards the centre of the earth, but neither of them exists because of this disposition. While there is room for new dispositions, newly acquired dispositions are normally not (and not automatically) new functions, for dispositions can be accidental properties, while functions are often seen as "integral to the entities in question" (Arp et al. 2015, 103), i.e., as essential and necessary properties (Röhl and Jansen 2014). This is what distinguishes functions from roles, which can be acquired and lost without threating the existence of their bearer (a view shared, e.g., by Arp et al. 2015; and by Spear et al. 2016).

In general, having a disposition to do something is not the same as having a function to do so. Of two things that are alike with respect to all of their dispositions, one may have a certain function that the other does not share. In the unlikely event that by sheer coincidence atoms assemble to a complex structure that resembles, atom for atom, a fine watch constructed by a Swiss watchmaker, there will be all dispositions necessary for performing the job of a watch, but there is no intention or evolutionary selection that would justify to ascribe a function to this incidental complex of atoms – while we have the proclaimed intention of the watchmaker that gives us any ground to ascribe a function to what we know as a watch. If this thought-experiment appears to be too far-fetched, consider the real-life example of the 'natural nuclear fission reactor' in Oklo, Gabun.[1] About 1.7 billion years ago, Oklo was the site of a long self-sustaining nuclear fission reaction, and parts of the

[1]Special thanks to Niels Grewe for pointing me to this example; for the details see Bentridi et al. (2011).

site participated in processes, which, in artificial nuclear power stations, would have been realisations of functions. But Oklo was not planned by engineers; it was a natural incident. The Oklo site had all necessary dispositions, but no functions. So the least thing to do is not to view any activity as a functioning and any disposition as a function, as the causal account does, but restrict this view to certain functions or enrich it by postulating that functions are dispositions plus something else. Hence we cannot stay content with the identification of functions with dispositions alone. More needs to be said.

7.5.2 Are All Functions Dispositions?

One strategy that has been suggested is to identify functions with not just any disposition, but with special dispositions, namely, with intended or evolutionary selected dispositions (Godfrey-Smith 1994; Arp et al. 2015, 102–103; Spear et al. 2016). Identifying functions with dispositions has the advantage to account for the causal efficacy of functions and to solve the "problem of the mad engineer": Not just any arbitrary function ascription by a mad engineer will bring about a corresponding function in the artefact created. On the downside, with such an identification of functions with special dispositions it becomes more difficult to account for malfunction – and type-malfunction is particularly problematic. The special-disposition account needs to treat all phantom functions as mere roles. This way, it becomes difficult for the classifier of artefacts to distinguish between artefacts with functions and artefacts with roles.

Another problem with the special-disposition account is, however, that there are seemingly "nonsensical" evolutionary survivals. For a long time it was assumed that the human appendix has long lost its function, although without doubt it has come along with a selective advantage. Today, there are new speculations about its present purpose (Spear et al. 2016 cite Laurin et al. 2011; cf. also Smith et al. 2017). In any case, the present positive effect needs not be the reason the appendix has been selected for in the distant evolutionary past. We can add to the examples. What, indeed, is the function of the human tailbone or the hindlimbs of whales? Both have come into existence "through evolution [...] in order to realize processes of a certain sort" (Spear and Smith 2015, 114) – but these processes were realised in early mammalian ancestors only, and I would not want to ascribe these functions to them as parts of recent whales and humans.

In general, for function ascription the recent history of a species seems to be far more relevant than its distant evolutionary origin (Godfrey-Smith 1994). This observation is in line with Tinbergen's distinction, among his famous four types of biological explanations, between functional and evolutionary explanations (Tinbergen 1963): Functional explanations ask for the present contribution of a certain feature of an organism to the animal's chances of survival and reproduction. Evolutionary explanations, in contrast, ask how a certain feature impacted on survival and reproduction in the past in comparison to other varieties, in order to explain why this feature evolved – even if it is no longer functional today.

Röhl and Jansen (2014) suggest considering a pair of sticks from the woods that incidentally have exactly the form of chopsticks as they are used in Chinese restaurants. Someone finding them can adopt them as chopsticks, i.e., use them like other chopsticks for eating purposes. When being adopted as chopsticks, the sticks have not acquired any new disposition, but they seem to have acquired a new function. Hence, or so the argument goes, functions cannot be dispositions, as the sticks have acquired a new function without acquiring a new disposition. The chopstick argument can easily be answered by the special-disposition account, for serving as a chopstick can be a role for these sticks and not a function (Spear et al. 2016, 120). Roles can be acquired and lost without any change in the physical make-up of the role bearer. Hence, if being used as a chopstick were a role of these sticks rather than a function, this would account for the fact that there is no intrinsic change in the role bearer and no new disposition acquired. On the downside, that means that among all the chopsticks in the world there may be some that do not have a chopstick function but only a chopstick role. This may be considered as an epistemic challenge, for sticks with a chopstick-function and sticks with a chopstick-role can only be distinguished by means of their histories, which might be a problem for, e.g., the users of a classification scheme that contains these two classes. However, if we want to circumvent this problem and ascribe a function to the sticks-adopted-as-chopsticks, we have a strong ontological implication: Sticks have no functions as long they are lying in the woods. When they are picked up and used as chopsticks, they are used intentionally for some purpose. But in order to say that the sticks acquired a function, we need to assume that they are thereby transformed to chopsticks: The function imposition creates new entities, i.e., new chopsticks, out of the sticks without any intrinsic change. While this can easily be modelled in an account of material constitution (Baker 2007), some may see this as an unnecessary multiplication of entities.

7.6 Possible Strategies for Special-Disposition Accounts

Spear et al. (2016) argue that the special-disposition account can not only deal with the malfunction challenge, but has also two advantages over the non-dispositional account by Röhl and Jansen (2014) in that it can account for the non-optionality of functions as well as for their causal efficacy. In this section, I will discuss how the special-disposition account is meant to answer the malfunction challenge, while I turn to the two supposed advantages of the special-disposition account in the following sections.

Spear and colleagues discuss at length how to account for malfunctions in the special-disposition view. They proceed in two steps. First, they argue that dispositions come in degrees, and that cases with a gradual change in disposition have to be distinguished from those cases in which the disposition has totally disappeared. For in many cases of malfunctioning (like hyper- or hypofunctioning) there is still an instance present of the disposition type in question; only it is not realised to the proper extent. This argument obviously exploits the fact that functions are often described in a rather generic way. Hearts have the function to pump blood, and they can still do so (i.e., they have the disposition to do so) even in cases of hyper- or hypofunctioning. However, the argument is threatened when we describe the function more precisely. If we say instead that human hearts have the function to pump blood with a blood pressure of something near to 120/80 mmHg, then even a hyperfunctioning heart lacks the matching disposition, i.e. the disposition to pump blood with a blood pressure of something near to 120/80 mmHg.

More important for the present context is, however, the second step of the argument. Assuming that they have solved the problem for gradual loss of disposition, Spear and colleagues continue to suggest three accounts for total loss of disposition (Spear et al. 2016, 122):

1. There is a heart with a pumping function but without a pumping disposition.
2. There is a heart without pumping disposition and hence without a pumping function; the ground for the statement "The heart is supposed to pump blood" is that the heart belongs to a kind whose instances are supposed to have the function and disposition to do so.
3. There is no longer a heart, because a "completely malfunctioning" heart no longer qualifies "as an instance of the kind" *Heart*.

Of these three, option 1 and 2 match to what Jespersen and Carrara (2011) call the "subsective interpretation" of malfunctioning, which entails that a malfunctioning *F* is an *F*; whereas option 3 matches to what they call the "privative interpretation", which entails that a malfunctioning *F* has ceased to be an *F* and thus now is not an *F*. Jespersen and Carrara do not only agree "that an artefact that is a malfunctioning *F* is characterized by lacking a certain disposition or capacity", they claim also that the privative interpretation goes well with a dispositional account of function, while the subsective account requires a non-dispositional account (2011, 126).

Option 1 is, of course, the view of Röhl and Jansen (2014) that presupposes that functions are distinct from dispositions. Both of the other options, however, are compatible with functions being dispositions. They have their problems, though. Option 2 manages to retain a ground for the normative statement that hearts are supposed to pump blood, but the ground is no longer the appropriate function but membership of a certain kind or type: "What grounds this *supposed to* is the kind of thing that the bearer is" (Spear et al. 2016, 122). However, if normativity is accounted for by kind membership anyway, we do not need to reserve this ground for the cases of absolute loss of disposition. Functions become superfluous in this respect, because kind membership can do the job in all cases – or rather, kind membership should be a constitutive element of having a function (see Krohs 2004, 2009, who, however, takes pain to avoid the term 'kind').

Option 3 is a viable option indeed, and I think that there are indeed cases that are properly described along this pattern. A cancerous lung (an example discussed in Röhl and Jansen 2014 and Spear et al. 2016) may indeed no longer be a lung.

Similarly, artefacts will often be irreversibly broken when they can no longer realise their function; we can assume that they have ceased to exist as a token of their kind. But I doubt that option 3 can be generalised. For this would imply that there never were, say, any absolutely malfunctioning hearts, or only in a loose sense: Even on this account, there is something that once *has been* a heart and that now has no disposition to pump blood. Nevertheless, this would not account for the normative implications of function ascriptions. Nor could we directly apply the strategy used in option 2, for there is no longer an organ that belongs to the kind *heart* whose instances are supposed to pump blood. We would need to retreat to the kind of the whole organism or artefact in question. For, say, humans belong to a kind whose members are supposed to have a heart that, in turn, is supposed to pump blood. In particular, in cases where an artefact or organ is only temporarily absolutely malfunctioning, it would be implausible to assume, say, that the function bearer went out of existence. For this would imply that a new artefact comes into existence after repair, or that the organism has a new heart after successful surgery. Because of this problem, option 3 should not be considered a solution for all cases.

There is also a technical problem for options 2 and 3, as they presuppose the metaphysical and grammatical possibility to ascribe predicates directly to kind terms. Metaphysically, this turns into a problem, once more nominalist leanings enter the scene, as not all philosophers will be happy with ascribing attributes on the level of universals. Grammatically, there is a problem when this option is to be formalised in first-order predicate logic without reference to universals, or in one of the more restricted dialects of description logic (Baader et al. 2007). In fact, in applied ontology the need to have a means of expressing statements like "Hearts are supposed to pump blood" without second-order predicates is one of the strongest motivations for a distinct category of functions.

If we do not want to identify functions with certain dispositions, we could try to retreat to weaker alternative statements about the relationship between functions and dispositions. We could state, e.g., that functions are not identical with dispositions, but that they depend on them for their existence. It is clear, however, that this version has similar problems with malfunctioning, as in the case of malfunctioning there seems to be a function but no disposition on which it could depend.

We could also try to retreat to the linguistic level and state that function ascriptions are disposition ascriptions (Houkes and Vermaas 2010, 84–94). Again, there is a problem with malfunctioning. Whoever reports a malfunctioning or a fault, claims that there is a function but not the matching disposition. Malfunction ascriptions imply, in fact, ascriptions of functions plus the denial of the matching dispositions. Similarly, the anthropologist reporting on functions of, say, amulets need not share any belief in the magic dispositions of these artefacts.

7.7 Non-optionality: Why Functions Are Essential

The first advantage that Spear and colleagues ascribe to the special-disposition account is that it is able to explain the non-optionality of functions (or, as Röhl and Jansen 2014 put it, their essentiality) "without appeal to social constructionism or some other variant of epiphenomenalism" (Spear and Smith 2015, 123).

It is true that the special-disposition account indeed delivers such an explanation. The point, however, is that this explanation does not at all rest on their tenet that functions are dispositions. The non-optionality of functions rather rests on those facts that are meant to be decisive for dispositions to be functions, namely, intentions or evolutionary selection. A non-dispositional account of function can point to the very same facts to explain non-optionality. An artefact, say, has a certain function, because its designer intended it to have this function. This is an external, historical and intentional fact about the artefact, but it is a fact that is deeply causally connected with the coming-to-be of this artefact. Similarly, someone holding the non-dispositional account can tell the evolutionary-selection story as well (as does Millikan 1984). Hence, the explanation of non-optionality that the special-disposition account gives is independent from the "disposition" aspect and relies on the "special" aspect alone.

In Sect. 7.2, I discussed Searle's remarks on functions as an example of an epiphenomenalist account of function. As I said, for Searle all functions are observer-relative; that is, a function report says as much about the observers and their values as about the function bearers. For Searle, functions are in a way created by the observer; hence, Searle's view can be seen as a constructionist account of functions. It should be clear, however, that Searle's view does not adequately reflect what we do when we ascribe functions. The anthropologists describing the amulets that are in use among the culture they study need not share the view that these amulets are any good; they need not even share the view that influencing one's fate is a value one should pursuit. Nor do the anthropologists create that function. Rather, the amulet function is grounded in the beliefs of the group they study (Searle 1995, 9–13). Hence, artefact functions are, in Searle's terms, rather observer-dependent than observer-relative. Similarly, biologists do not create biological functions by projecting values like survival and reproduction on the biological domain. Rather, survival and reproduction are necessary preconditions without which the whole fabric of life would not have gotten off the ground. Again, biological functions are not observer-relative, not even observer-dependent, but objective facts about biological entities.

Röhl and Jansen (2014) describe an as-if heuristic as a possible account of functions that is able to extend the intentionalist account for artefact functions to the biological domain. In a nutshell, the as-if account mirrors the kind of re-engineering that is often practised by biologists anyway: Imagine, in a thought experiment, that a given organism would be created by a smart bio-engineer in order to live in its normal biological environment. Body parts have those functions that this smart bio-engineer would ascribe to them. I want to underline that this is not meant to be a constructionist account of function. It might be suspicious that the as-if account involves not only someone conducting the thought experiment, but also a hypothetic bio-engineer. Nevertheless, the result of this thought experiment is not meant to be a psychological fact about those conducting these thought experiments. Rather, the thought experiment is instrumental to find out something about the organism and the functioning of his parts within its normal environment. The advantage of the thought experiment is that we can conduct it with the organism in its most recent state and environment – and thus we are able to stress the recent history of its kind without needing to deny the distant evolutionary roots.

One final note: In daily life, we sometimes say that a particular device "has lost its function", or "is now functionless". If functions are non-optional, phrases like these cannot be true in a literal way. What is probably meant by this phrase is that there is no longer a point in using the device. A floppy disk drive may still have the disposition to read and write data, but as nobody any longer wants to use floppy disks, it has become useless for most of us. Similarly, if one day nuclear disarmament is successful, there will be no longer any point in pressing the red button. The floppy disk and the red button continue to have their functions (and maybe all of their initial dispositions), but there is no longer any occasion to use them in the way they were intended to be used.

7.8 Causal Efficacy: A Dilemma and a Way Out

The second advantage that Spear and Smith (2015) claim for the special-disposition account is that they can account for the causal efficacy of functions, and it is to this that I turn now. How do the various accounts of function deal with malfunctioning? Causal theories, which equate functions with actual causal contributions, seem to be compelled to claim that there is nothing like having a function and not actualising it. In particular, malfunctioning seems to be impossible for terminological reasons. In contrast, etiological theories, which focus on the causal history of the function, can state that there was evolutionary success due to the realisation of a certain feature in past tokens, but there is no realisation of this feature in the present token. That is, in etiological theories having a function is mainly a historic feature. Intentional theories, too, can easily deal with malfunctioning (and maybe too easily, if it sounds odd to complain "My perpetuum mobile is malfunctioning; it comes to rest"). In a case of artefact malfunctioning, there is the intention that the malfunctioning device performs a certain activity, but the device lacks the disposition to do so, and hence does not perform the activity in question. In this perspective, the truthmaker of a function ascription is not only a historic fact but also extrinsic to those things to which it is ascribed. At this point, we seem to face a dilemma:

- Either the realm of functions is separated from the realm of dispositions; then functions seem to be causally inefficient.
- Or functions are identical with dispositions; but then malfunctioning seems to be conceptually impossible.

Spear and colleagues embrace the second horn (and then try to mollify the repercussions on malfunctioning with the strategies discussed in Sect. 7.5). My strategy is, in contrast, to argue that the dilemma is not well stated, because the first horn is ambiguous: two things can be connected or separated in more than one way. I will argue that while function types are not special types of dispositions, there are several important ways in which functions and dispositions are deeply connected with each other.

First, dispositions are important heuristically. Mostly, functions are ascribed after their realisation has been observed. That is, we observe instances of a kind of material objects *M* participate in instances of a certain process type *P*, and we infer that the instances of *M* are bearers of instances of a certain function type, whose realisations are instances of *P*. In this case, the observed instances of *M* need also bear instances of the matching disposition type *D*, for this is necessary in order for the process instances to happen. Alternatively, it can be argued that a material object has, in fact, a certain disposition to participate in instances of a process type *P* and that the object came about precisely because of this disposition. Then, on the etiological and intentional account, we can infer that this material object is bearer of a function to participate in instances of *P*.

In the biological domain, being replete with exceptions, it is of special importance that typical instances of a kind have the matching disposition in order to ascribe a certain function to a type. Typical hearts pump blood, only atypical hearts do not. In contrast, when human acrobats use their hands for walking (an example from Lord 2010), this is an atypical use of hands that does not imply that my hands are malfunctioning, although I am not able to use my hands in this way. In a biomedical perspective, the typical instances are normally identified with the physiological or healthy ones. Non-pumping hearts are to be treated; non-walking hands are not.

In order to discover functions of technical artefacts, we need not rely on the actual structure of an artefact and its dispositions alone. We can also refer to reports and testimonies of designers and users. Of course, any rational designer should aim at providing function bearers with the dispositions necessary to realise that function. Functions and dispositions also converge from an evolutionary perspective. In the past, enough tokens of a type must have born matching disposition instances, or else the organ in question would not have evolved to exercise the function it now has. (This is independent from the account of functions we want to adopt; but note that for the etiological account this is a constitutive ingredient of being a function, while on the other accounts it is a fact about evolution.)

There are, thus, important interrelations between functions and dispositions, (1) heuristically, (2) from a design perspective for artefact functions, (3) from an evolutionary perspective for types of biological functions. Any of these three suffices to constitute a close tie between functions and dispositions that is necessary to confer on them causal relevance, even if only indirectly via the dispositions which normally accompany them. Röhl and Jansen are indeed "committed to positing instances of function even in cases where these instances make no causal difference whatsoever" (Spear and Smith 2015, 123). Of course, this is not the typical case. However, as the discussion of malfunctioning in this paper shows, the admission of these cases is not a bug but a feature. It is needed to include the atypical cases where there are functions without matching dispositions, and to account for the normative connotations of function ascriptions when no second-order ascription of canonical features to kind terms are possible.

7.9 Conclusion

Malfunctional explanation, or so one could say, is the negative case of functional explanation. It turned out that the discussion of malfunction in this paper is relevant for both the ontology of functions and the theory of functional explanations. From the ontological point of view, the existence of malfunctions shows that function types are not identical with disposition types, nor do function types depend on dispositions. Nevertheless, dispositions are necessary both for the realisation of functions and for the discovery of functions. This leads to the next question: Why do functional explanations work? In part, we can say that functional explanations borrow from dispositional explanation: In most cases, things have the ability to perform their function and they do so. However, in an important respect they deviate from dispositional explanations. They do not primarily deal with what things can do, but what they should do. (Here, obviously, 'ought to' does not imply 'can'.) They add, that is, the standard of normal or physiological cases, which could be, depending on domain and theory, those cases that are evolutionary tested or, for artefacts, the intended cases. Again, the latter could be restricted to those cases that can be rationally intended. Given this standard, explanations that refer to a thing having a malfunction or a thing's malfunctioning work just like Goldschmidt's proof of the causal power of absences: The contrast to what is to be expected in normal cases leads to the light-bulb moment we aim at in an explanation. Such explanations can thus be considered to be special cases of contrastive explanation (Lipton 1990).

In this paper, I discussed what exactly accounts for the standard with which we compare a given case. I have argued that we should account for it by ascribing non-dispositional functions to the individuals in question. I have argued that the alternative solution to ascribe normative predicates not to the individuals but to the kinds in question is of no help when it comes to type-malfunction. In addition, though these two ways take a different direction in the beginning, it is only one short step from ascribing normative predicates to kinds to ascribing non-dispositional functions to instances. For having a non-dispositional function to realise in instances of a certain process type can easily be explained in terms of instantiating an object type that is supposed to participate in instances of that process type. Either way, functions are not dispositions; and malfunctioning occurs when there is a function but no matching disposition.

Acknowledgments Many thanks to Werner Ceusters, Manfred Drack, Georg Fuellen, Niels Grewe, Ulrich Krohs, Johannes Röhl, Petter Sandstad, Barry Smith and Andrew Spear for helpful comments on earlier versions of this paper.

References

Arp, R., B. Smith, and A.D. Spear. 2015. *Building ontologies with basic formal ontology*. Cambridge, MA: MIT Press.

Artiga, M. 2011. Re-organizing organizational accounts of function. *Applied Ontology* 6: 105–124.

Baader, F., et al. 2007. *The description logic handbook: Theory, implementation and applications*. 2nd ed. Cambridge: Cambridge University Press.

Baker, L.R. 2007. *The metaphysics of everyday life: An essay in practical realism*. Cambridge: Cambridge University Press.

Bentridi, S.-E., B. Gall, F. Gauthier-Lafaye, A. Seghour, and D.-E. Medjadi. 2011. Inception and evolution of Oklo natural reactors. *Comptes Rendus Geoscience* 343(11–12): 738–748.

Birnbacher, D., and D. Hommen. 2012. *Negative Kausalität*. Berlin/Boston: de Gruyter.

Burek, P. 2006. *Ontology of functions: A domain-independent framework for modeling functions*. Ph.D. thesis, University of Leipzig. http://www.onto-med.de/publications/2007/burek-p-2007-a.pdf

Cummins, R. 1975. Functional analysis. *Journal of philosophy* 72: 741–765.

del Frate, L. 2012. Preliminaries to a formal ontology of failure of engineering artifacts. In *Formal ontologies in information systems (FOIS 2012)*, ed. M. Donnelly and G. Guizzardi, 117–132. Amsterdam: IOS.

Godfrey-Smith, P. 1994. A modern history theory of functions. *Noûs* 28: 344–662.

Goldschmidt, T. 2016. A demonstration of the causal power of absences. *Dialectica* 70: 85.

Hart, H.L.A., and A.M. Honoré. 1959. *Causation in the law*. Oxford: Clarendon Press.

Houkes, W., and P. Vermaas. 2010. *Technical functions: On the use and design of artefacts*. New York: Springer.

Houkes, W., P. Vermaas, K. Dorst, and M.J. de Vries. 2002. Design and use as plans: An action-theoretical account. *Design Studies* 23: 303–320.

Jansen, L. 2008. Kategorien: Die top level Ontologie. In *Biomedizinische Ontologie. Wissen strukturieren für die Informatik-Praxis*, ed. Ludger Jansen and Barry Smith, 85–112. Zürich: vdf.

Jespersen, B., and M. Carrara. 2011. Two conceptions of malfunction. *Theoria* 77: 117–138.

Krohs, U. 2004. *Eine Theorie biologischer Theorien: Status und Gehalt von Funktionsaussagen und informationstheoretischen Modellen*. Berlin: Springer.

———. 2009. Functions as based on a concept of general design. *Synthese* 166: 69–89.

———. 2010. Dys-, mal- et non-: l'autre face de la fonctionnalité. In *Les Fonctions: Des Organismes aux Artefacts*, ed. J. Gayon, A. de Ricqlès, and M. Mossio, 337–351. Paris: PUF.

Laurin, M., M.L. Everett, and W. Parker. 2011. The cecal appendix: One more immune component with a function disturbed by post-industrial culture. *The Anatomical Record* 294 (4): 567–579.

Lipton, P. 1990. Contrastive explanations. In *Explanations and beyond*, ed. Dudley Knowles, 247–266. Cambridge: Cambridge University Press.

Lord, Ph. 2010. An evolutionary approach to functions. *Journal of Biomedical Semantics* 1 (Suppl 1): S4. https://doi.org/10.1186/2041-1480-1-S1-S4.

Merrill, G.H. 2010. Ontological realism: Methodology or misdirection? *Applied Ontology* 5: 79–108.

Millikan, R.G. 1984. *Language, thought, and other biological categories*. Cambridge, MA: Bradford Books/MIT Press.

Pahl, G., W. Beitz, J. Feldhusen, and K.-H. Grote. 2005. *Konstruktionslehre: Grundlagen erfolgreicher Produktentwicklung. Methoden und Anwendung*, 6. Auflage. Berlin/Heidelberg/New York: Springer.

Preston, B. 2013. *A philosophy of material culture: in Action, function, and mind*. New York: Routledge.

Röhl, J., and L. Jansen. 2014. Why functions are not special dispositions. An improved classification of realizables for top-level ontologies. *Journal of Biomedical Semantics* 5: 27. https://doi.org/10.1186/2041-1480-5-27.

Searle, J.R. 1995. *The construction of social reality*. New York: Free Press.

Smith, B. 2005. Against fantology. In *Experience and analysis*, ed. Johann C. Marek and Maria E. Reicher, 153–170. Wien: öbv & hpt.

Smith, H.F., W. Parker, S.H. Kotzéd, and M. Laurine. 2017. Morphological evolution of the mammalian cecum and cecal appendix. *Comptes Rendu Palevol* 16: 39–57.

Spear, A.D., and B. Smith. 2015. Defining 'Function'. In *Third International Workshop on Definitions in Ontologies (IWOOD 2015)*. Lisbon. 27 July 2015. https://sites.google.com/site/defsinontos2015/accepted-papers

Spear, A., W. Ceusters, and B. Smith. 2016. Functions in basic formal ontology. *Applied Ontology* 11 (2): 103–128.

Tinbergen, N. 1963. On the aims and methods of ethology. *Zeitschrift für Tierpsychologie* 20: 410–463.

Chapter 8
Disease Entities, Negative Causes, Multifactoriality, and the Naturalness of Disease Classifications. Remarks on Some Philosophical Misperceptions of Medical Pathology

Peter Hucklenbroich

Abstract In twentieth and twenty-first century medicine, the concept of disease entity has proven to be of key importance for pathology and the theory of diseases. Disease entities are kinds of complex clinical and etiopathogenetic processes that are triggered by specific primary causes and develop on anatomical, physiological, clinical and subjectively experienced levels. They are distinguished from healthy states of life by definite criteria of pathologicity. This paper sketches the prehistory as well as the central features of the current paradigm of disease entities. Since the 1970s, philosophical theories of disease tend to ignore or, at best, reject this concept. By examining the well-respected theories of H. Tristram Engelhardt, Jr. (The concepts of health and disease. In: Engelhardt HT, Spicker SF (eds) Evaluation and explanation in the biomedical sciences. Reidel, Dordrecht, pp 125–141, 1975 ff.), and Caroline Whitbeck (Philos Sci 44:619–637, 1977), it is shown that this defensive attitude results from a philosophical misconception of the concept. Engelhardt criticizes the concept of disease entity on account of historically obsolete, substantialistic or quasi-platonistic (essentialistic) interpretations of the concept. He erroneously assumes that explanations using this concept are inconsistent with explanations by laws of physiology. Caroline Whitbeck correctly refers to the modern, scientific version of the concept. But in her opinion, the concept "cause of disease" is defined according to certain "instrumental interests" that may differ between subjects and is, thus, neither objective nor unique and unequivocal. Hence, the concept of disease entity is ambiguous and not suited for establishing a unique, unambiguous, and unequivocal natural classification of diseases. It is shown that Whitbeck's objections rest upon misconceptions concerning the concept of "primary cause" resp. "etiological factor" and of the so-called "multifactorial" causation. By reference to a careful, medically and philosophically correct reconstruction of these concepts, it is shown that her objections do not apply.

P. Hucklenbroich (✉)
Institute for Ethics, History and Theory of Medicine, University of Münster, Münster, Germany
e-mail: Peter.Hucklenbroich@ukmuenster.de

A. Christian et al. (eds.), *Philosophy of Science*, European Studies in Philosophy of Science 9, https://doi.org/10.1007/978-3-319-72577-2_8

Keywords Disease entity · Negative causation · Multifactorial causation · Disease classification · Naturalness

8.1 Introduction

In the history of western medicine, two diametrically opposed conceptions concerning the nature of disease have been dominating the scientific approach (Temkin 1963): (i) the view that disease is *something* – a hostile or harmful "*entity*" – that befalls a healthy person from the *outside*, and (ii) the view that disease is an *endogenous deviation* from a normal, well-balanced condition, possibly influenced by some exogenous triggers and associated with suffering.

Only as recently as in the nineteenth and twentieth centuries, medical theorists realized that the apparent opposition between these both conceptions may be transcended and superseded by a unitary, integrative concept of disease entities (see Sect. 8.2). Since then, medicine possesses a general conceptual and theoretical basis for pathology – a genuine *concept of disease*. But contemporary philosophy of medicine, developing since the seventies of the last century, has not received, or even misperceived, this medical concept of disease. Thus, ever since a real overabundance of philosophical theories of disease has been developed that are starkly different, are even inconsistent with one another, and are incompatible with the medical concept of disease. This paper intends to reconstruct when, where, and why philosophy has lost its contact to medicine, and which authors and which philosophical mistakes have been responsible for the divide between philosophical analysis and medical reasoning. Since these mistakes have not been corrected in the philosophical literature until now, a renewed analysis of the conceptual background of medical pathology and the theory of disease is overdue.

8.2 The Concept of Disease Entity in Medicine

8.2.1 Prehistory of the Concept

Historians of medicine like to present the historical development of medical theory as being an age-long debate and controversy of two irreconcilable basic conceptions of disease: the ontological and the physiological concept (see Sigerist 1931, 1932; Temkin 1963; Engelhardt 1975).

The *ontological* concept characterizes diseases as being specific, exogenous, damaging kinds of entities (like demons, worms, miasms, bacteria) that are distinct from the human organism, that befall the body from the outside, and that become manifest in symptoms and natural courses of disease that are specific for each kind of entity. In history of medicine, this concept also has been called the *Platonic*, *realist*, *rationalist*, or *conventional* concept.

The *physiological* concept, in contrast, understands diseases as being individual, subjective conditions or states of health, particularly characterized by pain or suffering. States of this kind are interpreted as being *deviations* from a normal, balanced state of health and equilibrium that may be influenced by external factors but are primarily an expression and manifestation of the individual's particular constitution and biography. Thus, in history of medicine, this concept also has been called the *biographical*, *historical*, *Hippocratic*, *nominalist*, *empirical*, or *naturalistic* concept.

As indicated by the different designations given to these contrary basic ideas, there have been very different theoretical elaborations of them in history of medicine (Koch 1917; Sigerist 1931, 1932; Berghoff 1947; Riese 1953; Diepgen et al. 1969; Rothschuh 1978; Porter 1997). Theories adhering to the ontological idea preferred to neglect individual variations and to concentrate on invariant, typical patterns of diseases, that might be specific properties of the corresponding entity. They aimed at a *systematic* account and classification of all existing disease entities, enabling *subsumption* of individual cases. Theories sympathetic to the physiological view, in contrast, focused upon descriptions of individual patient *histories* (case histories) and the method of comparison and analogy (*casuistics*). They aimed at narratives of diseases instead of systematic classifications. (Interestingly enough, there have been several medical theories that described diseases as *deviations* from a medium, balanced and healthy state but aimed at systematic classifications, e.g., the theories of the ancient Methodic school of medicine (first century BC) speaking of *status strictus* and *status laxus* as deviations from the healthy *status mixtus*, or the theory of John Brown (1735–1788) describing the states *sthenia* (overexcitation) and *asthenia* (underexcitation) as deviations from the medium, equilibrated state of irritability of the nervous system.)

Both ontological and physiological theories suffer from specific theoretical difficulties: Ontological theories until the nineteenth century never succeeded in establishing and validating a complete, all-embracing systematic account and classification of disease entities that was theoretically valid and therapeutically successful. Physiological theories, in contrast, suffered from the unavailability of any generalized, systematic basis for therapy at all.

The battle between ontological and physiological theories remained undecided until the twentieth century. Only then, in the course of this century, the modern, synthetic theoretical conception emerged: the concept of clinically and etiopathogenetically defined disease entities.

8.2.2 The Modern Concept

The modern concept of disease entity forms a synthesis and a somewhat radical transformation of the traditional ideas concerning the nature of diseases. It unites and unifies theories from almost all fields of medical investigation, re-arranging and reconfiguring them into the new paradigm of disease entities. Thus, it could not emerge until these theories had been discovered and developed, a historical process that lasted more than four centuries. The theory of disease entities unified and integrated the modern theories of normal anatomy (Andreas Vesalius), pathological anatomy (Giovanni Battista Morgagni), histopathology (Xavier Bichat, Francois Broussais), cellular pathology (Rudolf Virchow), physiology and pathophysiology (Johannes Müller, Claude Bernard), and bacteriology (Louis Pasteur, Robert Koch) – to name only the most prominent fields and researchers (Ackerknecht 1992; Porter 1997, chs. 8–14). In the first half of the twentieth century, more and more pathological conditions, syndromes, and clinical pictures of diseases were explained as being manifestations of disease entities of a particular structure (Hucklenbroich 2014a, b, 2017a, b, c): They are characterized by

- a specific origin, forming a *transition* from (relative) health to a pathologically altered condition of life: the *etiological factor*;
- a specific *course* of pathological development, characterized by a chain of symptoms, clinical signs and findings forming a unique, identifiable and diagnosable pattern, called *natural course* or *natural history* of the disease;
- an underlying chain or cascade of *causally connected* pathological states and processes, called *pathogenesis*;
- and a typical pattern of possible outcomes, from spontaneous recovery over chronic illness or disability to early, premature death, called *exitus*.

By discovering and detecting the etiological factors and the causal processes of pathogenesis, the twentieth century also established the systematic foundations of therapeutic interventions, thus creating the scientific basis of medical therapeutics.

There are four features of the modern concept that deserve closer attention from philosophy of science:

- the exact *conceptual nature* of etiological factors,
- the relationship between *laws* of physiology and pathophysiology, and typical *patterns* of disease entities,
- how the emergence and development of the modern concept is documented and mirrored in medical *textbooks* of the twentieth century,
- and the way how modern, disease entity based medicine deals with the *individual* aspects of disease.

8.2.2.1 The Concept of Etiological Factors

Etiological factors are not things, physical objects, or events in the environment of the organism; they are not exogenous, external entities that befall a human body. Instead, they are specific *interactions* between organisms and their environments, and they are *alterations* of the physiological processes that form specific *transitions* from health to disease. Thus far, etiological factors do not resemble the disease entities of previous ontological theories. To say more: etiological factors are not bound to be single events at a single, unitary space-time location, but may be combinations of several events of this localized kind together forming *the* etiological factor. And even more: the general etiological factor of a disease entity may be logically complex, it may be an alternative or disjunction of several alternatively possible, equivalent events, or even conditional events: Event e is an etiological factor of disease entity D *provided that* condition c is present. They may be even "negative" events, in the sense of lack, absence, loss, or deficit of something in the interaction of organism and environment (for examples, see below). However, etiological factors cannot and must not be construed at will or arbitrarily. Rather, they have to be identified in order to allow the identification, demarcation, and definition of a unique pattern of natural course and pathogenesis that results from the etiological factor and forms a specific disease entity. Thus, etiological factors may be very complex without loss of uniqueness and specificity. Frequently, one disease entity A is part of the etiological factor of a second disease entity B. In such cases, A is called *predispositional* disease of B, or B is called *secondary* disease or *sequela* of A. E.g., disease entity *coronary heart disease* (CHD) is a predispositional disease of disease entity *myocardial infarction*; or disease entity *rheumatic fever* may be a sequela of disease entity *streptococcal tonsillitis* ("*strep throat*"). The logical complexity of many etiological factors has led to irritations and shortcomings in theoretical and philosophical debates regarding causes of disease, particularly concerning the so-called *multifactorial* causation (see below).

8.2.2.2 Physiological Laws and Disease Entities

Contrary to the opinion of some philosophers (e.g., H. T. Engelhardt, Jr.), there is not any conflict or incompatibility between the existence of *laws* of physiology and pathophysiology, and the existence and possibility of discrete, regular, law-like disease entities. Closer analysis shows that laws of the kind established in physiology and pathophysiology refer to discrete, localizable regularities and cause-effect relationships: E.g., that there is a high chemical affinity between the hemoglobin of the erythrocytes and molecular oxygen in the air, or that percussing the patellar ligament triggers the patellar reflex, aka knee jerk (laws of physiology); or, that gaseous CO (carbon monoxide) may *replace* molecular oxygen in erythrocytes because of its much higher chemical affinity, or that a *missing* patellar reflex indicates *damage* of the corresponding neural substratum (laws of pathophysiology). Laws of this type are established and investigated by experimental research, segregating, dissecting and isolating local, discrete cause-effect relations in the laboratory. But disease entities are sort of "experiments of nature", they are complex chains, concatenations, and cascades of *multiple* physiological and pathological processes that result in *unique natural kinds* but that, nevertheless, obey all the laws of physiology and pathophysiology. Thus, the existence of (patho-)physiological laws as well as the existence of lawful disease entities are not only compatible but even necessary, regarding the complexity of human nature.

8.2.2.3 Disease Entities and Medical Textbooks

The discovery in twentieth century medicine that the apparently unmanageable and perplexing multitude of pathological phenomena may be clarified, ordered, explained, and classified by a system of disease entities in the modern sense, was reflected and mirrored by the emergence of two novel types of medical textbooks. The first type is represented by the "Wörterbuch der klinischen Syndrome" (Dictionary of Clinical Syndromes), as published by Bernfried Leiber and Gertrud Olbrich 1957 (1st edition). The second type is represented by the books "Differentialdiagnose innerer Krankheiten" (Differential Diagnosis in Internal Medicine) by Robert Hegglin (1952) and "Vom Symptom zur Diagnose" (From Symptom to Diagnosis) by Walter Hadorn (1960).

Leiber's and Olbrich's book presented a comprehensive collection of those syndromes and complexes of symptoms that had been described in the medical literature but had not yet been explained by any disease entity. In the introduction of this book, Leiber provided a conceptual explication of the concepts "syndrome" and "disease entity", a typology of different kinds of syndromes and their differences from disease entities, and a short sketch of the historical pattern that leads from the discovery of syndromes to, ultimately, the discovery of underlying disease entities (Leiber 1990/1997, XXIII–XXIX). Particularly, Leiber's clear and precise definition and exposition of the concept of disease entity marked a unique milestone in the history of medical theory. This huge handbook underwent 8 editions until 1997, when it was transformed into a digital database. But recognition of Leiber's achievement seems to have been restricted to the specialized community of expert medical researchers and scientists, and the handbook has not been recognized by theorists outside this community or outside theoretical medicine; particularly, philosophy of science and philosophy of medicine have totally neglected this important conceptual contribution.

In some sense, the books of Hegglin and Hadorn constitute the counterpart of Leiber's handbook: they aimed at a comprehensive collection of all symptoms and syndromes that are known to be part of an already established and known disease entity. Thus, they aimed to trace the way "from symptom to diagnosis" in order to support the practical task of medical differential diagnosis. These books underwent several new editions until this day and saw several co-editors (Battegay 2017, 21st ed.; Siegenthaler 2007, 1st engl. ed.; Zöllner and Hadorn 1986, 8th ed.). Because of their high practical value, these books partially are in clinical use till now. Particularly, they have triggered the appearance of many similar books and computer-aided tools that support the way "from symptom to diagnosis" in nearly all special fields and subdisciplines of clinical medicine.

Apart from their practical value, the emergence and success of books of the Leiber and Hegglin/Hadorn type demonstrate the success of the disease entity model as well as its widespread recognition in twentieth century medicine.

8.2.2.4 Disease Entities and Individual Sickness

Regarding the enormous success of the disease entity model, the question arises of how the individual aspects of disease – the patient's experience of being ill and sick – are taken into account in modern medical theory: Where is the legacy of the physiological idea to be found? The answer might be twofold: First, there is a *theoretical* response to this challenge that consists in the development of psychological and psychosomatic medicine. Modern medical theory of the human organism comprises very detailed theories about psychological, psychosocial, and psychosomatic aspects of human life and the subjective experience of diseases and pathological conditions. This scientific knowledge covers a great proportion of what is counted as individual experience in the previous physiological conception of disease.

However, there remains a residuum of the *genuine individual*, *unique* life and experience of human beings that also shapes experience of disease. This truth is respected, and accounted for, in modern *clinical methodology*: Because these highly individual, personal experiences and conditions may only be perceived and understood by other human persons and may escape any objectifying theoretical access, *here* is the methodological role and function of the physician as individual subject and fellow human being. Medical scientific diagnosis of disease entities can only cover those aspects of the individual's sickness that are, in principle, generalizable features. But the uniqueness of the sick individual calls for the uniqueness of the physician, as its counterpart and mirror image. Thus, medical theory is complemented by a clinical methodology that acknowledges the irreplaceability of the physician as a human subject capable of insight, understanding, and empathy (Hahn 1988). This theoretical and methodological insight, elaborated already in the methodology of psychoanalysis and psychosomatic medicine (Loch 1986; Greenson 2007), forms the second, most important proportion of the legacy of the physiological schools in history of medicine.

8.3 Disease Entities in Philosophy of Medicine

Around the year 1970, the concept of disease entity had been established in all clinical subdisciplines of medicine, with the sole exception of *psychiatry*. Psychiatry was still suffering from the antipsychiatry debate, and there were considerable doubts concerning the existence of genuine psychiatric *diseases* – as distinct from mere *disorders* – at all. But *psychiatry* was the field that attracted the attention of most philosophers of medicine – not ordinary, somatic medicine. Maybe that was the reason only two eminent philosophical authors paid attention to the concept of disease entity at all – and both of them rejected the notion! Thus, in the following decades, the question "What is disease?" generated an ever-increasing philosophical debate and dozens of philosophical theories of disease, but the medical concept of disease entity was almost totally ignored (cf. the overviews in Boorse 2011; Simon 2011; Cooper 2017; Kingma 2017; Sisti and Caplan 2017). Only the book "The Nature of Disease", written by the psychiatrist Lawrie Reznek, attempted to reconstruct the concept of disease entities but denied the existence of objective criteria of pathologicity (Reznek 1987); regrettably, this remarkable reconstruction gathered very little attention. Until now (2017), this situation has not changed. Ironically, in the meantime even psychiatric theory is on the move to genuine disease entities (Kendler and Parnas 2008, 2012; Walter 2013; Hucklenbroich 2017b), whereas philosophy continues to ignore the concept. Therefore, in the following sections I am going to scrutinize the criticisms H. T. Engelhardt and C. Whitbeck launched against the concept of disease entity; in order to show that there are serious shortcomings in both of them, and that ignorance of this concept may be the greatest obstacle to progress in philosophical theory of disease.

8.3.1 H. Tristram Engelhardt's Misguided Interpretation of Disease Entities

In his influential 1975 paper on "The Concepts of Health and Disease", Engelhardt sketched an interpretation of disease entities that differs considerably from the concept established in modern medicine: According to Engelhardt, disease entities either are "things", physical objects, or "ideas", invariant "platonic" types of disease. Evidently, Engelhardt here refers to the antiquated, obsolete "ontological" interpretation of disease entities, as distinct from (pathologically altered) physiological processes obeying the laws of ordinary physiology and pathophysiology. This interpretation transpires from Engelhardt's statements in the paper:

> Diseases were not things, nor were they perduring types of pathology. Rather, for the physiological or functional nosologists, diseases were more contextual than substantial, more the resultant of individual constitutions, the laws of physiology and the peculiarities of environment, than the result of disease entities (Engelhardt 1975, 131).

> Disease entities offered a level of abstraction that could bind together the signs and symptoms in an etiological matrix. In particular, reifying medical ontological theories could treat diseases as the substances which bear the signs and symptoms (Engelhardt 1975, 132).

> Diseases are, in fact, not only multifactorial, but also multidimensional, involving genetic, physiological, psychological, and sociological components. [...] The result is a multidimensional concept of disease with each dimension – genetic, infectious, metabolic, psychological, and social – containing a nexus of causes [...] [T]he disease could be alternatively construed as a genetic, metabolic, anatomic, psychological, or sociological disease, depending on whether one was a geneticist, an internist, a surgeon, a psychiatrist, or a public health official. The construal would depend upon the particular scientist's appraisal of which etiological variables were most amenable to his manipulations. [...] [E]pidemiological studies of disease [...] indicate a pattern-pattern analysis of disease within which the pattern of signs and symptoms clustering in a syndrome is bound to a pattern of causal variables (Engelhardt 1975, 133–134).

> Diseases such as cancer, tuberculosis, and schizophrenia thus exist, but as patterns of explanation, not as things in themselves or as eidetic types of phenomena (Engelhardt 1975, 136).

These statements demonstrate Engelhardt's misrepresentation of the modern, twentieth century concept of disease entity: Engelhardt simply does not see and recognize that, in the modern version, disease entities are, indeed, "resultants of individual constitutions, the laws of physiology and the peculiarities of environment", *as well as* "perduring, eidetic types of pathological phenomena" (in their clinical pictures). He just misses the crucial point: The laws of physiology and pathophysiology, combined to specific etiological factors that rely on the individual constitutions as well as on the peculiarities of environment, just result in typical courses of diseases – *called disease entities* (see Sect. 8.2.2.2.)! But in Engelhardt's analysis, disease entities dissolve into two kinds of patterns that may be correlated with each other according to epidemiological probabilities of coincidence, discipline-specific perspectives of investigators, and instrumental interests in manipulation and intervention – thus forming a multifactorial and multidimensional nexus of probabilistic correlations between heterogeneous, remote or unspecific variables instead of patterns of causally connected chains that form unique pathogeneses and specific natural courses. Apparently, the concept of an etiological factor in the sense of a unique primary cause or primary lesion that forms a transition to a pathological state has got lost and completely disappeared in Engelhardt's analysis. This view was probably influenced by the impressive success of great *epidemiological* studies, like the famous Framingham study, in the 1960 and 1970 decades, that helped to recognize the importance of risk factors, i.e., *indirect*, *remote* pathogenic factors, but did not consider *direct*, genuine etiological factors.

As opposed to Engelhardt's philosophical analysis, it has to be stressed that, in contemporary medicine, neither all causally relevant factors are considered to be *genuine etiological factors*, nor are disease entities *substances* or *things*, or even platonistic *ideal types*. Engelhardt's criticism hits a target that was already antiquated in medical theory when he wrote his 1975 article.

8.3.2 Caroline Whitbeck's Misrepresentation of Disease Entities

Two years after the publication of Engelhardt's paper (and in the same issue of *Philosophy of Science* as Christopher Boorse's – now famous – article on "Health as a Theoretical Concept" was published), Caroline Whitbeck published her paper titled "Causation in Medicine: The Disease Entity Model" (Whitbeck 1977). This paper may be viewed as a kind of correction and rejoinder to Engelhardt's. Whitbeck points out that, in modern medicine, disease entities are neither physical objects nor bodies but types of processes:

> I would emphasize that the question of whether we are to accept the ontological conception of disease is a question of whether an individual case of disease can be construed as a situation in which the person has something like a physical object lodged in his or her body. This conception is not shown to be inadequate just because a disease, in the sense of a disease entity (or disease type) could not be [a] very much like a body. Compare the situation with that which exists with regard to biological species. A horse is a body, but the species Equus caballus is not. Unfortunately, a number of discussion [sic] of the nature of disease either mistakenly fault the ontological conception for construing a disease entity on the model of a body, or conversely maintain there are special problems about disease entities because they fail to be bodies. For example, after discussing the problems surrounding the identification and differentiation of disease entities, Lester King raises the question of whether diseases exist prior to their discovery and asserts "Diseases are not things in the same sense as rocks, or trees or rivers. Diseases represent patterns or relationships which are not material. The problem then becomes, how real is a pattern?" [...] Diseases are not particular physical objects, but this does not prevent their existence being as objective as types of rock or species of trees (Whitbeck 1977, 623).

Whitbeck's argumentation in this paper consists of two parts: in the first part, she correctly underscores several central truths about modern medicine and its concept of disease entity. In the second part, she erroneously launches three objections against this concept. I am going to analyze and reject her criticism in the following sections. First, she correctly stresses the following truths:

- In contemporary medicine, the disease entity model is the dominant and universally accepted model of disease.
- Disease entities are *types*, or typical *patterns*, of complex pathological processes.
- Disease entities become manifest in clinical and subjective signs and symptoms as well as in pathomorphological and pathophysiological alterations of normal structure, function, and behavior.
- Disease entities are caused by *specific etiological factors* and exhibit *specific natural courses* or *natural histories*.

By these statements, Whitbeck directly contradicts and corrects Engelhardt's position. But then, she raises three objections and concludes that the disease entity model is inappropriate: "[...] there is much evidence to which our present model is not adequate. Thus it must be discarded or at least radically revised" (Whitbeck 1977, 635).

These objections are:

- Identification of the causes (etiological factors) of disease entities is not objective but depends on "instrumental interests", i.e., interests regarding medical interventions.
- Identification of etiological factors and of disease entities is, at least in some cases, not unique and unambiguous, e.g., in the case of "multifactorial" diseases.
- Therefore, the co-called *natural*, etiological *classification* of diseases is neither objective nor unambiguous.

8.3.2.1 Whitbeck's Argument Against Objectivity of Etiological Factors

Whitbeck summarizes her objections in the following passages:

> The focus upon a particular cause (whether contributing or indispensable) and the identification of it as *the* cause is frequently simply a matter of what factors the speaker is at least studying to control. (Whitbeck 1977, 630)

> I wish to argue that there are several criteria which govern the choice of the causal factor to be regarded as *the* etiologic agent. There is, first, a preference for a proximate as opposed to a remote cause, and second, a preference for a factor which exists in the environment prior to contact with the patient's body and which may then act upon it. The first preference seems natural enough, for if we begin to consider remote causes we may never stop. The rationale for the second preference is that it leads us to fix upon something in the environment which can be recognized and removed before it acts on the body (Whitbeck 1977, 631).

This argument relies on a misguided reconstruction of how the unique, specific etiological factor of a disease entity is defined and identified. First, the etiological factor of a disease entity *never* is an object or process outside the organism and located solely in the environment. Instead,

- it is an event that refers to an interaction or interactive relationship involving both organism and environment (except in the case of genetic diseases, where the etiological factor is identical with just the origin of the individual organism itself);
- it may be an event that is composite in substance and logically complex;
- it is *the* event that is distinguished and marked as being the *first pathological event* in a causal chain of *pathological* events and interactions of the organism – the *pathogenesis* of a specific disease entity.

Note that the pathologicity of an event is defined by a system of natural criteria of pathologicity, spelled out by general medical pathology. The theory of pathologicity and the system of criteria cannot be presented in this paper (cf. the account in Hucklenbroich 2013, 2014a, b, 2016, 2017c).

Thus, there is no freedom of choice regarding identification of etiological factors; particularly, "instrumental interests" are not allowed to influence and, in fact, to falsify this identification. Whitbeck's error may be explained and excused, however, by a particular linguistic usage of "causal therapy" that is common and widespread in clinical medicine; clinicians speak of "causal therapy" when they treat any causal factor in the causal chain of pathogenesis, provided that this treatment will terminate the pathological processes and restore health – but the selected factor is not bound to be, and often cannot even be, the *first* pathological event, i.e., *the* etiological factor.

8.3.2.2 Whitbeck's Argument Against the Uniqueness and Unambiguousness of Etiological Factors

Whitbeck's objection relies on a misguided reconstruction of "multifactorial" causation, as stated in the following passages:

> However, there is much evidence to which our present model is not adequate. Thus it must be discarded or at least radically revised. This is because the assumption which underlies our preferred mode of classifying diseases, viz., that unique etiologic agents can be found for each complex process that we wish to call a disease entity, seems false.
>
> Individuals whose immune mechanisms fail to work properly may be subject to many different diseases and each of these disease processes may be precipitated by any number of agents. Thus according to what was said in section four about the criteria which determine etiologic agents there will be no answer to the question "What is the etiologic agent for this type of cancer." There will be no answer, first, because in these cases there is no factor in the environment which is indispensible for precipitation of the disease process, and second, because the factor which is crucial in determining whether the process will develop to the point of being a threat to life and health, is the reaction of the host's body. (Such a disease is termed "multi-factorial.") (Whitbeck 1977, 635)

This argument is misguided because etiological factors may be logically abstract and complex, and may be composite in substance. E.g., the clinical disease entity "common cold" or "influenzal infection" (as distinct from "influenza" itself) may be caused by several types of virus that are clinically indistinguishable but can be distinguished by methods of molecular biology. Must we conclude that the etiological factor of common cold is not unique? Of course we do not, since the correct etiological factor is "infection by one of the n different types of virus v1 or v2 or... or vn". This is a case of a logically complex etiological factor, i.e., a disjunction (connection by "or"). A second example for so-called multifactorial causation is given by etiological factors that are *conjunctions* of n distinguishable events that trigger the disease entity only if they *coincide*. By analyzing this coincidence (or conjunction) of events as being *the* etiological factor, uniqueness of causation is preserved.

By the way: note that the etiological factor may be *abstract* in the sense of being *negative*. To die of thirst, or to die of starvation, means to die of the negative (pathological) conditions of *absence* of water, resp. absence of food. Even better examples are deficiency diseases like vitamin deficiencies, e.g., scurvy (vitamin C) or rachitis (vitamin D). In the case of scurvy, the exact etiological factor is the event that the balance between intake and loss of vitamin C changes to negative values: At this moment, the individual suffers from "latent scurvy", that may develop into "manifest scurvy" showing severe signs and symptoms – even death, or may revert to health if the intake of vitamin C increases again, thus changing the intake-loss balance back into positive values. Thus, *absences* may be genuine etiological factors of disease entities.

The case of immune deficiency followed by cancer, mentioned by Whitbeck in the above argument, is to be analyzed as follows: Immune deficiency is a part or manifestation of one of several possible disease entities, e.g., AIDS, leukemia, pancytopenia, or immunosuppression by medical treatment. This is disease entity

A. Disease entity A is part of a complex, disjunctive etiological factor of a possible second and secondary disease entity B, e.g., any kind of cancer. The etiological factor of B thus is complex and composite, consisting of A conjunctively combined with the disjunction of all possible triggers of cancer B. This is a typical case of multifactorial causation, as it is found in many fields of clinical medicine, and it is by no means a counterexample to the disease entity model.

8.3.2.3 Whitbeck's Argument Against the Naturalness of Disease Classifications

Whitbeck states her objections against the assumed naturalness of classifying diseases according to their etiology, or etiological factors, in the following passages (the first passage was already quoted in Sect. 8.3.2.2. above):

> [...] the assumption which underlies our preferred mode of classifying diseases, viz., that unique etiologic agents can be found for each complex process that we wish to call a disease entity, seems false (Whitbeck 1977, 635).

> [Sydenham] clearly states that just as there are zoological and botanical species so there are species of disease, "species morbi." What is entailed in this (or any) claim that some classification is natural? I take it that to claim that some classification is natural is to claim that sorting of individuals in accord with the criteria set down in the proposed classification scheme makes possible the maximal number of correct inferences (of the sort one is interested in) concerning these individuals. Thus classification of chemical substances on the basis of their color and appearance would result in the sorting of rhombic and amorphous sulfur into different classes and thus would not maximize the number of correct inferences regarding the chemical reactions it would undergo. Therefore, it would not be natural in the requisite sense, even if such a classification would be a (psychologically) natural one to try. (Whitbeck 1977, 620)

This is an argument that consists of two objections:

1. There is no unique, unambiguous correlation between diseases and etiological factors, thus the required classification by etiological factors cannot be constructed.
2. In order to be *natural*, any classification must enable the maximal number of correct inferences concerning the individuals classified.

The first objection, concerning the impossibility of unique etiological factors, was rejected in the sections above; thus, it may be neglected here. The second objection, concerning the number of inferences "made possible" by any classification, seems to be somewhat obscure or confused: There is no unique, unambiguous and necessary correlation between criteria of classifications and methods of logical inference. Additionally, measuring and comparing numerically sets of logical inferences may be difficult and, particularly, not helpful at all, because sets of this type usually are nonfinite (countably infinite) and, thus, of *equal* cardinal number.

But Whitbeck's own example, of crystallographic and chemical classifications, shows that the crucial point in her argument refers to theoretical assumptions underlying any classification that can be empirically true or false: If it is theoretically assumed that classifying chemical substances (or elements) according to crystallographic properties, i.e., according to color and appearance, will coincide with classifying them according to chemical structure and reactivity, then this assumption will be falsified by chemical experience. Apparently, this is the crux of Whitbeck's argument.

Turning now back to medical classification, we can outline the following counter-argument to Whitbeck: if there is a unique, unambiguous correlation between diseases and etiological factors – as was shown above – then we can construct an etiological classification of diseases. (This is done, by the way, in medical *nosology*.) The theoretical assumption underlying this classification states that this classification coincides with classifying diseases according to their natural courses, or better to the patterns of their natural courses, namely to clinical picture and pathogenesis. This is, indeed, the general theoretical assumption underlying medical theory, and is the *theoretical* justification for the disease entity model. The *empirical* justification of the model has been provided by the impressive success of its application in the twentieth and twenty-first centuries (Ackerknecht 1992; Groß and Winckelmann 2008; Kiple 1993). To sum up:

1. Classifying disease entities by their etiological factors is *possible*, and it yields a unique, *unambiguous* and *unequivocal* nosological system correlating etiological factors, natural courses, and pathogeneses.
2. Proposition 1 is not an *a priori*, logical or conceptual truth (and may not be refuted by *a priori* objections), but has to be proven *empirically*.
3. To date, there is compelling empirical evidence that proposition 1 is *true*.

8.4 Epilogue

After publication of Engelhardt's 1975 and Whitbeck's 1977 papers, the concept of disease entity lost attention of philosophical authors and was not any longer considered in philosophical theories of disease. Instead, philosophical theories focused features that are alien to the medical concept, e.g., isolated *statistical differences* in physiological function (Boorse 1977, 1997, 2014), or even (negative) *moral values* like evil or harm (Clouser et al. 1981, 1997; Wakefield 1992), as representing the conceptual foundations of pathology. Meanwhile, the situation in philosophy of medicine is summarized by two of the most renowned experts in the field by "[. . .] *dozens of philosophical theories paint starkly different pictures of the essential nature of disease*." (Sisti and Caplan 2017; cf. also Boorse 2011; Cooper 2017; Kingma 2017). Apparently, the missing contact to medical theory is responsible for this uncontrolled proliferation of theories, similar to a *Kuhnian crisis* of a paradigm.

In this paper, I have sought to argue that the arguments (by Engelhardt and Whitbeck) against the disease entity model are misguided and rely upon shortcomings in the analysis of modern medical theory. Thus, I am convinced the multitude of "starkly different", mutually contradictory philosophical theories will be overcome only by a renewed, more careful and profound analysis of the modern medical model of disease entities. Evidently, progress in this direction will require close cooperation of both medical and philosophical experts.

References

Ackerknecht, E.H. 1992. *Geschichte der Medizin*. 7th ed. Stuttgart: Enke.

Battegay, E., Hrsg. 2017. *Differenzialdiagnose innerer Krankheiten*. 21st ed. Stuttgart: Thieme.

Berghoff, E. 1947. *Entwicklungsgeschichte des Krankheitsbegriffs*. 2nd ed. Wien: Maudrich.

Boorse, C. 1977. Health as a theoretical concept. *Philosophy of Science* 44: 542–573.

———. 1997. A rebuttal on health. In *What is disease?* ed. J.M. Humber and R.F. Almeder, 1–134. Totowa: Humana Press.

———. 2011. Concepts of health and disease. In *Handbook of philosophy of science*, *Philosophy of medicine*, ed. F. Gifford, vol. 16, 13–64. Amsterdam: Elsevier.

———. 2014. A second rebuttal on health. *Journal of Medicine and Philosophy* 39: 683–724.

Clouser, K.D., C.M. Culver, and B. Gert. 1981. Malady: A new treatment of disease. *Hastings Center Report* 11: 29–37.

———. 1997. Malady. In *What is disease?* ed. J.M. Humber and R.F. Almeder, 175–217. Totowa: Humana Press.

Cooper, R. 2017. Health and disease. In *Bloomsbury companion to contemporary philosophy of medicine*, ed. J.A. Marcum, 275–296. London/New York: Bloomsbury Academic.

Diepgen, P., G.B. Gruber, and H. Schadewaldt. 1969. Der Krankheitsbegriff, seine Geschichte und Problematik. In *Prolegomena einer Allgemeinen Pathologie*, Hrsg. F. Büchner, E. Letterer, and F. Roulet, 1–50. Berlin: Springer.

Engelhardt, H.T. 1975. The concepts of health and disease. In *Evaluation and explanation in the biomedical sciences*, ed. H.T. Engelhardt and S.F. Spicker, 125–141. Dordrecht: Reidel.

Greenson, R.R. 2007. *Technik und Praxis der Psychoanalyse*. Stuttgart: Klett-Cotta.

Gross, D., and H.J. Winckelmann, eds. 2008. *Medizin im 20. Jahrhundert: Fortschritte und Grenzen der Heilkunde seit 1900*. München: Reed Business Information.

Hadorn, W. 1960. *Vom Symptom zur Diagnose*. Basel: Karger.

Hahn, P. 1988. *Ärztliche Propädeutik. Gespräch, Anamnese, Interview*. Berlin: Springer.

Hegglin, R. 1952. *Differentialdiagnose innerer Krankheiten*. Stuttgart: Thieme.

Hucklenbroich, P. 2013. Die wissenschaftstheoretische Struktur der medizinischen Krankheitslehre. In *Wissenschaftstheoretische Aspekte des Krankheitsbegriffs*, Hrsg. P. Hucklenbroich and A. M. Buyx, 13–83. Münster: Mentis.

———. 2014a. ‚Disease entity" as the key theoretical concept of medicine. *Journal of Medicine and Philosophy* 39: 609–633.

———. 2014b. Medical criteria of Pathologicity and their role in scientific psychiatry – Comments on the articles of Henrik Walter and Marco Stier. *Frontiers in Psychology* 5: 128. https://doi.org/10.3389/fpsyg.2014.00128.

———. 2016. Die Normativität des Krankheitsbegriffs: Zur Genese und Geltung von Kriterien der Krankhaftigkeit. *Analyse & Kritik* 38 (2): 459–496. https://doi.org/10.1515/auk-2016-0122.

———. 2017a. Medical theory and its notions of definition and explanation. In *Handbook of the philosophy of medicine*, ed. T. Schramme and S. Edwards, vol. 2, 791–801. Dordrecht: Springer. https://doi.org/10.1007/978-94-017-8688-1_44.

———. 2017b. Disease entities and the borderline between health and disease: Where is the place of gradations? In *Vagueness in psychiatry*, ed. G. Keil, L. Keuck, and R. Hauswald, 75–92. Oxford: Oxford University Press.

———. 2017c. ‚Krankheit" als theoretischer Begriff der Medizin: Unterschiede zwischen lebensweltlichem und wissenschaftlichem Krankheitsbegriff. *Journal for General Philosophy of Science 48*.https://doi.org/10.1007/s10838-017-9367-y. http://rdcu.be/tQIV.

Kendler, K.S., and J. Parnas, eds. 2008. *Philosophical issues in psychiatry*. Baltimore: Johns Hopkins.

———, eds. 2012. *Philosophical issues in psychiatry II: Nosology*. Oxford: Oxford University Press.

Kingma, E. 2017. Disease as scientific and as value-laden concept. In *Handbook of the philosophy of medicine*, ed. T. Schramme and S. Edwards, vol. 1, 45–63. Dordrecht: Springer.

Kiple, K.F., ed. 1993. *The Cambridge world history of human disease*. Cambridge: University Press.

Koch, R. 1917. *Die ärztliche Diagnose*. Wiesbaden: Bergmann.

Leiber, B., Hrsg. 1990/1997. *Die klinischen Syndrome*. 7th/8th ed. München: Urban & Schwarzenberg.

Leiber, B., and G. Olbrich. 1957. *Wörterbuch der klinischen Syndrome*. München: Urban & Schwarzenberg.

Loch, W. 1986. *Zur Theorie, Technik und Therapie der Psychoanalyse*. Frankfurt/M: S. Fischer.

Porter, R. 1997. *The greatest benefit to mankind. A medical history of humanity from antiquity to the present*. London: Harper Collins.

Reznek, L. 1987. *The nature of disease*. London: Routledge.

Riese, W. 1953. *The conception of disease, its history, its versions and its nature*. New York: Philosophical Library.

Rothschuh, K.E. 1978. *Konzepte der Medizin in Vergangenheit und Gegenwart*. Stuttgart: Hippokrates.

Siegenthaler, W., ed. 2007. *Differential Diagnosis in Internal Medicine*. New York: Thieme.

Sigerist, H.E. 1931. *Einführung in die Medizin*. Leipzig: Thieme.

———. 1932. *Man and medicine. An introduction to medical knowledge*. New York: Norton.

Simon, J.R. 2011. Medical ontology. In *Handbook of philosophy of science, Philosophy of medicine*, ed. F. Gifford, vol. 16, 65–114. Amsterdam: Elsevier.

Sisti, D., and A.L. Caplan. 2017. The concept of disease. In *The Routledge companion to philosophy of medicine*, ed. M. Solomon, J.R. Simon, and H. Kincaid, 5–15. New York/London: Routledge.

Temkin, O. 1963. The scientific approach to disease: Specific entity and individual sickness. In *Scientific change*, ed. A.C. Crombie, 629–647. London: Heinemann.

Wakefield, J.C. 1992. The concept of mental disorder. *American Psychologist* 47: 373–388.

Walter, H. 2013. The third wave of biological psychiatry. *Frontiers in Psychology* 4: 582. https://doi.org/10.3389/fpsyg.2013.00582.

Whitbeck, C. 1977. Causation in medicine: The disease entity model. *Philosophy of Science* 44: 619–637.

Zöllner, N., and W. Hadorn, Hrsg. 1986. *Vom Symptom zur Diagnose*. 8th ed. Basel: Karger.

Part III
Philosophy of Social Sciences and Values in Science

Chapter 9
Identifying Agnotological Ploys: How to Stay Clear of Unjustified Dissent

Martin Carrier

Abstract Agnotology concerns the creation and preservation of confusion and ignorance. Certain positions are advocated in science in order to promote sociopolitical interests with the result of launching mock controversies or epistemically unjustified dissent. I propose to identify agnotological ploys by the discrepancy between the conclusions suggested by the design of a study and the conclusions actually drawn or intimated. This mechanism of "false advertising" serves to implement agnotological endeavors and helps identify them without having to invoke the intentions of the relevant agents. I discuss three agnotological cases, i.e., studies on bisphenol A, Bt-maize/Roundup, and Oslo's airport Gardermoen. Pinpointing agnotological endeavors is a means for weeding out approaches that look fitting at first glance, but are blatantly inappropriate, in fact.

Keywords Agnotology · Ignorance · Unjustified dissent · Bias · Epistemic and non-epistemic values

9.1 Delineating Agnotological Endeavors

It goes without saying that doing research illuminates certain questions. It is not so obvious, however, that doing research also eclipses certain questions. The latter effect is brought to the fore by addressing matters of "agnotology," a notion introduced and a subject matter highlighted by Robert Proctor in 1992. The designation derives from the Greek "agnosia" for "non-knowledge" or "ignorance," and it is intended to refer to the production or preservation of ignorance. Agnotology is supposed to represent the downside of epistemology and to refer to the deliberate suppression or inadvertent neglect of information (Proctor 2008, 27–28). Proctor's own work examines purposeful attempts to dismiss certain pieces of information

M. Carrier (✉)
Department of Philosophy, Bielefeld University, Bielefeld, Germany
e-mail: martin.carrier@uni-bielefeld.de

A. Christian et al. (eds.), *Philosophy of Science*, European Studies in Philosophy of Science 9, https://doi.org/10.1007/978-3-319-72577-2_9

for economic or political reasons. He revealed that the hazard involved in smoking tobacco had been deliberately obscured by the pertinent industry (Proctor 2012). Naomi Oreskes and Erik Conway expanded this approach to global warming. They diagnosed a wanton opposition to the well-founded consensus view, an opposition driven by political motives and intended to confuse the human-made character of climate change (Oreskes and Conway 2010; Oreskes 2015). The procedure employed in both cases was generating doubt by placing the threshold of acceptance for unwelcome claims at exceedingly high levels. With respect to smoking, epidemiological studies were accused of not being controlled laboratory inquiries and, therefore, of not being trustworthy. Laboratory experiments with rats, however, were declared irrelevant because the effects might be different in humans. In the end, nothing would ever convince the critics. Doubt is manufactured, the sole intention being to prevent political bodies from taking action (Proctor 2008, 11–18; Michaels 2008, 91).

Proctor himself emphasizes that agnotology is rich in meaning and not meant to be derogatory in each context. In particular, he briefly alludes to two other kinds of ignorance, namely, virtuous ignorance and ignorance as a passive construct. Virtuous ignorance is actively constructed or respected. Science is required to stay away from examining certain issues. Traditionally, the realm of the sacred was exempt from critical scrutiny, but Robert Merton's ethos of organized skepticism committed science to not bowing its head before the revered and venerated (Merton 1942). However, as Proctor emphasized, there are many things that we rightly do not want to know and that we rightly do not want science to explore. Virtuous ignorance is the ignorance that results when "not knowing" is deliberately accepted in research as a consequence of adopting certain values. Such intentional ignorance arises, for example, when knowledge would be procured by improper means (e.g., involving serious risks to human or animal subjects), and it should be respected when knowledge would be too harmful or dangerous (such as the know-how involved in weapons of mass destruction). Moreover, the right to privacy is a legitimate shield of ignorance, and not knowing one's own genetic diseases is considered bliss by many (Proctor 2008, 20–21).

Ignorance as a passive construct is the by-product of choices made in the research process (Proctor 2008, 7–8). Scientists are free to ask certain questions while leaving others unaddressed, and they can choose principles of testing and confirming assumptions in science. Such choices can be made differently and have been made differently in the history of science. We may look at nature from a variety of angles and pursue diverse goals in our quest for knowledge. Ian Hacking argues that styles of scientific reasoning pave the way toward problems of a certain sort while discouraging or preventing scientists from taking up others. Hacking maintains that one obvious reason for this limitative character of styles of reasoning is the restricted amount of resources available. In a world of limited funds, devoting efforts to a particular issue withdraws the means for tackling a different issue. Conceptual constraints are yet another influential factor. If research has moved in a certain direction, other avenues of theory development may remain closed (Hacking 1999, 2000). Look at the contrast between evidence-based medicine and personalized medicine for illustration. The former emphasizes randomized controlled trials in which averages across large numbers of patients are considered. The personalized approach is based on the observation that different drugs work differently in different patients so that medication should be customized to the individual at hand. As a result, no large numbers of users are available, and physiological mechanisms or small-scale studies are accepted as sufficient information. The more general point is that scientific research operates like a searchlight. Each such endeavor illuminates certain aspects of experience and thereby leaves other features in the dark. Seeking knowledge in a particular way or seeking particular forms of knowledge inevitably affects what is left unexplored.

I focus on the first of Proctor's notions, i.e., the intentional production of misleading information or the deliberate creation of epistemically detrimental dissent. Several non-trivial epistemological problems require clarification. First, the purpose of generating confusion is typically difficult to ascertain. Accordingly, identifying a publicly accessible mistake would be helpful for pinpointing agnotological ploys. Second, the idea underlying Proctor's notion is that sociopolitical motives have trumped or outplayed the quest for knowledge. However, implementing this idea demands the distinction between epistemic and non-epistemic values. The former appreciate knowledge and understanding, while the latter refer to sociopolitical interests and utility. Many philosophers of science do not acknowledge an in-principle distinction between the two. At the same time, they are committed to scientific pluralism. Both considerations come together in raising the problem which methodological standards are violated in the production and maintenance of ignorance. What may intuitively appear as inappropriately biased could, in a framework of value pluralism, be passed off easily as merely being guided by a particular choice of values. Without a clear set of epistemic values and methodological rules, there is no basis for rejecting any such approach as being actively distorting. Everybody may feel free to draw on one's favorite values for designing and interpreting a study; a variety of different such endeavors can be taken to enrich the cognitive landscape of science. Yet, this seems counterintuitive for the promotion of smoking and the denial that climate change is caused by humans. In these cases, sociopolitical values seem to have overridden knowledge claims.

My goal in this article is to suggest a methodological standard that is violated in some agnotological endeavors. There are various procedures to this effect, and one has just been outlined: raising the threshold of acceptance beyond what can be achieved in practice. The difference between being rightly demanding and being unreasonably strict, however, is not that clear. Look at another case for illustrating this difficulty. The precautionary principle was introduced into international environmental regulations by the Rio Declaration of 1992. The principle roughly says that activities which are potentially hazardous should not proceed unless proven safe. This shifted the burden of proof to those parties who sought to launch some enterprise (such as building a plant). The principle was later adopted by non-governmental organizations (NGOs) in their fight against the market release of genetically modified organisms (GMOs). NGOs strove to rule out GMOs without any proven damage because their safety was claimed to be not confirmed beyond reasonable doubt. David Magnus (2008, 258) commented on this strategy to the effect that NGOs mirrored the procedures used by corporations in the tobacco case: they opposed a scientific consensus and appealed to minority views. The precautionary principle was used as a lever for raising the threshold of acceptance to unreasonable heights.

There is no universal rule as to what amount of evidence warrants acceptance. It is, however, worthwhile having a criterion for distinguishing between objections that should be taken seriously and criticism that can be discarded as being misleading and illegitimate. My goal is to outline a strategy for identifying agnotological machinations. There are various strategies for obscuring the public; I wish to draw attention to a mechanism that has so far not been identified, a mechanism I call "false advertising" and which capitalizes on the discrepancy between the design of a study and its use. My claim is that this mechanism provides a sufficient indicator of agnotological ploys without the need to invoke obscurantist intentions (that are often hard to establish).

9.2 Characterizing Agnotological Maneuvers

In this section I develop my narrative more extensively. Epistemically detrimental dissent is usually understood such that economic and political interests have deprived scientists of their neutrality or impartiality in weighing the evidence. What seems to be lacking at first sight in such unjustified dissent is an appropriate response to data and opposing factions in the scientific community. Agnotological agents just dismiss counterevidence and objections. However, it is not that clear what scientific method demands. Thomas Kuhn and Imre Lakatos consider firmness in the face of epistemic challenges as a scientific virtue. Accordingly, staunchly following an approach, even one-sidedly and lacking the open-mindedness often attributed to a scientific attitude, is not generally viewed as a violation of scientific standards and therefore cannot serve to identify agnotological ploys. There is no methodological rule that specifies the right balance of responsiveness or resistance to evidence and critics (see Carrier [forthcoming] for additional details).

In addition, pluralism has been defended by various authors as an apt epistemic tool. A social notion of objectivity is advocated which highlights the benefit of reciprocal criticism and control. Scientific method demands that scientists from different perspectives criticize each other. A variety of contrasting accounts serves this critical spirit best. Thus, dissent and opposition is among the core features of scientific method (Popper 1966, 412–413; Longino 1990, 66–80; Carrier 2013, 2548–2549). Such a social approach has been extended to include non-epistemic considerations. Philip Kitcher has famously argued that the desire of scientists to receive recognition and fame is likely to produce a split-up of the scientific community such that it distributes itself across various competing approaches. The diversification of risks underlying such behavior represents the epistemic optimum

(Kitcher 2001, 112–113). Judged on this basis, there is nothing wrong with pursuing idiosyncratic research objectives and invoking non-standard procedures. Employing a variety of goals, methods, study designs, and criteria of judgment is suited to boost conceptual pluralism in science which is a major precondition of its objectivity. Agnotology appears to be a misnomer.

However, all the advocates of a social notion of objectivity assume the importance of epistemic ambitions. By contrast, agnotological endeavors seem to be characterized by the prevalence of non-epistemic, sociopolitical values. I agree with this characterization; agnotological maneuvers lack an epistemic attitude (Carrier 2013, 2563–2564). They are driven by a sociopolitical agenda and their advocates do not really want to know. However, bringing this idea to bear on an analysis of agnotological challenges requires additional assumptions which are not always granted. First, such an analysis demands the distinction between epistemic (or knowledge-oriented) values and non-epistemic (or sociopolitical) values, which is not generally admitted. Values are highly ambivalent. They can play a fruitful or indispensable role in science, but also create bias. They may provide a research endeavor with cognitive orientation and determine the relevance of given pieces of evidence, but they may also overrule epistemic aspirations. Without the distinction between epistemic and non-epistemic commitments, it is difficult to identify the latter case. Such an identification requires separating the painstaking quest for knowledge from the profit-driven, one-sided pursuit of a project that promises rich remuneration. Otherwise, different research communities could just select different goals and criteria, and no basis for abandoning some such endeavors as positively misleading would be available.

Second, the charge that the adversary's position is tainted by non-epistemic interests is leveled by either side. Climate change deniers are accused of pursuing a political agenda and of being market fundamentalists who oppose government regulation and refute every single environmental challenge. Their primary concern is said to obstructing certain policies based on scientific conclusions (Kitcher 2011, 162; Oreskes 2015, 44–46). This strategy, however, works in both directions. Climate change deniers point to economic interests that are served by combating allegedly urgent climate change (Kitcher 2011, 30). Companies involved in renewable energy and insurance or re-insurance business are charged to pursue vested interests in underscoring the risks of climate change and warning against its impact (Driessen 2009). Third, even given the distinction between epistemic and non-epistemic standards, we still need to elaborate which epistemic requirement has been transgressed by an agnotological maneuver. Such maneuvers are usually not implemented by fabricating evidence, but in a more subtle way by designing studies in a one-sided fashion. The challenge is what the epistemic nature of agnotological maneuvers precisely is.

Advocates of a ban of agnotological endeavors point to the hazards for public policies if unjustified dissent is allowed to undermine scientific consensus. This applies to the two typical examples of smoking and climate change, in which interest-guided, scientifically unfounded intervention succeeded in delaying action considerably. There is prima-facie evidence for cases in which deliberately sapping a standard view in science does damage to the state of knowledge and has a detrimental impact on public policies.

9.3 The Impact Centered Approach: Agnotology as the Shift of Inductive Risks

Justin Biddle and Anna Leuschner have recently suggested an approach that seeks to pinpoint epistemically detrimental dissent on methodological grounds alone and without having to immerse in the underlying motives and intentions (which they rightly take to be difficult to determine). In their view, epistemically detrimental dissent is identified by a deviation from "well-entrenched conventional standards" that lead to a shift in inductive risks from producer risks to public risks. More specifically, they suggest four conditions that are supposed to be jointly sufficient for identifying epistemically detrimental dissent. These conditions state that the "non-epistemic consequences of wrongly rejecting [hypothesis] *H* are likely to be severe," the dissenting research "violates established conventional standards" and "involves intolerance for producer risks at the expense of public risks," assuming that the two "risks fall largely upon different parties" (Biddle and Leuschner 2015, 273–274). In short, risks of error are shifted toward the public by violating an established methodological rule.

One of the examples they mention is bisphenol A, a substance used in many plastics. Bisphenol A chemically resembles estrogen and is feared to interfere with human hormonal balance for this reason. Tests with this substance performed by the relevant industry employed rats, which were said to be particularly insensitive to estrogen. This test design of bisphenol A breached the recognized methodological rule to choose animals that respond to the substance in question. In addition, the corresponding risks were shifted from the producers to the public. Barring the substance erroneously is a producer risk, since the effort invested in its development would be invalidated without justification; releasing the substance mistakenly to the market is a public risk since unrecognized health hazards may emerge. As a result, the agnotological character of the test design can be demonstrated by exclusively relying on the relation between the risks involved and the violation of a well-entrenched conventional standard (ibid.). It is sufficient to take the impact of fabricating a test design into account. There is no need for speculating about the motives behind the ploy.

While this impact-centered approach may look plausible, it seems to include major uncertainties regarding how risks are individuated and what counts as a relevant risk. Take the 2009/10 swine flu epidemic which eventually turned out to be minor. When it was feared that the figures of infected people would soar, two vaccines were admitted to the European market that had merely run through a test procedure of reduced severity and whose safety had not been ascertained accordingly. The motive behind this lowering of standards was the fear of imminent danger which required immediate action. At the same time, German authorities relieved these companies from their liability due to the possibly premature administration of new agent substances. The companies were liable only for the damage caused by gross negligence while damage resulting from side-effects of the vaccination itself was borne by the government (Kuhlen 2009; TAZ 2011; Schlitt 2013). In other words, standards of accepting vaccines as safe were lowered. Such a shift certainly involves non-negligible risks, and assuming government liability amounted to shifting these risks to the public. However, the situation looked different from the perspective of the relevant authorities at the time. They were convinced that lowering the standards meant speeding up the test procedure and that this acceleration actually prevented risks caused by the epidemic. As a matter of fact, however, hardly any such risk existed, and the fear was largely overdrawn. The converse risk associated with suspending the established safety protocol, however, was real and may have led to increased narcolepsy incidence. Good intentions do not mean anything in the impact-centered approach; it is the result that counts. Yet, this suggests that the two first conditions of the Biddle-Leuschner account are satisfied in this example: Erroneously accepting a vaccine as safe without approval by the standard protocol (this is the dissenting view in their terminology) involved incurring possibly severe risks. The two last conditions apply as well: These risks were shifted away from the producers toward the public.

However, it seems implausible to consider this case an agnotological maneuver. The reason is that the authorities acted in good faith and their objective was to prevent public risks. Thus, this is not an agnotological case in terms of motives and intentions. But it looks like one in terms of impact. As a result, this example is a false positive of the impact-centered approach: It qualifies as detrimental dissent in its light, but it seems significantly different from the paradigmatic agnotological cases.

9.4 Agnotology and False Advertising

I conclude that shifting risks by violating established standards is not plausible as a sufficient criterion for identifying agnotological ploys. Thus, what else is a methodological recipe for misleadingly casting doubt on certain claims? One is arguing for raising the threshold of acceptance with respect to unwelcome hypotheses (as mentioned before). However, this scheme has obviously not been applied in the case of bisphenol A which looks like a typical agnotological ploy. The critical feature here is not the location of the threshold of acceptance but rather the one-sided blueprint of the study. The methodological flaw involved in adopting misleading designs needs to be understood differently.

This additional strategy proceeds by what I call "false advertising." The pertinent blunder is that such studies actually bypass the issues they pretend to address. In the case of bisphenol A, the issue supposedly attended to was whether human health was in jeopardy. The expectation is that the substance is established as being safe. This issue was avoided by employing a strain of rat that was rather insensitive to the health risks at hand. Due to this lack of sensitivity, the study fails to achieve this objective and rules out, instead, that the use of bisphenol A is prematurely barred as being unsafe. The issue allegedly tackled is the harmlessness of bisphenol A, but this issue is bypassed by the design of the study. This design is neither flawed nor inappropriate in general. The procedure is suited to ensure that bisphenol A is not dismissed for the mistaken reason that its use is unsafe. Employing more sensitive creatures could have made the alarm bell ring falsely and too early and would have made us overestimate the health risks involved. The use of less sensitive strains of rat reduces this risk of error and avoids overly strict regulation. However, the question the study intimated to answer is the converse one, namely, to make sure that health risks are not underestimated. This discrepancy was not recognized but passed over tacitly. False advertising is the methodological flaw involved here.

False advertising of this sort underlies the illicit intrusion of non-epistemic interests in the outcome. In general, researchers may feel free to pursue non-epistemic research objectives. Accordingly, one might take apparently biased test designs as the legitimate expression of value commitments that we do not share ourselves (see Sect. 9.1). Sociopolitical values invoked by industry demand not to outlaw bisphenol A unless health risks are revealed beyond doubt. Different communities are free to pursue different research objectives; there is no reason to complain. In fact, however, we rather consider such designs as being one-sided in an unjustified fashion (Wilholt 2009, 97–98). We take such experiments to be illegitimate and misleading and regard them as symptoms of agnotological machinations. Supporting such an assessment, however, requires the identification of a methodological shortcoming which, in my view, consists in an unrecognized discrepancy between the design of a study and how the study is used. Its design makes it suitable to rule out the overestimation of health risks, but it is used for suggesting that health risks have not been underestimated. The discrepancy between design and use means that the setup of the experiment makes it rather insensitive to the issue it purportedly addresses. This incongruity is concealed and glossed over in the interpretation of the results (Carrier forthcoming).

9.5 Generalizing the False Advertising Account

In this section I try out the false-advertising account by considering two additional prima-facie instances of agnotological maneuvering and exploring how this account and the alternative impact-centered approach come to grips with them. The first such instance is the 2012 experiment of Gilles-Eric Séralini who fed rats with genetically modified Bt-maize which is resistant to the herbicide Roundup. Séralini reported an increased rate of cancer in rats that had been fed for 2 years with low doses of the two substances (Séralini et al. 2012). However, the study was heavily criticized in methodological respect: it was accused of having employed too small a sample of rats for obtaining significant results and of having chosen a strain of rat that is predisposed to contract cancer sooner or later anyway (de Souza and Oda 2013). Consequently, the study was retracted against the will of the authors, but republished in a revised version in 2014 in a different journal. The editor of the journal justified the withdrawal of the paper by stating that "there is a legitimate cause for concern regarding both the number of animals in each study group and the particular strain selected." This is said to cast doubt on the reality of the claimed difference between study group and control group (http://www.sciencedirect.com/science/article/pii/S0278691512005637, accessed February 11, 2017). The salient point in the present context is that Séralini et al. (2012) announce in their Introduction that they perform an experiment on potential toxic effects of genetically modified Bt-maize and Roundup, and for toxicological studies the chosen group size and strain of rat used would comply with usual standards. However, in fact, the health effect most extensively discussed is tumor growth. Cancer development is then attributed to toxic effects (ibid. 4229), but the bottom-line is that the standard protocol for cancer studies would have required a fivefold group size and a less cancer-prone strain of rat. The health damage chiefly reported in the paper is cancer, and it is this effect that actually underlies their recommendation. Séralini et al. implicitly recognize the justification of this objection by confining the republished study Séralini et al. (2014) on toxic effects in a narrow sense. Cancer is rarely mentioned. This shift in emphasis confirms that the original paper is a piece of false advertising: The survey was designed as a toxicity study, but interpreted as a carcinogenicity study. Its design made it unfit for this latter purpose so that we are faced with a striking discrepancy between design and use and, consequently, with an agnotological endeavor.

In contradistinction, how is this case to be assessed in light of the impact-centered approach? There are two critical items: first, the violation of a conventional methodological standard, which, second, is apt to shift risks from producers to the public. Can these two features be identified in the case at hand? Regarding the violation of standards, the core objection was that Séralini et al. (2012) had passed off a study of cancerous effects as a study of toxic damage and had thereby taken advantage of less demanding methodological conventions. There is some plausibility to this charge which is underlined by the tacit correction of this feature in the republished paper. We should, however, bear in mind that the authors deny this charge and claim to explore the emergence of cancer as the result of toxic influences. Thus, although there is reason to assume that a methodological rule has been broken, this case also conveys that the notion of "established conventional standards" is sometimes ambivalent. Whether standards have been violated depends on whether the experiment is viewed as a study on toxic or on carcinogenic effects. The impact-centered approach has difficulty establishing which conventional standards are the relevant ones.

Second, granting that methodological requirements have been breached, what are the risks involved? On the face of it, the assumed violation serves to keep Bt-maize and Roundup away from public use even if they did not cause harm. Given the design of the study, the outcome is likely to be alarming even if the two substances did not, in fact, cause cancer. The risk-related reasoning underlying the interpretation of the experiment might be something like the following. A mistaken acceptance of Bt-maize and Roundup as innocuous would cause tremendous damage and should be avoided at any rate. Thus, seeking to keep the damage done in the event of error at a minimum requires raising the threshold for releasing genetically modified substances to the market. This framework appears to have guided the interpretation of the experiment. In this understanding, the interpretation given by Séralini et al. (2012) involves a shift from public to producer risks. Accordingly, this strategy apparently does not qualify as agnotological by Biddle-Leuschner's lights. I take this to be a false negative of their account (see Carrier [forthcoming] for additional details).

This speculative reconstruction of the train of thought underlying Séralini et al.'s (2012) interpretation suggests that their driving motive was to promote the common good. In the false-advertisement account, this assumed good cause does not change the assessment of the corresponding maneuver as being agnotological. Even if risks are in fact or believed to be shifted away from the public, the discrepancy between design and use exploited for this purpose still makes the move illicit in methodological respect. In the impact-centered approach, benefiting the public by breaking a methodological rule never qualifies as agnotological, whereas it may do so in the false-advertising account. What speaks in favor of the latter account is the intuition that methodological blunder should not be justified by appeal to the good political cause thereby supported. Deceiving the public in order to promote the common good is still an agnotological strategy.

The Bt-maize/Roundup example emphasizes once again that agnotology is not exclusively tied up with the commercialization of science. Agnotological maneuvers do not always serve economic interests. They rather may also be invoked for furthering political goals. Agnotological agents of the latter kind tend to be activists; they are driven by a mission. Séralini has a long record of fighting GMOs. He is the author of the popular anti-GMO and anti-pesticide book *Tous cobayes*! (Paris: Flammarion 2013), i.e., we are all guinea pigs, in which industry is accused of exposing the public to unknown effects of GMOs. Being driven by a mission may make scientists amenable to the allure of agnotological moves.

Consider another example of false advertising that concerns the assessment of the environmental risks involved in the construction of Oslo's airport Gardermoen which was planned in the mid-1990s. Environmental concerns were raised because the future airport was located in the vicinity of groundwater supplies that could be spoiled by the de-icing fluids employed in wintertime. An expert committee was set up, its task being to make the airport "100% environmentally safe." In particular, it was supposed to be ruled out in any event that de-icing fluids would eventually contaminate the groundwater. However, due to time restrictions, measurements and explorations were done in a rather superficial manner so that a lot of uncertainty about the geological and chemical conditions remained. The committee listed all these risks in their report, but they disappeared as the report was condensed into an executive summary. After the airport had been built on the basis of the truncated report, it quickly became clear that the pollution propagated much deeper into the soil than anticipated.

The experts had been asked to guarantee that the de-icing procedures would have no adverse effects on the environment. There are two types of risk involved which arise from underregulation and overregulation, respectively. The former means that damage is done by procedures falsely supposed to be harmless. The political commission to make the airport 100% environmentally safe is tantamount to excluding underregulation. The experts had been asked to certify that all procedures employed were sufficient for protecting the groundwater. The uncertainty admitted in the original report revealed that this goal was unattainable on the basis of the available evidence. The crucial item, however, is that the exclusive commitment to not jeopardizing the environment had lost its salience when the executive summary had finally reached the political decision-makers. When the report approached the political level, the opposite risk of overregulation had gained in importance. Overregulation means that a certain procedure is gratuitous. It is claimed to be indispensable for keeping the water clean, whereas, in fact, even without the measure in question, no damage would have occurred. Accordingly, what gained in significance here is the cost of protection and the need to remain within the planned schedule. Avoiding possibly superfluous measures whose deployment would be expensive and time-consuming had become a vital goal. In the decision-making process, emphasis shifted from avoiding underregulation to preventing overregulation (Droste-Franke et al. 2015, 49–51).

It is worth noting that no struggle obtains here between scientific and social objectives. Both commitments to environmental safety and to budget discipline and the schedule are non-epistemic and of a social nature. Therefore, the violation of standards was not so much the shift in itself but letting it occur implicitly. When the original report was processed in moving upward to the political ranks, the weight attached to different risks was quietly readjusted. Rearranging the balance between the two kinds of hazards without notice is tantamount to being guilty of false advertising. That is, the executive summary was used for demonstrating environmental safety, whereas, in fact, emphasis had been shifted to economic considerations. In light of the false-advertising account, we are faced with an example of agnotology.

This consideration reveals, in addition, that bringing the false-advertisement account to bear on singling out agnotological moves does not require the distinction between epistemic and non-epistemic values. This distinction is indispensable for pinpointing the methodological challenges posed by such moves. If the distinction is abandoned between seeking knowledge or understanding and striving for sociopolitical goals, we lose the explanandum (see Sects. 9.1 and 9.2). But the distinction is not part of the explanation. Unrecognized shifts may occur between non-epistemic values and yet represent an agnotological ploy. Accordingly, in no way does the false-advertisement account suggest that scientific research should be confined to pursuing epistemic ends. The idea underlying Proctor's notion of agnotology as active neglect and deliberate suppression of information can be interpreted such that epistemic concerns have been overruled by non-epistemic aspirations. However, in no way does this entail that pursuing the latter aspirations in general interferes with the process of knowledge-gain. No sin of agnotology is committed by seeking to make research socially relevant. In sum, the distinction between epistemic and non-epistemic values is presupposed for characterizing agnotological maneuvers, but is not employed in their identification. The distinction resembles Wittgenstein's ladder in that it may be dispensed with after it has served to move us one level up.

The conclusion to be drawn in light of the false-advertising account is that the cases of Bt-maize/Roundup and Gardermoen airport involve agnotological maneuvers. This account can also be used for accommodating Proctor's smoking example. The relevant industry raised objections to the design of the studies on health risks of tobacco smoking to the effect that these studies had either been not controlled, if humans were involved, or not involving humans, if they were controlled (see Sect. 9.1). These objections amount to demanding "proofs beyond a reasonable doubt" from the studies at hand and thus import a legal category into science. However, scientific investigations always come with a limited degree of reliability (as evidenced by the use of error probabilities). The design of a scientific study is never suited to prove a point with certainty. In demanding certainty, industry interprets possible epistemic achievements of science in a way not justified by the design of the pertinent studies. Accordingly, Proctor's chief agnotological case also comes out as agnotological in the false-advertising account. This example illustrates, in addition, that the illegitimate use of a study can be effected by parties that did not participate in its design. The same lesson accrues from the account of some NGOs' fights against GMOs given in Sect. 9.1. Economically driven strategies employed by industry may sometimes not be completely different from politically motivated moves of activists.

9.6 Identifying and Coping with Agnotological Machinations

It is one thing to uncover inappropriate bias, but it is a different thing to remove such bias from science. Given the lack of methodological rules specifying the neutral or even-handed judgment of a hypothesis, one of the more convincing counter-strategies for correcting inappropriate one-sidedness is conducting contrasting studies. A study of the defective kind can be supplemented with another one addressing the neglected perspective. In fact, given that there is no methodological rule for judging a hypothesis unambiguously, pluralism is the only means for approaching a balanced assessment. For instance, in the early 2000s, scientists debated worries that the anti-clotting efficacy of aspirin would decrease over the years. Some years later it was revealed by a journalist that the whole debate had been instigated by a company competing with the chief manufacturer of the drug. The competitor produced alternative anti-clotting agents and intended to boost its products via a scientific controversy. Conversely, some of the leading scientists opposing this alleged drop in effectiveness had been funded by the aspirin manufacturer (Wise 2011, 288). This mock controversy can be taken as revealing the unsavory effects of commercialization but also as supporting the claim that the one-sided focus of certain studies can be offset by an approach biased in the opposite way. In this case, the economic interests in play could have created an agnotological situation. They did not, however, because competing interests were at work, and this competition exerted a corrective influence. The ensuing pluralism served to promote objectivity.

The same lesson emerges if the agnotological paper by Séralini et al. (2012) is placed in its wider context. In a response to its forced retraction, a large number of French scientists, in an article published in the periodical *Le Monde*, called upon a re-examination of the earlier industrial studies that had not found adverse effects under similar conditions. They argued that it would have been better to assign already the original studies done before market release to a variety of researchers (Andalo et al. 2012). This response shows that in spite of its dubious value Séralini's study served to highlight the importance of pluralism for obtaining trustworthy results. In addition, the republished version Séralini et al. (2014) corrected the methodological blunder of the original paper and thus improved on its predecessor. Séralini received the whistleblower award of the Federation of German Scientists for 2015 in support of "independent science" in recognition of his demonstration of "toxic effects" of Roundup herbicide on rats. Accordingly, it is widely perceived that studies pointing in different directions can prompt more painstaking research that eventually produces more reliable outcome.

Speaking more generally, what helps to meet agnotological challenges is transparency and plurality. Transparency means that the partial and incomplete character of judging hypotheses is recognized and laid open. Furthermore, understanding that there is no uniquely correct way of judgment leads to acknowledging plurality. Unrecognized incompleteness and unacknowledged one-sidedness can best be revealed by pursuing research projects in which the questions addressed and the criteria of judgment employed differ from the studies under debate. This plurality makes it conspicuous that the agnotological judgments in question offer a partial view of the matter, but fail to acknowledge this partisan character.

However, we now seem to be faced with quite the same quandary as initially. Pluralism seems to suggest that it is quite legitimate to keep a variety of competing approaches in play and that there may be no way to single out one account in a justified manner. This could mean that we end up with being overwhelmed by the multiplicity of different points of view. Yet, in fact, we are in a better position to reduce the spectrum of approaches. Agnotological machinations can be identified by looking for indications of false advertising. Such approaches can be dismissed as not being serious epistemic endeavors. We are left with a qualified pluralism, and this means some progress at least. Pluralism is ambivalent. It is epistemically beneficial in enabling demanding procedures of test and confirmation, but it is also epistemically detrimental in leaving us without a definite system of knowledge. In this latter respect, the major challenge is to curb the manifold of contrasting assumptions and to focus on serious competitors. Identifying agnotological endeavors is among the means to achieving such a reduction and to produce a manageable range of alternatives. Some such accounts can be dropped as lacking the necessary credentials for being considered in earnest. They are disqualified as serious contenders for entering the system of knowledge. Such a reduction in the range of challenges to be taken into account justifies ignoring some dissenting approaches and thus helps to manage pluralism.[1]

9.7 Conclusion

The goal of this article is to point out one of the mechanisms underlying the deliberate creation and maintenance of ignorance. The challenge is to pinpoint agnotological patterns that are able to create confusion in a prima facie persuasive manner and to achieve this identification without invoking intentions. I seek to elucidate one such agnotological strategy, namely, false advertising. By this I mean the unrecognized discrepancy between the conclusions buttressed by the design of a study and the conclusions actually drawn or suggested. Agnotological challenges are best met by transparency and plurality. Transparency requires acknowledging the partial character of a study while plurality encourages conducting a different study so as to produce a more balanced picture. Pinpointing agnotological ploys is a means for weeding out approaches that look fitting at first glance, but which are, in fact, blatantly inappropriate. Identifying such endeavors serves to reduce the range of studies under consideration and thus contributes to making pluralism manageable.

References

Andalo, Ch., et al. 2012. *Science et Conscience*. Le Monde, Nov. 16, 2012. http://www.lemonde.fr/idees/article/2012/11/14/science-et-conscience_1790174_3232.html. Accessed 8 Apr 2016.

Biddle, J.B., and A. Leuschner. 2015. Climate skepticism and the manufacture of doubt: Can dissent in science be epistemically detrimental? *European Journal for Philosophy of Science* 5: 261–278.

Carrier, M. 2013. Values and objectivity in science: Value-Ladenness, pluralism and the epistemic attitude. *Science & Education* 22: 2547–2568.

———. forthcoming. *Agnotological challenges. How to capture the production of ignorance in science*. Under review.

de Souza, L., and L.M. Oda. 2013. Letter to the editor. *Food and Chemical Toxicology* 53: 440.

Driessen, P. 2009. *Disclosing climate change risks*. http://www.cdfe.org/center-projects/the-climate-lobby/disclosing-climate-change-risks/. Accessed 13 Apr 2015.

Droste-Franke, B., M. Carrier, M. Kaiser, M. Schreurs, Ch. Weber, and Th. Ziesemer. 2015. *Improving energy decisions. Towards better scientific policy advice for a safe and secure future energy system*. Heidelberg: Springer.

[1]It is true, even the agnotological study by Séralini et al. (2012) had the beneficial effect of stimulating pluralism. But appreciating the importance of contrasting studies in examining effects and side-effects of GMOs does not have to resort to an agnotological study. That is to say, the goal of stimulating pluralism is not a good reason for conducting and debating agnotological studies.

Hacking, I. 1999. Weapons research. In *The social construction of what?* 163–185. Cambridge, MA: Harvard University Press.

———. 2000. How inevitable are the results of successful science? *Philosophy of Science 67 (Proceedings)*: S58–S71.

Kitcher, Ph. 2001. *Science, truth, democracy*. Oxford: Oxford University Press.

———. 2011. *Science in a democratic society*. Amherst: Prometheus.

Kuhlen, I. 2009. Die Haftungsfrage, *Deutsches Ärzteblatt* 106(49), 4 Dezember 2009, 103. http://www.aerzteblatt.de/archiv/67026. Accessed 5 Apr 2016.

Longino, H. 1990. *Science as social knowledge: Values and objectivity in scientific inquiry*. Princeton: Princeton University Press.

Magnus, D. 2008. Risk management versus the precautionary principle. Agnotology as a strategy in the debate over genetically engineered organisms. In *Agnotology: The making and unmaking of ignorance*, ed. R.N. Proctor and L. Schiebinger, 250–265. Stanford: Stanford University Press.

Merton, R.K. 1942. The normative structure of science. In *The sociology of science. Theoretical and empirical investigations*, 267–278. Chicago: University of Chicago Press. 1973.

Michaels, D. 2008. Manufactured uncertainty: Contested science and the protection of the Public's health and environment. In *Agnotology: The making and unmaking of ignorance*, ed. R.N. Proctor and L. Schiebinger, 90–107. Stanford: Stanford University Press.

Oreskes, N. 2015. From nuclear winter to climate change: The political uses of scientific dissent. In *Wissenschaftliche Politikberatung im Praxistest*, ed. P. Weingart and G.G. Wagner, 35–50. Weilerswist: Velbrück.

Oreskes, N., and E.M. Conway. 2010. *Merchants of doubt: How a handful of scientists obscured the truth on issues from tobacco smoke to global warming*. New York: Bloomsbury Press.

Popper, K.R. 1966. *The open society and its enemies II. The high tide of prophecy*. 5th rev ed. London: Routledge.

Proctor, R.N. 2008. Agnotology: A missing term to describe the cultural production of ignorance (and its study). In *Agnotology: The making and unmaking of ignorance*, ed. R.N. Proctor and L. Schiebinger, 1–33. Stanford: Stanford University Press.

Proctor, R. 2012. *Golden holocaust. Origins of the cigarette catastrophe and the case for abolition*. Berkeley: University of California Press.

Schlitt, R. 2013. Ein Piks mit finanziellen Folgen. *Deutsches Ärzteblatt* 110 (8): 22. Februar 2013, A 304. http://www.aerzteblatt.de/archiv/134836. Accessed 5 Apr 2016.

Séralini, G.-E., et al. 2012. Long term toxicity of a roundup herbicide and a roundup-tolerant genetically modified maize. *Food and Chemical Toxicology* 50: 4221–4231. Retracted by the editor. http://www.sciencedirect.com/science/article/pii/S0278691512005637

Séralini, G.-E., et al. 2014. Republished study: Long-term toxicity of a roundup herbicide and a roundup-tolerant genetically modified maize. *Environmental Sciences Europe* 26: 1–17.

TAZ. 2011. *‚Riskant für Kinder und Jugendliche"*, 9.9.2011. http://www.taz.de/!5112434/. Accessed 5 Apr 2016.

Wilholt, T. 2009. Bias and values in scientific research. *Studies in History and Philosophy of Science* 40: 92–101.

Wise, N.M. 2011. Thoughts on politicization of science through commercialization. In *Science in the context of application. Methodological change, conceptual transformation, cultural reorientation*, ed. M. Carrier and A. Nordmann, 283–299. Dordrecht: Springer.

Chapter 10
The "Ought"-Dimension in Value Theory: The Concept of the Desirable in John Dewey's Definition of Value and Its Significance for the Social Sciences

Elizaveta Kostrova

Abstract The concept of "value" is widely used in various fields, and it has recently become the subject of empirical research. However, there is no common understanding of what it is. From the very start, the scope of value has been part of the opposition of what "is" to what "ought to be," and the fact that value judgments contained a normative element seemed to make the exclusion of value from the area of scientific analysis inevitable. John Dewey offers a different way of reasoning about values, which would allow scientists to keep the normativity in a way of saving the specificity of the concept. In order to do this, Dewey links the source of value with the evaluation process and introduces the concept of the "desirable" drawing the line between the "desirable" and the "desired". Clyde Kluckhohn later borrowed this concept from Dewey, while formulating the concept of values within Parsons' theory of action. Thanks to him, the "desirable" has become a favorite part of value definition among different researchers. As a result of this development, the concept of "desirability" has been transformed: for example, in social psychology, the "desirable" has moved closer to the "important", and the significance of the normative aspect has diminished, evolving to a more descriptive understanding, while the social dimension, though present already in Dewey, has greatly increased.

Keywords Definition of value · Normativity · Theory of action · Pragmatism · Desirability · Fact/value dichotomy · Value research in social psychology and sociology

---

E. Kostrova (✉)
Saint Tikhon's Orthodox University, Moscow, Russia
e-mail: elizakos@mail.ru

A. Christian et al. (eds.), *Philosophy of Science*, European Studies
in Philosophy of Science 9, https://doi.org/10.1007/978-3-319-72577-2_10

10.1 Introduction

The term *value* is used in a wide range of fields. Even putting aside its economic meaning, it can be found in the works of philosophers, psychologists and sociologists, not to mention its use in oral speech and public discourse. It seems that in all these cases the term implies some kind of a common hunch. At the end of the day, no one questions the existence of such utterances as “X is good” or “Doing X is right” which in themselves contain something that makes them different from simple statements of facts and descriptions. However, determining what it is exactly is not an easy matter. At first, *value* used to be a purely theoretical concept, but the second half of the twentieth century brought in numerous and rather informative empirical studies of this field in the context of social sciences (for a review, see Hitlin and Piliavin 2004). It, too, failed to provide a precise and unambiguous definition of value, though. In 1951 Clyde Kluckhohn wrote,

> Reading the voluminous, and often vague and diffuse, literature on the subject in the various fields of learning, one finds values considered as attitudes, motivations, objects, measureable quantities, substantive areas of behavior, affect-laden customs or traditions, and relationships such as those between individuals, groups, objects, events (Kluckhohn 1962, 390).

In 2000, Meg Rohan quotes him stating that since then the situation has hardly become any clearer (Rohan 2000, 255–256). To see the vast variety of concepts that can be understood as a value, take one of the existing lists where we can find such heterogeneous concepts as the nation, thriftiness and the respect for the parents (Krobath 2009, 549–551).

From the very beginning *value* has been firmly embedded in the opposition of entity and duty originated from David Hume’s theory on the logical specifics of the *ought* link verb as opposed to the *is* link. Since then the opposition has developed into establishing a clear and crucial distinction between the two concepts where one is related to causality, fact-finding, existential judgments and natural sciences; and the other is connected to teleology, the world of moral principles and values, normative and value judgments, and conforms to the principles of philosophy and humanities. The idea of a distinct border between that which *is* and that which *ought to be* has firmly taken root in philosophy. Despite all the attempts to eliminate it or to make it less prominent,[1] this kind of division has become an axiom (for example, Krobath 2009, 41).[2] In the abovementioned article, Clyde Kluckhohn points out that among all the various points of view on the nature of *value*

[1] One of the most famous attempts was made by John Searle; see (Searle 1964).

[2] A close connection between *value* and *norm* can also be proved by their inclusion in one and the same encyclopedia article: (Marini 2000, pp. 2828–2840). The correlation of the evaluative and normative is a separate topic; although both have the meaning of *should*, they are of dramatically different characters (e.g. see: Marini 2000, 2829–2830; Habermas 1992).

> [t]he only general agreement is that values somehow have to do with normative as opposed to existential propositions (Kluckhohn 1962, 390),

and in 1981 Hilary Putnam states that the fact and value opposition

> has assumed the status of a cultural institution (Putnam 1981, 127).

It seems that the normative element found in the statements of value inevitably prevented values from being subject to a scientific analysis as the true/false criteria obviously did not work well with them. So, in science there were two main ways of treating values. On the one hand, they could be treated as independent ideal objects that require a special approach and method, different from those of the natural sciences. The illustration of this approach can be found in the works of Heinrich Rickert who distinguishes between the natural science (Naturwissenschaften) and humanities (Geisteswissenschaften): According to him, values famously neither exist nor do not exist but remain valid (gelten) (e.g., Rickert 1915, 265). On the other hand, the empirical human science denied the existence of values, considering them a manifestation of *desire*[3] or *approval* (e.g., Ayer 1983), thus discarding *value* as a research subject altogether.

John Dewey suggested another way of treating values which allowed researchers to provide a foundation for their scientific empirical study – but as a *sui generis* reality not reduced to any other, say, psychological phenomena. First, they should not be considered metaphysical entities and instead some correlate should be found that belongs to the world of things; and the said correlate should lie outside the subject's inner world to enable objective observation. That is why Dewey preferred to interpret value as a property of a thing that was attributed to the latter by a subject, not as an independent entity, and related it to activity.

Moreover, he needed to get round the normative element found in the concept of value and connect values to statements of fact. The paradox was that he at the same time needed to keep some of the normative to preserve the value's very specifics. For this purpose, Dewey introduced the concept of the *desirable*.

In this article, I am going to consider the appearance of the term in the definition of value as well as its role in it and its further application in the study of values.

10.2 The Invention of the "Desirable": John Dewey

As has already been mentioned, Dewey's position[4] focuses on the two extreme points in the interpretation of values with which he is equally unsatisfied: on the one hand, a strict division into the field of facts and metaphysical field of values adopted

---

[3]See C. Ehrenfels's understanding of value as "being desired" (Begehrbarkeit) (Ehrenfels 1897, 53) or R.B Perry's view on value as "any object of any interest" (Perry 1950, 115).

[4]Dewey has several works on *values* (Dewey 1929, 1949, 1972); in this article I will stick to the latest one (Theory of valuation, 1939).

in the philosophical discourse on values, and on the other hand, a reductionist approach typical of psychology. Both of these make a scientific study of values impossible: in the first case, the language of science cannot be applied to them; in the second, *value* is nothing but a word and has no real substance of its own. That's why Dewey suggests a third option which will firstly preserve the specific character of values, and secondly make them available for a scientific research.

Dewey manages to avoid the metaphysical interpretation of values by establishing a connection between value and activity. To emphasize this connection he even prefers to speak about *valuation* or *something valuable* instead of *value*. It is the close observation of how human activity proceeds that cures us from the belief in any kind of end-in-itself – and that is exactly what values are usually considered to be in metaphysical theories, being valuable in themselves and serving as ultimate ends. Dewey completely denies the existence of that kind of objects. Firstly, he finds it unreasonable that an end would prevail whatever the means are. He claims in a rather straightforward manner that,

> ends framed in separation from consideration of things as means are foolish to the point of irrationality (Dewey 1972, p. 26).

Secondly, he argues that it is impossible to separate means from ends once and for all and suggests that we understand the relation between the purpose and the means not as a permanent constellation but as a continuum in which they are constantly swapping places (I exercise my voice a lot to improve my singing; I want to sing better in order to become a star; I want to become a star in order to be envied by my neighbor etc.). The absolutization of intrinsic properties, according to Dewey, results from the lack of empirical grounding. Thus, Dewey makes a crucial (in terms of the problem of values in empirical science) turn from the problem of values to the problem of a person's evaluative activity, making values descend from a high place and to the earthly world open for scientific analysis.

Now one should see to it that the specific nature of valuation is not reduced to any other psychological phenomena and thus prove value as an objective phenomenon available for observation and recording, as opposed to manifestations of a subject's inner world where true/false criteria are not applicable. Dewey achieves this by separating such notions as *wish*, *impulse*, *desire* and *desirable*.

The first difference, the one between wish and desire, lets him explain what distinguishes his own position from the one defended by, for example, Alfred Ayer (Ayer 1983). From Ayer's point of view, value expressions reflect the speaker's inner world akin to exclamations or blushing from embarrassment, so they cannot be considered logical utterances but mere manifestations of emotions or means of getting a certain reaction from the other. This approach excludes the possibility of empirical trial, as one's inner reality is not available for observation. Dewey, in his turn, insists on the objectiveness of the phenomena of values. That is why he distinguishes between wish and desire: while the former can be understood as an expression of fondness, that something is liked, the latter, desire, will necessarily presuppose certain efforts made to achieve something that is declared worthy,

> Effort, instead of being something that comes after desire, is seen to be of the very essence of the tension involved in desire. For the latter, instead of being merely personal, is an active relation of the organism to the environment (as is obvious in the case of hunger), a factor that makes the difference between genuine desire and mere wish and fantasy (Dewey 1972, 16).

Desire can well be scientifically recorded as valuation accompanied by activity: a person's behavior, their efforts make it clear for the observer what is considered valuable by this person, and the utterance by which it is recorded will contain a description of facts available for trial and true/false valuation (Dewey 1972, 51):

> In empirical fact, the measure of the value a person attaches to a given end is not what he says about its preciousness but the care he devotes to obtaining and using the means without which it cannot be attained (Dewey 1972, 27).

Dewey defines an act of valuation expressed by human activity, e.g. a mother considers her child valuable and makes obvious efforts to ensure its wellbeing. However, the science of values, according to Dewey, should not stop here but go deeper. Saying this, he makes a step forward leaving behind those theorists who considered any object of desire or interest (as Chr. Ehrenfels or R. Perry did) a value, and introduces a yet more subtle borderline between desire, impulse and the desirable. The difference between the said notions lies in the role the end and the means play in the activity.

Impulsive activity will be on the lowest level; here an action happens so fast, closely following necessity, so that there is almost no clear understanding of the end of an action. A hungry person grabs food hastily and a scared one runs from danger. Unlike the impulse, desire features some kind of a hunch of an end, an understanding of what the results are expected to be like – something that Dewey called *end-in-view*. While, according to the reductionist theories, even an impulse action contains a judgment of value (as a person running from a hazard obviously values safety), according to Dewey, even desire does not include it yet.

Only when we compare, for example, two ways of acting (e.g. two treatment programs suggested by different doctors) and determine some of them better for achieving some goal we want to reach (e.g. a patient's health), in this case our judgment will be evaluative and not just a statement of fact. If we have such a goal, its achievement can be imagined as a sequence of trials and errors or as a comparison and evaluation of actions and things from the point of view of their usefulness in terms of achieving the said goal, which means some of the actions may be considered *stupid* – not helping to get the desired or even hindering it. The alternative then would be the *desirable* option. The same procedure is applicable to comparing of our ends (as they are never ultimate).

Figure 10.1 illustrates Dewey's distinctions.

In this way, Dewey defines the thin border between the *desired* and the *desirable:* the former describes the object of my endeavor, whereas the latter introduces the normative element suggesting what I *should* choose. According to Dewey, the origins of norms are also found in activity, as the source of evaluation lies in the past experience,

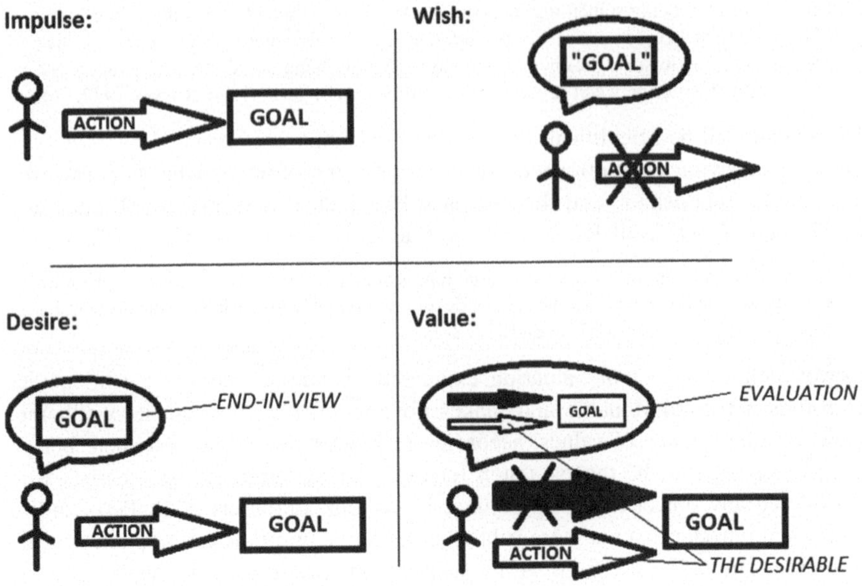

Fig. 10.1 Dewey's distinction between impulse, wish, desire and value

> The "desirable," or the object which should be desired (valued), does not descend out of the *a priori* blue nor descend as an imperative from a moral Mount Sinai. It presents itself because past experience has shown that hasty action upon uncriticized desire leads to defeat and possibly to catastrophe (Dewey 1972, 32).

Growing up and remembering all their unfortunate actions in the past, one learns to analyze their impulses and not to follow them blindly, take into consideration the current conditions[5] and foreseeable consequences. That is why a statement of value (i.e. a statement of what is desirable or not), unlike a statement of fact, can be interpreted as a norm describing the conditions for future actions.

The idea of the desirable allows Dewey to find a connection between fact and norm, preserving the division itself, building bridges between what used to belong to the *sphere of eternal values* and actual circumstances of human existence, highlighting the border separating *is* and *ought*. He shows how normativity arises from the human action itself. As Hans Joas points out,

> Dewey does not intend with these deliberations to eliminate the distinction between "is" and "ought"; instead, in good pragmatist fashion, he opposes all dualistic tendencies and emphasizes the continuity between the terms which have been differentiated (Joas 2000, 108).

[5]Including social conditions (Dewey 1972, 32).

Thus, Dewey manages to demonstrate the existence of statements of value that

> may rest upon scientifically warranted empirical propositions and are themselves capable of being tested by observation of results actually attained as compared with those intended (Dewey 1972, 24).

10.3 The Definition of Value for the Social Science: Clyde Kluckhohn

A key role in the discussion of the problem of value in social sciences belongs to Talcott Parsons (for a review, see e.g. Spates 1983). His idea was that a value is something that connects a person and the social system, which is the foundation of modern sociology understanding (Krobath 2009, 511–512). From Parsons' point of view, the cultural system, that, of course, encompasses values, too, provides a person with the standards and norms, which play the role of a benchmark in certain situations where a choice between several courses of action ought to be made (see Parsons and Shils 1962, 3–243). In his article "Values and Value-Orientations" American anthropologist Clyde Kluckhohn sets an objective to make Parsons' definition of value more precise, turning to Dewey's philosophy among others. Kluckhohn does not claim his definition to be universal, limiting it to a definition that can constitute a part of an action theory and will not, if possible, contradict the other uses of the word.

Kluckhohn's now classical definition of value is preceded by an analysis of the opposition of fact and value, existential and evaluative statements. The difficulty of the division seems to be connected with the fact that a real utterance can contain many various elements; the difference between evaluative and existential statements turns out to be no more than a difference in the degree, where one shade of meaning prevails over the other. It is a two-way principle. That is, for a scientist, in his phrase "It is true", there is an additional value attributed by him to the notion of truth (Kluckhohn 1962, 392). The reflection of values in the statements of fact can be indirect: as pointed out by Kluckhohn,

> In scientific discourse, at least, our propositions relate to matters we consider important (Kluckhohn 1962, 394).

Besides, the values deeply integrated in society become for its members nearly indistinguishable from the facts, so

> beliefs about "what is" are often disguised assumptions of "what ought to be" (Kluckhohn 1962, 410).

On the other hand, one cannot say that ideas of values do not depend on facts: as a rule, the ideas of a desired state of things are limited by some unshakable basic principles that characterize reality, e.g. the laws of physics, and create a frame within which the ideas of values can develop.

However, Kluckhohn does not agree with those who intend to discard the value-fact opposition. Although this borderline cannot be easily noticed in real life, it should be preserved on the theoretical level. That is why he agrees with Dewey on identifying three "fundamental types of experience": something that exists – the field of facts and existential utterances; something that someone wants – the field of desire; something that ought to be wanted – the field of the desirable (Kluckhohn 1962, 395). The last one is a kind of superstructure over the first two; calling something a value, we, according to Kluckhohn, mean approximately the following,

> This appears to be naturally possible [the frame of facts – E.K.]. It does not exist or does not fully exist, but we want to move toward it, or, it already exists but we want to preserve and maintain it [that expresses our subjective desire – E.K.]. Moreover, we aver that this is a proper or appropriate or justified want [that establishes norms and the aspect of values proper – E.K.] (Kluckhohn 1962, 393f.).

According to Kluckhohn,

> A value is a conception, explicit or implicit, distinctive of an individual or characteristic of a group, of the desirable which influences the selection from available modes, means, and ends of action. (Kluckhohn 1962, 395.)

It should be mentioned that Kluckhohn makes a direct reference to Dewey as a source of the key difference between the *desired* and the *desirable*, between *preference* and *justified preference* and dwells upon the said difference, which he claims to be of great importance. Its importance can be easily demonstrated by the well-known situation when a person suppresses and condemns his desires. Moreover, this word allows Kluckhohn to name an important connection between the affective and cognitive elements of a value: the mere feeling of affection accompanied by a moment of the reflection on the said affection, a simple *wish* accompanied by an *appraisal* (Kluckhohn 1962, 400). Thus, spinach eating becomes a value only if it has some rational reasoning, e.g. spinach's special nutritious properties (Kluckhohn 1962, 397, fn. 16). That means, *valuable* and *desirable* are different from *desired* and *preferred*. The difference lies mainly in the temporal aspect as well as the specificity of an object, being subject to natural impulses, possible sanctions for making a wrong choice, or the existence of conscious reasoning (Kluckhohn 1962, 397).

While Dewey introduces the *desirable* first of all due to one's own experience and only vaguely outlines the social dimension, for Kluckhohn, an anthropologist and Parsons' follower, sociology is the starting point for reasoning. At the very beginning of his article, he states that

> [h]uman life is – and has to be – a moral life precisely because it is a social life (Kluckhohn 1962, 388.).

Values constitute a logical construction, the result of the abstraction from many acts of social interactions (Kluckhohn 1962, 388). Dewey emphasized the process of summing up my past experience to hold me back from unreasonable choices; for Kluckhohn, the social reality grows in significance and becomes the source for my ideas of the desirable. Accordingly, the sanctions imposed in case of wrong behavior are assumed to influence my decision along with the perspective of failure. The restrictions placed by values on desires are connected to the preservation of a person's and society's integrity; that is

> [v]alues define the limits of permissible cost of impulse satisfaction (Kluckhohn 1962, 399).

The necessity in such restrictions results from the situation where society must ensure the existence of people – and that is a guarantee of some kind of stability and predictability of the behavior of separate individuals and the social system as a whole. It is the social aspect that defines the notion of the desirable: as it has already been mentioned above, it contains a subtle promise of sanctions that may follow in case of a violation.

10.4 The Convergence of the "Desirable" and the "Important": Social Psychology

Kluckhohn's definition, alongside with the concept of the desirable, played a crucial role in the further study of values by empirical science. He managed to formulate the main characteristics of values and their differences from the related concepts. However, the results varied depending on each science's specifics. The psychological approach focuses on an individual conscience, thus putting the social aspect aside. Moreover, there is an opinion that theoretical issues may hinder a scientist from observing what is actually going on at the empirical level and that science should develop *from bottom to top.* As a result, social psychology adopts a looser definition than the one aimed at by Dewey and Kluckhohn: in spite of preserving the connection with their works, the psychologists dim the normative element related to the concept of the *desirable*.

Psychologist Robert Kilby makes a rare attempt to generalize and sum up everything that has been achieved in the study of value so far (Kilby 1993). Of course, he touches upon Kluckhohn's definition and while he acknowledges its undeniable advantages, yet, Kilby finds it too narrow. Kluckhohn virtually reduces values to evaluative dispositions which allows him to give a very clear definition of values that, however, does not completely correspond to the entirety of phenomena. Kilby decides that for the purpose of a psychological, not anthropological, study of values the clarity and precision of the definition can be loosened a bit by adding the *worth values,* i.e. the belief in "the worthy, the significant, and the important", to the *evaluative values*, i.e. the belief in the "right", the "good", the "should", where the former can be relatively independent from the latter (Kilby 1993, 34). Kilby is talking about something that can be called a value for me, which is opposed to something that is right, when I think that one ought to possess such a value; such likings do not come from the desirable but from what I get from it: for me those activities or things are of *intrinsic worth.* It is notable that to prove his words he chooses an example from the field which is virtually excluded by Kluckhohn and Dewey, the followers of action theory: Kilby speaks of the admiration of art which becomes a value for me because of the aesthetic delight it brings – although I do not believe that other people ought to share my point of view.

According to Kilby, such a broadening of the definition is justified as when one analyzes some real cases of human interactions it is hardly possible to separate *valuable* from simply *important:* these two notions tend to get mixed up. Both of them are subject to selectivity: one thing or activity is chosen over the others; in both cases we can use the words *good*, *right*, etc.; *valuable* and *important* can be combined and interdependent in many ways: very often something at first considered a *value*, becomes then *important*, so it can be rather complicated to separate one from the other (Kilby 1993, 34–35). That is why Kilby suggests that we imagine these two as a continuum and not as two different categories (Kilby 1993, 35).

In the end, he defines values as the "conceptions of the desirable or the worthwhile" and is ready to reduce it to the "conceptions of the desirable", as the latter, according to him, unites the meaning of *valuation* and *importance* (Kilby 1993, 36). Furthermore, though, this denial of the connection between value and duty – even if only in the veiled form of the *desirable* – affects the clarity of borders: it is hardly clear how Kilby is going to separate values from interests (Kilby 1993, 44). Preserving the term, Kilby removes the difference between preference and desire, on the one hand, and values, on the other, while it was so important for Dewey and Kluckhohn.

Shalom Schwartz, one of the most significant figures in the empirical study of values, follows the same trajectory in his work. He mainly uses two definitions, which can be found in his various publications. In some cases (Schwartz 1999, 23–47) he uses explicit definitions, in others (Schwartz 1992, 2012) – implicit, defining the values through their properties. In both cases, he mentions the notion of the *desirable*. So, in 1999 he suggested the following definition of values,

> I define values as conceptions of the desirable that guide the way social actors (e.g. organizational leaders, policy-makers, individual persons) select actions, evaluate people and events, and explain their actions and evaluations (Schwartz 1999, pp. 24–25).

In another article he says,

> When we think of our values, we think of what is important to us in life (Schwartz 2012, p. 3).

The *desirable* appears several lines further when Schwartz enumerates the six characteristics of values, summing up his predecessors' works,

> *Values refer to desirable goals* that motivate action (Schwartz 2012, 3).

Then he concludes,

> Values are the socially desirable concepts used to represent these goals[6] mentally and the vocabulary used to express them in social interaction (Schwartz 2012, 4).

[6]Meaning the ten purposes realized in the ten types of values which serve to handle "three universal requirements of human existence", namely, "needs of individuals as biological organisms,

That is, in his formal definition Schwartz seems to be following Kluckhohn's reasoning (only taking into consideration that psychologist Schwartz is more interested in a person rather than a group) and uses his normative concept of the *desirable*. However, in his broader definition he refuses to understand value in the same way as Kluckhohn did. Just like Kilby, Schwartz wants to broaden the concept of value, making the *desirable* equivalent to the *important*. In his surveys, he uses them as synonyms: his questionnaires contain mostly the notion of *important* to establish the respondents' values (Schwartz 2012, 56–60). The *egoistic* values, e.g. those of self-direction, are the best to demonstrate how the normative and objective elements disappear and separating the *desired* from the *desirable* becomes impossible[7].

10.5 Conclusion: The Normative Aspect of Desirability

An attempt to combine Dewey's and Parsons' theories made by Kluckhohn can be considered one of the most important starting points of the empirical study of values by twentieth century social scientists. Parsons' conception creates the main course of understanding the value according to the action theory, and Dewey's philosophy allows researchers to clarify how exactly this connection works, while the typically pragmatic approach makes this union possible. Unlike many of his contemporaries, Dewey insists on the knowability of values. This implies that there are certain objective phenomena corresponding to the statements of value. Our idea of the *good* and the *right* does not belong to one's subjective inner world nor to the field of metaphysical entities but to the field of human activity. However, Dewey does not completely eliminate the normative aspect which is one of the most important characteristics of a value, but rather shows, by introducing the concept of the *desirable,* how the property of normativeness emerges from the notion of an end which is most crucial for any action. The concept is borrowed by Kluckhohn who manages to popularize the term, so that the element of the *desirable* becomes widely used by various scientists in the definition of values.

This approach has some great advantages: Most importantly, it allows social scientists to connect the study of values with objective reality, something that can be observed and recorded. However, it also has some flaws. For example, it hardly allows working with aesthetic values (compare Boyer 1956). It is no surprise that Dewey's followers also failed to study them (Kluckhohn 1962, 389). Thus, in Schwartz's theory of values they can only be included in the category of hedonism, which is at best controversial.

---

requisites of coordinated social interaction, and survival and welfare needs of groups" (Schwartz 2012, 4).

[7]See the categories of *hedonism* and *power* (Schwartz 2012, 56–60): here we can hardly imagine a conflict between the *desired* and the *desirable* in the form of "I do not want to, but I will".

The concept of the *desirable* creates yet other difficulties. Dewey tries to show that normativeness is not a special ontological space found beyond the objective world but can be explained through activity and the human ability to get experience by remembering and correlating actions and results as well as the ability to foresee the expected outcomes of their future actions. In this case, normativeness is nothing more than a usage of the past experience in the present situation. "I ought to do X" can be developed into "Maybe, I do not want to do X. But if I don't do X, then (as imposed by my own or historical experience) Y will happen, and I don't want Y"[8]. In Dewey's terms, "I ought not to do X" can be substituted by "X is undesirable" – which is drastically different from "I do not want X" or "I condemn X", the way value is understood by naturalists and emotivists (for example, the abovementioned Ehrenfels and Ayer). As the result, the highly problematic *ought to* appears to be explicated into the non-normative elements.

However, it only works as long as you keep in mind the whole set of Dewey's distinctions. Meanwhile, they rather gradually retreat into the background until nothing is left except for the word *desirable*; and the latter only seems to have solved the problem, as stated by Rüdiger Lautmann in his profound study (Lautmann 1969). First of all, its meaning is ambiguous as it can mean both something that is worth striving for and – descriptively – something that can be desired, that is actually desired, that is synonymous to interesting, useful, etc.[9] Thus, the difference between the *desired* and the *desirable* is not as prominent as Dewey wanted.

What is yet more important is that the clarification of the concept of the desirable makes us again turn to the notions that contain the normative and value aspects (Lautmann 1969, 37–38).[10] Even the explication of the action situation suggested by Dewey turns out to be imperfect and fails to avoid an impasse in the definition, and that is, for it to become operative, we need to evaluate the possible results of an action (Y from the example) as negative, *bad*. This scheme can function only in cases of relatively primitive situations connected to, say, survival, where the valuation of the results can be considered intuitively obvious. However, it is evident that this will not work in cases that are more complex.

Nevertheless, introducing the concept of the *desirable*, Dewey seems to have found a rather witty solution to the problem and suggests a way of concretizing the nature of the *ought*-element found in values, particularly the aspect connected with the social nature of a value, which is so important for Humanities. This particular shade of meaning can be noticed in the word itself: the *desirable* is something that the others want from me, thus influencing my choice.

---

[8]Or vice versa: "I ought not to do X" can be interpreted as "I probably want X. But if I do X, Y will happen, and I do not want Y".

[9]One can say, "this position is highly desirable" (i.e. "Everyone wants this position"), and "it is *desirable* to distinguish between these concepts" (i.e. "one needs/ought to/should distinguish between these concepts").

[10]E.g. the dictionary definition of *desirable* refers to *worth*, while the latter is defined through *value* (see e.g. Oxford English Dictionary).

Dewey says little about the social sources of normativeness that are so important for the modern scientists. He focuses on a person and valuations produced by separate people. However, he realizes that some of the goals are not set by a person themselves but rather by the society in which they live.[11] According to Dewey, though, such goals, unless they are thought over, are of no interest to the theory of valuation as they are no more than habits: they create the space within which a person will determine their own goals. Kluckhohn probably means the same, talking of institutionalized values, which become part of a situation alongside existential beliefs.[12]

Dewey must understand these cultural values as the result of a certain generalization mechanism. The generalized ideas of values are the result of accumulating experience from similar repetitive situations and thus are abstract but not at all *a priori*; despite not being connected to some separate case, they, however, depend on the sum of empirical situations. In addition, their application to new cases, in its part, serves as an empirical trial (Dewey 1972, 44).

Empirical scientists following Dewey's idea of the *desirable* give more prominence to the social aspect: the focus is now on the second, less normative meaning of the word ("everyone wants it"). It is obvious for both Kluckhohn and Schwartz, that values are what connects a person with society. Kluckhohn says at the very beginning of his article,

> ...all group values... are the products of social interaction as embodied in culture (Kluckhohn 1962, 388),

while Schwartz substitutes *desirable* with *socially desirable* (Schwartz 2012, 4). Here the word *desirable* is a very happy one, as *desirable* normally lacks the subject of desire but contains a positive valuation and a normative element, a distance between the object and myself. That very much fits in with the sociological and psychological idea of values: a person's idea of what should be desired is nothing less than an internalized idea of what everyone (or many) desires (see Darity 2008, 581–583).

The concept of the *desirable* successfully conveys the understanding of a subjective-objective character of values typical of pragmatism ("Wert" 2004, 572). On the one hand, the word's composition refers to the subjective, but also hints at the generalization of the opinions of many: as if other people share my valuation (or disagree with it), thus proving it. The idea of the connection between the values and people's social existence has developed into the most advanced strategy of overcoming the split between facts and values. Working on this topic, Hilary Putnam and Richard Rorty will suggest discarding the classic division altogether (Putnam 2004, Rorty 1982).[13]

---

[11]"At any given time an adult person in a social group has certain ends which are so standardized by custom that they are taken for granted without examination, so that the only problems arising concern the best means for attaining them" (Dewey 1972, 43).

[12]"By institutionalization value *is* part of the situation" (Kluckhohn 1962, 410).

[13]Actually, John Searle's criticism of the fact-value opposition is also based upon the existence of social institutions.

Despite everything that has been said, in social sciences there has been no absolute rejection of the normative element. Even when *desirable* was closely associated to *important*, there was still the idea that any value originates from a conflict: evaluative words *right/wrong*, *better/worse*, etc.

> are certainly used whenever it is felt that there is an incomplete matching between an existent state of affairs and what is possible in nature (Kluckhohn 1962, 398).

Strictly speaking, unless the ideas of values confront the reality or each other, they are hardly perceived and exist as a background of human activity (Schwartz 2012, 4). Thus, the origin of values still has a close connection with a person's refusal to put up with what actually exists and their pursuit of what only exists in their imagination.

References

Ayer, A.J. 1983. *Language, truth and logic*. Harmondsworth: Penguin.

Boyer, M.W. 1956, September. An expansion of Dewey's groundwork for a general theory of value. *The Journal of Aesthetics and Art Criticism* 15 (1): 100–105.

Darity, W.A., ed. 2008. *International encyclopedia of the social sciences*. Vol. 8. 2nd ed. Detroit: Macmillan Reference USA.

Dewey, J. 1929. *Experience and nature*. The Paul Carus lectures. London: Geroge Allen & Unwin, Ltd.

———. 1949. The field of value. In *Value: A cooperative inquiry*, ed. R. Lepley. New York: Columbia University Press.

———. 1972. Theory of valuation. In *International encyclopedia of unified science: Foundations of the unity of science, 2*, 4. Chicago: University of Chicago Press.

Ehrenfels, Ch. von. 1897. *System der Werttheorie*. Vol. 1. Leipzig: O.R. Reisland.

Habermas, J. 1992. *Faktizität und Geltung: Beiträge zur Diskurstheorie des Rechts und des demokratischen Rechtsstaats*. Frankfurt am Main: Suhrkamp.

Hitlin, S., and J.A. Piliavin. 2004, August. Values: Reviving a dormant concept. *Annual Review of Sociology* 30: 359–393.

Hügli, A. 2004. Wert. In *Historisches Wörterbuch der Philosophie*, Völlig neubearb. Ausg. des "Wörterbuchs der philosophischen Begriffe" von Rudolf Eisler., 12: W-Z:1556. Basel: Schwabe.

Joas, H. 2000. *The genesis of values*. Chicago: University of Chicago Press.

Kilby, R.W. 1993. *The study of human values*. Lanham/New York/London: University Press of America.

Kluckhohn, C. 1962. Values and value-orientations in the theory of action: An exploration in definition and classification. In *Toward a general theory of action*, ed. T. Parsons and E. Shils, 388–433. Cambridge, MA: Harvard University Press.

Krobath, H.T. 2009. *Werte: Ein Streifzug durch Philosophie und Wissenschaft*. Würzburg: Königshausen & Neumann.

Lautmann, R. 1969. *Wert und Norm: Begriffsanalysen für die Soziologie*. Köln und Opladen: Westdeitscher Verlag.

Marini, M.M. 2000. Social values and norms. In *Encyclopedia of sociology*. 2nd ed. 4: 2828–2840. New York: Macmillan Reference USA.

Parsons, T., and E. Shils, eds. 1962. *Toward a general theory of action*. Cambridge: Harvard University Press.

Perry, R.B. 1950. *General theory of value: Its meaning and basic principles construed in terms of interest*. Cambridge, MA: Harvard University Press.
Putnam, H. 1981. *Reason, truth, and history*. Cambridge/New York: Cambridge University Press.
———. 2004. *The collapse of the fact/value dichotomy and other essays: [Including the Rosenthal Lectures]/Hilary Putnam*. 3. Print. Cambridge, MA: Harvard University Press.
Rickert, H. 1915. *Der Gegenstand der Erkenntnis. Einführung in die Transzendentalphilosophie*. Tübingen: JCB Mohr (Paul Siebeck).
Rohan, M.J. 2000. A rose by any name? The values construct. *Personality and Social Psychology Review* 4 (3): 255–277.
Rorty, R. 1982. *Consequences of pragmatism: Essays, 1972–1980*. Minneapolis: University of Minnesota Press.
Schwartz, S.H. 1992, December 1. Universals in the content and structure of values: Theoretical advances and empirical tests in 20 countries. In *Advances in experimental social psychology* 25: 1–65. Elsevier.
———. 1999, January. A theory of cultural values and some implications for work. *Applied Psychology* 48 (1): 23–47.
———. 2012, December 1. An overview of the Schwartz theory of basic values. *Online Readings in Psychology and Culture* 2 (1).
Searle, J.R. 1964, January. How to derive 'ought' from 'is'. *The Philosophical Review* 73 (1): 43–58.
Spates, J.L. 1983. The sociology of values. *Annual Review of Sociology* 9: 27–49.

Chapter 11
From Stability to Validity: How Standards Serve Epistemic Ends

Lara Huber

Abstract The paper explores standardisation from the perspective of epistemology. Its aim is to enquire into the reality of standards as being very specific tools with defined uses, but at the same time sharing general suppositions about which ends they serve within the realm of science. The paper focuses on the questions how standards relate to ends that facilitate and/or allow for knowledge claims in the sciences. Therefore, scientific practices in different fields of research are assessed, ranging from measurement to experimental trial design in medicine and psychology.

Keywords Experiment · Robustness · Homogeneity · Validity · Standardisation · Epistemology

11.1 Introduction

Standards are said to provide trust in scientific methodology in general and measuring devices in particular. To standardise means to formalise and regulate scientific practices, to prioritise instrumental and methodological prerequisites of research: Standardisation impacts on the design of experiments, concerns the reporting of outcomes and the assessment of research (e.g. peer review process).

Studies in the history of science and technology have shown that standards contribute significantly to the evolution and validation of scientific practices (e.g. Chang 2004). The philosophy of science is as yet only beginning to analyse systematic challenges posed by standardisation (e.g. Tal 2011; Schlaudt and Huber 2015). Currently, scholars almost exclusively focus on measuring in science. Take, for instance, Allan Franklin's book *Experiments in Particle Physics in the Twentieth*

L. Huber (✉)
Kiel University, Kiel, Germany
e-mail: huber@philsem.uni-kiel.de

A. Christian et al. (eds.), *Philosophy of Science*, European Studies in Philosophy of Science 9, https://doi.org/10.1007/978-3-319-72577-2_11

Century, in which he illustrates how standards of measurement that determine scientific values, especially significance and credibility, are under threat due to shifting practices in science (Franklin 2013).

The main interest of this paper is to elaborate on the question how standards relate to ends that facilitate and/or allow for knowledge claims in experimental sciences in general. In the following, the paper intends to inform about scientific practices in different fields of research that address given ends of standardisation. First of all, the paper presents three examples of standards in science.

11.2 What Are Standards?

Standards are ubiquitous in our life in general. Just consider plugs and power adapters. Standards are also common tools in the scientific laboratory in particular: think about gauges and reference samples. Standards are either objects, for example measuring devices that have been standardized, or they are employed to describe a formal practice (e.g. protocol). The former often serve as tools that allow for standardisation (e.g. prototype, gauge). The latter formalises and guides experimental practices. For instance, they inform about technical preconditions and, wherever required, about inclusion and exclusion parameters regarding research subjects.

Def.1: A standard is ....

.... an item, that is commonly acknowledged for measuring length, for example a metallic bar, an artefact standard (e.g. the former "standard meter"); or

.... a living organism, for example a mouse or a rat that has been transformed through breeding and/or genetic engineering; or

.... a formalised practice, for instance a protocol, that defines key features of an experimental design.

All the examples mentioned in this definition represent prominent cases of standardisation in the natural sciences. Still, their potency might not be restricted to scientific purposes only. One might even say that all three examples present us with "kinds" of standards or at least "typical" ways of standardisation. In the following, I shall not aim to develop a typology of scientific standards.[1] Instead, my paper will focus on scientific practices that refer to these exemplary cases of standards. Are there common ends of standardisation in science from an epistemological point of view? The analysis begins with some preliminaries on this question.

[1] I had recently suggested how standards for scientific purposes could be differentiated (Huber 2016). By introducing "kinds of scientific norms", my aim, clearly, was not to defend a strict "typology" of standards, but to address the need to systematically discuss the status of standards and to reflect on their scope from an epistemological point of view. According to the paper's systematic ratio, the first and second examples, as mentioned above, are "technical norms" (technische Normen), whereas the third one is an example of a "methodological norm" (methodische Norm). For an introduction, see ibid., pp. 400–407.

11.3 How Do Standards Relate to Ends?

As concerns scientific practices, standards are commonly attributed a given operational value. Accordingly, this value explains why certain objects or protocols are regarded as "standards". In other words, a standard is a prioritised tool/instrument to serve a given end (e.g. measuring gauge).

First of all, in order to address a given operational value of a standard you have to define an end towards which this standard is working. Accordingly, a standard is a *means* that is attributed an operational value if it allows to ascertaining a given *end*. In an earlier paper I have elaborated on how standards rise, that is to say, how they become "scientific norms" (Huber 2015). I differentiated two ways of how standards gain momentum and are implemented as a "norm/rule" that is widely acknowledged: "in-progress-stabilisation" and "explicit fixation". The latter addresses the need to set standards explicitly to securing the interests of a scientific community (e.g. stable classification system and nomenclature). The former acknowledges mere "bottom-up" processes. One main driving force of "in-progress-stabilisation" is what I call the "operational value" or "functional capability" of a designated standard to serve a given end (ibid.). Regarding this process of stabilisation, a further analysis is obviously required. For instance, concerning the query of how norms/rules come into existence in the sciences.[2] In this paper I shall focus on the operational value of means to *certain* ends. Especially, I am interested in the question of how standards serve ends that facilitate and/or allow for knowledge claims in the sciences, and therefore might be called "epistemic ends".

> Def. 2a: A standard is a scientific norm/rule because of its operational value to ascertain ends that facilitate and/or allow for knowledge claims.

Standards differ from mere "conventions", in the sense of pragmatic rules. In the case of conventions we are seeking to apply a rule (rule "a") that could be equally substituted (rule "b"). According to this reading, conventions do not change the outcome of actions, whereas standards do, because they are "founded rules".[3] In other words, standards are mandatory means to allow for a certain end. I would even claim that they are *constitutive* for scientific practices—as long as they are *constitutive* for epistemic ends.[4]

> Def. 2b: Standards are founded norms/rules because they reliably serve epistemic ends.

---

[2] A general account of the emergence of (social) norms/rules provide Ullmann-Margalit (1977) and Raz (1990).

[3] Bunge, for instance, distinguishes "groundless or conventional rules" from "founded rules" (1967, Vol. 2, 132): "Many rules of conduct, work, and sign, are *conventional*, in the sense that they are adopted with no definite reasons and might be exchanged for alternative rules with little or no concomitant change in the desired result." According to Bunge, "founded rules" are "norms satisfying the following *Definition*: A rule is grounded if and only if it is based on a set of law formulas capable of accounting for its effectiveness."

[4] For my reading of "constitutive", see Searle (1995).

In the following, I shall present three case studies on standardisation in different fields of scientific research, ranging from physics and measurement science to population-based trial design in psychology and medicine. My concern here is to explore general assumptions about standardisation from the perspective of the philosophy of science. More explicitly, my aim is to enquire into the reality of standards as being very specific tools with defined uses, but at the same time sharing general suppositions about which ends they serve within the realm of science.

The paper focuses on three ends purported to serve epistemic needs in different fields of scientific inquiry. These are (1) *stability*, (2) *homogeneity*, and (3) *internal validity*. First of all, I shall introduce each of these ends with regard to general assumptions about prospects of standardisation from an epistemological point of view and name in a rather exemplary manner, additional methodological challenges (see 2.1 to 2.3). In the following I shall present further arguments that support the claim that standards are necessary tools to ascertain the epistemic status of measurement procedures and allow for knowledge claims in reference to experimental trial design.

11.3.1 Stability as Epistemic End of Standardisation

Precision measurement presupposes *stability*: Units and other standards of measurement are said to be of "constant value". Therefore, in the nineteenth century scientists Charles Sanders Peirce and Charles Babbage addressed the weakness of artefact standards, given that prototypes of measurement such as metallic bars (here "standard meter") might change significantly due to time and use. Peirce's scientific and practical interest in measurement was not only restricted to measurement as such, but also included the design of instruments, for example for measuring gravity using pendulums of his own design.[5] Peirce is also known for his research on the constancy of standards of measurement, such as the matching of a wavelength of light to the length of a rod, in this case, the rigid body of the artefact standard of the metre (cf. Hacking 1990, 202f.).

Physical constants, such as the redefinition of the metre in terms of wavelengths of light, are physical quantities that are generally believed to be both "universal in nature" and having "constant value" in time.[6] In his scientific report on the *Comparison of the Metre with a Wave-Length of Light*, written in 1882, Peirce

[5]Accordingly, Peirce's research was strongly oriented towards issues of verifying measurement results and ascertaining the accuracy of standards and balances. He also addressed the need for scientific authority and scientific control for all operations of weighting and measuring (cf. *Testimony on the Organization of the Coast Survey*, 1885. In: Peirce 1982–1999, Writings. Vol. 5: 1884–1886, pp. 149–161).

[6]For an introduction to the evolution of measurement units and further information on the current reform of the *International System of Units* (SI), visit the Homepage of the *Bureau International des Poids et Mesures* (BIPM) <http://bipm.org> (24.01.2017).

reports that metallic bars used as standards of length have more than once been found to have changed their lengths in the course of years.[7] In his text on *Six Reasons for the Prosecution of Pendulum Experiments*, also written in 1882, Peirce expresses his views on the matter by stating that recent investigations support the supposition "to attach increased importance to experiments with the pendulum in their connection with metrology."[8] Peirce explicitly states his trust in pendulum experiments as follows:

> I will even go so far as to say that a physicist in any remote station could ascertain the length of the metre accurately to a one hundred thousandth part more safely and easily by experiments with an invariable reversible pendulum than by the transportation of an ordinary metallic bar.[9]

In this matter, Peirce was also supported by his contemporaries. About around the turn of the nineteenth century assessments such as these were combined with concrete demands for dematerialization of measurement units and the establishment of "absolute standards" (Maxwell) and "natural constants" (Babbage) respectively.[10]

As the critical evaluation of artefact standards highlights, stability reflects an internal quality of standards, given that an ideal standard should provide a baseline or a reference (constancy of value) that does not change over time. In the case of artefact standards, *constancy of value* coincides with *constancy through time (and space)*. Therefore, stability is a quality in itself that is associated with a given epistemic gain, especially regarding measurement procedures.

In the empirical and, above all, the experimental sciences, stability is also read as "the consequence of compelling evidence" (Hacking 1999, 33). In theses cases, stability reflects the epistemic quality of data generated by a given experimental trial design. Regarding the latter, stability is said to be a concept "richer than mere lack of change".[11] In the scientific literature the epistemic quality of (empirical) data is also addressed as "robustness": Data that are said to be robust show a *stable* representation, a "suggestive pattern" or "structure".[12] Given that a stable pattern or structure might also reflect a mere artefact (e.g. disturbances in the data),[13] robustness is secured by repeated measurement or, if possible, with evidence

[7]Peirce 1982–1999, Writings, Vol. 4: 1879–1884, pp. 269–298; here: p. 269.

[8]Ibid., pp. 356–360; here: p. 357.

[9]Ibid., p. 358.

[10]Cf. Schaffer (1992, 25f.).

[11]See Radder in his introduction to *The Philosophy of Scientific Experimentation* (2003, 2f.). Feest gives the following account (2011, 59): "Roughly, this notion is taken to refer to the processes whereby scientist (a) empirically identify a given phenomenon or entity, and (b) gradually come to agree that the phenomenon is indeed a stable and robust feature of the world, rather than being (say) an artifact of any particular experiment or instrument."

[12]Bechtel, for instance, refers to a "suggestive structure in the results" (1994, 174): "Robustness is, for many scientists, the strongest criterion of what is real in or about the world."

[13]For further challenges see, for instance, Bechtel, who uses the case of research techniques in cell biology (1994, 174f.).

generated by multiple techniques.[14] It is only in these cases that a *stable* pattern is truly informative, that is to say, that it represents a real phenomenon. Most prominently, Ian Hacking has referred to the stabilisation of phenomena on the basis of experimental practices and experimental knowhow. In *Representing and Intervening*, he states (1983, 231):

> To experiment is to create, produce, refine and stabilize phenomena. If phenomena were plentiful in nature, summer blackberries there just for the picking, it would be remarkable if experiments didn't work. But phenomena are hard to produce in any stable way. That is why I spoke of creating and not merely discovering phenomena. That is a long hard task. Or rather there are endless different tasks. There is designing an experiment that might work. There is learning how to make the experiment work. But perhaps the real knack is getting to know when the experiment is working.

To sum up, the first step of my analysis has shown that "stability" is a possible feature of different entities, of material objects, of formalized practices as much as of patterns of data. As an epistemic end, it refers to different observations of constancy and/or robustness, for example, the material constancy of objects through time and/or space, the formal constancy of practices as much as the constancy of patterns of data through repeated measurement. Against this background, two readings of stability as an epistemic end could be differentiated: First of all, stability reflects a quality that comes with an essential epistemic gain, especially for the purpose of measurement (*constancy* of a reference unit). Secondly, there are cases, where stability as a quality (e.g. stable pattern of data) is additionally associated with epistemic values (*evidence* in measurement data, *validity* of experimental results). The latter illustrates that different epistemic ends of standardisation might coevolve or even coincide from time to time. I shall elaborate on this topic more broadly, especially with regard to the concept of "internal validity" as an epistemic end of population-based trial design (see below 2.3). First of all, the paper introduces a second quality of material artefacts that is realised and maintained by standardisation, namely "homogeneity".

11.3.2 Homogeneity as Epistemic End of Standardisation

In engineering and the technical sciences, homogeneity or uniformity as an ideal of standardisation corresponds to the paradigm of *interchangeability*. As driving force of mechanisation and industrialisation, standardisation permits rather large

[14] Franklin introduces robustness in his study on *Experiments in Particle Physics* as follows (2013, 231): "Robustness may [...] be provided by [...] experiments in which the same quantity is measured using different experimental apparatuses and in which the systematic uncertainties, the backgrounds, and the selection criteria are different." Besides, robustness might also be challenged, as other scholars point out: For instance, if there is only one technique available (Gandenberger 2010). In other cases, where different techniques are available, the question might remain whether their outcomes could be compared (Stegenga 2009).

production runs of component parts that can be readily fitted to other parts without adjustment. Think, for instance, about bolts and nuts. Accordingly, homogeneity or uniformity describes the specific quality of an item ("material artefact"), namely a standardized part or product, with respect to its *design* and/or *performing function.*

Then again, homogeneity as an ideal of standardisation is no longer restricted to the realm of engineering. In the life sciences, different biotechnical practices address homogeneity as an epistemic end of standardisation: In recent years, synthetic biology adopted the industrial and engineering concept of a part. Current research promotes strategies of engineering biology in creating, sharing and using standardized biological "parts", namely "BioBricks".[15] *BioBricks* are DNA sequences, for example "portions of proteins", "RNA sequences", "plasmids" and the like. Databases, most prominently, the *Registry of Standard Biological Parts*, provide the resources to construct new "parts" and "devices" that could be used for measurement purposes or the creation of simplified biological systems (e.g. the minimal synthetic genome of a bacterial cell). The Registry also communicates what is called "assembly standards", which inform about how biological "parts" could be put together.[16]

The standardized biological "part", actually, is only one of many examples of how homogeneity in biology and biomedicine reflects the ideal of "pure culture" and practices of "purification".[17] Chemists isolate and purify chemical compounds by repeated dissolution, distillation and crystallization. By the end of the nineteenth century medical bacteriologists adopted this ideal by referring to the "pure nature" of bacterial organisms. Robert Koch, for instance, addressed a "pure culture" of bacterial organisms as a culture "which bred true and displayed no alterations in morphology."[18]

Early practices of genetic engineering extended this ideal of pure culture to encompass more complex organisms (e.g. Wilhelm Johannsen's work on "pure

---

[15]The give and get-policy of sharing standardized biological parts, which was implemented by the *Registry*, plays an essential role in community building and might also contribute to defining valid targets of research (cf. <http://parts.igem.org>; 24.01.2017).

[16]Cf. Mackenzie (2012, 158): "The BioBrick standard says nothing about the specific function of the biological parts. It really only addresses how parts can be put together. The standardisation concerns only those aspects of the biological part that pertain to assembly, or linking together. Obviously, parts that cannot be connected or interfaced easily are not engineerable." He especially problematizes the underlying concept of a "part" (ibid., 157): "DNA comes nowhere near complying with the form-matter, or coding-coded distinctions that are layered into most industrial and engineering concepts of a *part.*"

[17]In genetics, for example, specially designed plasmids (small, closed, circular pieces of DNA) are used in experimental research. Here, the "plasmid prep" is utilized for plasmid purification and isolation and could be regarded as a commonplace tool in recombinant DNA research.

[18]Cf. Gossel (1992, 292): "A new or unusual form he labelled a "contaminant"." She demonstrates how "pure culture-techniques" invented by Koch and others, were adopted throughout bacteriology. Gradmann (2014) describes how "standard" methods of nineteenth century bacteriology ("Koch's postulates") are depicted in current bacteriological and microbiological research.

lines" of beans).[19] Since the beginning of the twentieth century, practices of purification inform about the creation and maintenance of what are said to be "homogeneous" animal populations, "pure strains"–an issue Bonnie Tocher Clause outlines in her historical study on the *American Wistar Institute* and the rise of the "WISTARAT" around 1910:

> [T]rust toward uniformity and apparent perfection was realized through [...] breeding experiments; as a result of the process of close inbreeding and continuous selection for desirable traits, the gene pool was narrowed down so that the probability of "breeding true," or achieving reproducible genetic uniformity, was maximized, and the possibility of the appearance of undesirable or new traits was minimized. [...] Thus the Wistar Institute was able to produce rats of known traits, both anatomic and genetic, biometrically quantifiable within a known range of variation and comparable to a set of known growth curves. The inbred Wistar albinos, then, were initially defined and promoted as "homogeneous," meaning that they were thought to be predictably uniform in both measureable and hidden traits.[20]

What allowed for the standardisation of animals in the early experimental sciences, especially in physiology, was selective breeding. As the case of "Wistarat" illustrates, the new "life form" of inbred albinos also affected assumptions about the generality of outcomes of experimental trial design.[21] Accordingly, "homogenous" or "pure" strains became "ideal tools" for experimental research in physiology (Kirk 2012, 61). And they were soon regarded as important resources for research into human diseases. In this context, "homogeneity", literally, meant different things–as Ilana Löwy and Jean-Paul Gaudillière have pointed out in their 1998 paper on *Disciplining cancer.* Standardisation with the end of providing "homogeneity" of strains in animal-based cancer research, could either be "targeted at the production of animals with the 'same genes'" or at "the selection of strains showing the same incidence of cancer."[22] This reflects different targets of biomedical research into cancer. Accordingly, if at all, different human populations might profit from these trial designs. In the following I shall address some challenges that either arise with regard to the transfer of pathological markers into animal organism, or with regard to the translation of experimental outcomes of animal-based research into clinical medicine. I shall restrict my analysis to philosophical questions and refrain from addressing queries with regard to defining aims of research into human diseases that have been critically discussed by scholars of sociology and epidemiology.

Today, "homogeneous" strains of animals are especially important resources for model-based approaches into human diseases: In biomedicine homogeneity is presupposed to establish demands of "similarity" within experimental settings. With regard to explanatory inference it has been stated that model and target

[19]For an introduction see Müller-Wille (2007).

[20]Cf. Clause (1993, 337 or 345f., respectively).

[21]In her historical study, Logan (2002) elaborates on the changing role of test animals in physiology due to the standardisation of tools and the selective breeding of animals. She also focuses on the question in how far the scientist's assumptions about diversity and generality were transformed.

[22]Cf. Löwy and Gaudillière (1998, 221).

systems should share *relevant similarities* (e.g. Parker 2009). Whereas what is considered "relevant" depends on the particular question an experimental system wants to provide answers for. As concerns animal-based approaches into human diseases, "similarity" is instantiated through genetic engineering: For this purpose bioscientists, for instance, create transgenic mice that express a human pathogen.[23] First of all, a human pathogen has to be identified. Secondly, a standardized strain of animals has to be chosen or invented from scratch to create a transgenic strain of test animals. Ideally, every individual transgenic animal should show a given set of pathological features that affect human subjects. Further prerequisites of research include standardized tools, for instance staining techniques in immunohistopathology. In the case of research into Alzheimer's Disease, a disease that does not spontaneously occur in mice, "similarity" is only regarded as being achieved if findings in the brains of transgenic mice display the same features as those found in human brains (cf. Huber and Keuck 2013).

It is noticeable that both ideals associated with standardization, the ideal of "purity" as much as "similarity", are challenged by the biological nature of experimental organisms (e.g. *Mus musculus, Drosophila melanogaster*). One of the fundamental issues would be, under what circumstances a given biological system could be reliably used as a human stand-in? One specific application challenge would be to resolve the question of how to cope with variability on an individual level, for instance, if an animal's behaviour poses threats to a given experimental protocol. Obviously, systematic and methodological challenges do not only arise with regard to the means of modelling but also as concerns the targets of research (e.g. patients with cancer). Therapeutic aims of biomedicine are concentrated mainly to certain groups of patients. Patients, who share a given pathology and display quite similar clinical symptoms, for instance regarding disease progression, form the focus of research. Actually, the variety of symptoms is far more diverse than populations of clinical and pharmaceutical research would lead us to suppose. On the other hand, different forms and states of progression of diseases present a de facto obstacle regarding the recruitment of "homogeneous" patient groups for clinical studies. Neither the laboratory set-up in biomedicine nor clinical studies (phase 3 trials), which are clearly favoured in evidence-based-medicine, respond sufficiently to the heterogeneity of clinical symptoms and states of diseases that are actually seen in clinical day-to-day treatment. This challenge lies beyond the more general debate on the priority of methodological tools and the validity of experimental trial designs that I will address in the following.

[23]In 1984 the first transgenic strain of mice that expressed a human pathogen for tumour growth ('myc'-protein) was created at Harvard University ("OncoMouse"). Harvard's three US OncoMouse patents, exclusively licensed to DuPont, lasted for 17 years. The last one expired in 2016. But "OncoMouse" is still a registered trademark owned by DuPont.

11.3.3 Internal Validity as Epistemic End of Standardisation

My analysis of stability as an end of standardisation has already highlighted that a given quality of measurement data ("stable pattern", "robustness") might be said to reflect the validity of experimental outcomes ("compelling evidence of a phenomenon"). In this special case, "stability" and "validity" as ends of standardisation coincide. In empirical methodology, two readings of validity are commonly distinguished, namely "internal" and "external validity" (cf. Campbell and Stanley 1963). I shall briefly introduce both concepts, but restrict my analysis in the following to "internal validity" as a primary epistemic end of standardisation.

Internal validity addresses the fact that all experimental errors and biases of a test system are known and could be successfully excluded.[24] According to Francesco Guala, internal validity is achieved "when the structure and behavior of a laboratory system (its main causal factors, the ways we interact, and the phenomena they bring about) have been properly understood by the experimenter. For example: the result of an experiment E is internally valid if the experimenter attributes the production of an effect B to a factor (or set of factors) A, and A really is the (or a) cause of B in E." (2003, 1198) When are scientific results considered *externally* valid? *External validity* addresses the need to ensure that scientific results could be reproduced outside the scientific laboratory, at least in a more real-life setting (e.g. in a clinical study). Guala states, for example, that the result of an experiment E is externally valid if "A causes B not only in E, but also in a set of other circumstances of interest, F, G, H, etc.." (ibid.). For example, a new drug does not only have certain positive effects on animals that have been tested in the laboratory, but also on humans who are affected with a given disease.[25]

In the following, I shall introduce internal validity as an epistemic end of standardisation that often, but not always, precedes further inquiries into the external validity of a given trial design, taking the case of population dependent settings. Most experimental protocols in psychology as much as in clinical and pharmaceutical trials contains at least one "control condition". Most importantly, a *control condition* provides a baseline against which some variable of the experiment can be compared.[26] Accordingly, it allows to minimizing experimental error and excluding known biases.

Control conditions are key features of "randomized controlled trials" (RCTs), a highly standardized experimental trial design in medicine. RCTs are regarded as the "gold standard" of evidence-based medicine and said to reveal the effectiveness of

[24]Campbell and Stanley give the following account (1963, 5): "*Internal validity* is the basic minimum without which any experiment is uninterpretable".

[25]Cf. Campbell and Stanley (Ibid.): "*External validity* asks the question of *generalizability*: To what populations, settings, treatment variables, and measurement variables can this effect be generalized?"

[26]Boring (1954, 577): "The term control experiment, as a standard of comparison, has now got into most of the dictionaries. Its synonym is test experiment."

a given treatment (cf. Timmermans and Berg 2003). In clinical and pharmaceutical trials the control condition is, for instance, referred to as "baseline phase".[27] Here, a set of data found at the beginning of a study is used for comparison with later data in order to establish a relative, rather than an absolute, meaning to data. Pharmaceutical trials often try to achieve statistical significance by introducing a placebo baseline period at the beginning of a given clinical trial. Accordingly, the *baseline phase* represents the period in which the intervention to be evaluated is not offered to the subject, whereas the "treatment phase" represents the time period during which the intervention is implemented. When repeated measurements are taken during the baseline phase, several threats to *internal validity* are controlled, purportedly.[28] Thus, internal validity refers to the extent that the outcome for a trial can be attributed to the experimental treatment, instead of any alternative explanation, such as the natural course of the target problem.

In general, between-subjects designs in psychology and medicine include at least one "control group". In its simplest form, the *control group* is the group of patients or other research subjects that does not receive the levels of interest, for example the new treatment that is to be evaluated. Today RCTs are commonly blinded, so as not to allow subjects to discover that they are receiving placebos, given the knowledge that one is not receiving treatment affects outcomes (hence is a known possible "bias"). Ideally, control conditions are said to be constant during a given experimental trial design. For example, how a control group is composed should not change over the entire term of an experimental setting.[29]

As the latter shows, internal validity of population-based trial design and stability of empirical data ("robustness") coevolves, or even coincides. The same could be said about *internal validity* and *experimental control*, i. e. the value of an experimental design.[30] Concerning clinical trial design, internal validity is "controlled" through repeated measurement and graphical rendering of data in the baseline phase. In this case, validity, actually, is said to correspond to a graphical ideal, namely a *stable* line that ideally is said to emerge from the data, and to allow for visual control.[31]

---

[27]Cf. Nestor and Schutt (2012).

[28]For a critical assessment, see, for instance, Mohr et al. (2009).

[29]Control conditions in the biomedical sciences can require large sample sizes to be adequately powered with regard to statistical analysis. Sample sizes that are too small are not able to indicate the effect of interest, i.e., of a given therapeutic intervention. Hence, with regard to clinical trials it is rather common to involve several medical centers in the study (multicentre study). This comes with several epistemic as much as medical challenges (cf. Levine 2008).

[30]Guala, for instance, states: "Roughly, a good experimental design is one that maximises control." (2003, 1199).

[31]Cf. Nestor and Schutt (2012, 282f.): "In the baseline phase, measurements are taken until a pattern emerges. One type of pattern is a *stable line*. A stable line is a line that is relatively flat, with little variability in the scores so that the scores fall in a narrow band. This kind of line is desirable because changes can easily be detected, and it is likely that there are few problems of testing, instrumentation, statistical regression, and maturation in the data."

11.4 Is There an *Epistemic* Gain of Standardisation?

In the beginning I stated that standards in science are attributed a given operational value and that this operational value corresponds to an end, an end towards which the standard is working. Ends that facilitate and/or allow for knowledge claims in the experimental sciences were said to be "epistemic ends".[32] Epistemic ends, such as stability or robustness, homogeneity or purity, internal validity or experimental control, share special affinity to practices that prepare, manipulate and intervene into objects. Seemingly, they correspond to strategies of "making sure", in the way John Dewey has described the specific character of scientific endeavour in his collection of papers titled *Experience and Nature*:

> [I]n the practice of science, knowledge is an affair of *making* sure, not of grasping antecedently given sureties. What is already known, what is accepted as truth, is of immense importance; inquiry could not proceed a step without it. But it is held subject to use, and is at the mercy of the discoveries which it makes possible. It has to be adjusted to the latter and not the latter to it. (1981 [1925], 123)

After this reading, scientists refer to standards with the prospect of ascertaining that the behaviour of an organism is caused by a given experimental intervention and does not reflect the natural course of events. The case studies have provided us with some examples of how standards (e.g. prototype, pure strain, control condition) serve ends that facilitate or allow for knowledge claims.

> Def. 3: Standards are *founded* norms/rules that allow for epistemic ends. Therefore, they could be regarded as necessary (but not sufficient) preconditions of knowledge claims in science.

As highlighted in the beginning of this paper (see Def. 2a and 2b), a means that is chosen to serve an end is not a random tool, but an established and acknowledged norm/rule (= standard) with an exceptional functionality in allowing to a given end (= operational value). When it comes to a specific interest, for instance, defining tumour growth in oncology, ideally only one means is prioritised as standard to serve this interest. Any physician with the purpose such as this will rely on the TMN-classification of tumours. If oncologists, on the other hand, would respectively prefer their own protocols, the epistemic rank of the TMN-classification in oncology – as the standard tool for accessing tumour growth – would be challenged. This serves to further exemplify that stability reflects an internal quality of all standards–including material artefacts, formalised procedures, such as protocols as much as classification systems. Still, by stressing this quality, I shall not imply that standards ever cease to be the objects of improvement or scientific progress.

Stability, homogeneity, and internal validity are common epistemic ends in very different fields of research. They facilitate and/or allow for knowledge claims

[32] Presumably, some scholars might prefer to speak of "cognitive ends"–in restricting "epistemic" to truth-indicative ends. Laudan, for instance, differentiates between "cognitive" and "epistemic values" in science (1984). For an introduction, also see Douglas (2013).

in metrology, in physics, in biology, in psychology, in medicine and beyond. According to the instrumental and methodological prerequisites of research in these fields of inquiry, very different standards (means) actually serve these ends. Epistemic ends, therefore, could not be regarded as specific, but rather as general ends in the realm of science, given that they respond to a limited set of basic needs that, if at all, might be addressed via standardisation. As a consequence, ends such as "homogeneity" or "stability" differ significantly, depending on the objects and research materials they target: "Homogeneity", for instance, is achieved through different strategies of "purification". This includes DNA and plasmid purification in genetics as much as the inbreeding and genetic engineering of "homogeneous" strains of experimental organisms in biomedicine.

We can conclude that epistemic ends of standardisation are especially significant from an epistemological point of view, wherever science is concerned with fluid and mutable processes. Take, for instance, the case of Hacking's "phenomena": If the stabilisation of a given phenomenon is successful then this will simultaneously amount to a proof of evidence (this is not an artefact, but a real phenomenon), given all mandatory prerequisites of experimental trial design are fulfilled. This invites coming to terms with the fact that, especially in the experimental sciences, the resulting "stability" of a phenomenon can not be understood as the natural property of the phenomenon (to manifest itself in this way and none other), but rather as criterion pertaining to the operative value regarding the proof of evidence. In a nutshell, this proof of evidence is reached via acknowledged and established regimes, i.e., "standard procedures".

11.5 Conclusions

Stability, homogeneity and internal validity were introduced as epistemic ends of standardisation in the experimental sciences. As the case studies have highlighted, epistemic ends often correspond to mandatory preparatory procedures in the sciences, for instance, the standardisation of resources of research. But also, further procedures, such as the reporting of outcomes, are addressed. This can be traced back to the fact that technical designs in these sectors are bound up to a high degree. We can also safely state, that the significance of standards is to be seen in a direct relation to a completely altered understanding of how we need to deal with research objects and resources in experimental settings, in order to achieve reliable results.

My analysis does not imply that the practical aspect of scientific research is more important than the theoretical assumptions. I merely wish to highlight that we need to place a special emphasis on the practical reality of experiments in the sciences. One indispensible aspect regarding this reality is the presence and significance of "scientific norms/rules". This includes all those in particular, that are employed in the context of *securing* knowledge in science.

Acknowledgments This paper is based on a larger systematic research project on scientific norms (i.e., standards), which is unpublished to date. An earlier version of this paper was presented at the *Second International Conference* of the *German Society for Philosophy of Science* (GWP) at the University of Düsseldorf, Germany, in 2016. I would like to thank the audience and an anonymous reviewer for their valuable comments.

References

Bechtel, W. 1994. Deciding on the data: Epistemological problems surrounding instruments and research techniques in cell biology. *Philosophy of Science* (Proceedings of the Biennial Meeting of the Philosophy of Science Association, Volume Two: Symposia and Invited Papers) 2: 167–178.

Boring, E.G. 1954. The nature and history of experimental control. *The American Journal of Psychology* 67 (4): 573–589.

Bunge, M. 1967. *Scientific research*. Vol. 2. Berlin/Heidelberg/New York: Springer.

Campbell, D.T., and J.C. Stanley. 1963. *Experimental and quasi-experimental designs for research*. Boston: Houghton Mifflin Company.

Chang, H. 2004. *Inventing temperature: Measurement and scientific progress*. Oxford/New York: Oxford University Press.

Clause, B.T. 1993. The Wistar rat as a right choice: Establishing mammalian standards and the ideal of a standardized mammal. *Journal of the History of Biology* 26: 329–349.

Dewey, J. 1981. *The later works, 1925–1953*. Vol. 1: 1925, ed. Jo Ann Boydston. Carbondale: Southern Illinois University Press.

Douglas, H. 2013. The value of cognitive values. *Philosophy of Science* 80: 796–806.

Feest, I. 2011. What exactly is stabilized when phenomena are stabilized? *Synthese* 182: 57–71.

Franklin, A. 2013. *Shifting standards. Experiments in particle physics in the twentieth century*. Pittsburgh: University of Pittsburgh Press.

Gandenberger, G.S. 2010. Producing a robust body of data with a single technique. *Philosophy of Science* 77: 381–399.

Gossel, P.P. 1992. A need for standard methods: The case of American bacteriology. In *The right tools for the job. At work in twentieth-century life sciences*, ed. A. Clarke and J.H. Fujimura, 287–311. Princeton: Princeton University Press.

Gradmann, C. 2014. A spirit of scientific rigour: Koch's postulates in twentieth-century medicine. *Microbes and Infection* 16: 885–892.

Guala, F. 2003. Experimental localism and external validity. *Philosophy of Science* 70: 1195–1205.

Hacking, I. 1983. *Representing and intervening. Introductory topics in the philosophy of natural science*. New York: Cambridge University Press.

———. 1990. *The taming of chance*. Cambridge/New York/Melbourne: Cambridge University Press.

———. 1999. *The social construction of what?* Cambridge, MA/London: Harvard University Press.

Huber, L. 2015. Measuring by which standard? How plurality challenges the ideal of epistemic singularity. In Schlaudt & Huber, 207–215.

———. 2016. Gentlemen in, genuine knowledge out? Zum Status wissenschaftlicher Normen für die Erkenntnissicherung. *Analyse & Kritik* 38: 391–415.

Huber, L., and L. Keuck. 2013. Mutant mice: Experimental organisms as materialised models in biomedicine. *Studies in History and Philosophy of Biological and Biomedical Sciences* 44: 385–391.

Kirk, R.G.W. 2012. Standardization through mechanization: Germ-free life and the engineering in the ideal laboratory animal. *Technology and Culture* 53: 61–93.

Laudan, L. 1984. *Science and values. The aims of science and their role in scientific debate*. Berkeley/Los Angeles/London: University of California Press.

Levine, R.J. 2008. The nature, scope, and justification of clinical research. What is research? Who is a subject? In *Oxford textbook of clinical ethics*, ed. E.J. Emanuel et al., 211–221. Oxford/New York: Oxford University Press.

Löwy, I., and J.-P. Gaudillière. 1998. Disciplining cancer: Mice and the practice of genetic purity. In *The invisible industrialist. Manufactures and the production of scientific knowledge*, ed. J.-P. Gaudillière and I. Löwy, 209–249. Ipswich: Macmillan Press.

Logan, C.A. 2002. Before there were standards: The role of test animals in the production of empirical generality in physiology. *Journal of the History of Biology* 35: 329–363.

Mackenzie, A. 2012. Technical objects in the biological century. *Zeitschrift für Medien- und Kulturforschung* 1: 151–168.

Mohr, D.C., B. Spring, K.E. Freedland, V. Beckner, P. Arean, S.D. Hollon, J. Ockene, and R. Kaplan. 2009. The selection and design of control conditions for randomized controlled trials of psychological intervention. *Psychotherapy and Psychosomatics* 78: 275–284.

Müller-Wille, S. 2007. Hybrids, pure cultures, and pure lines: from nineteenth-century biology to twentieth-century genetics. *Studies in History and Philosophy of Biology and Biomedical Sciences* 38: 796–806.

Nestor, P., and R.K. Schutt. 2012. *Research methods in psychology. Investigating human behavior*. Los Angeles: SAGE.

Parker, W. 2009. Does matter really matter? Computer simulations, experiments, and materiality. *Synthese* 169: 483–496.

Peirce, C.S. 1982–1999. *Writings of Charles S. Peirce. A chronological edition*. In N. Houser et al. 6 Vol. Bloomington/Indianapolis: Indiana University Press.

Radder, H., ed. 2003. *The philosophy of scientific experimentation*. Pittsburgh: University of Pittsburgh Press.

Raz, J. 1990. *Practical reason and norms*. 2nd ed. Oxford/New York: Oxford University Press.

Schaffer, S. 1992. Late Victorian metrology and its instrumentation: A manufactory of ohms. In *Invisible connections. Instruments, institutions, and science*, ed. R. Bud and S.E. Cozzens, 23–56. Bellingham/Washington, DC: SPIE Optical Engineering Press.

Schlaudt, O., and L. Huber, eds. 2015. *Standardization in measurement. Philosophical, historical and sociological issues*. London: Pickering & Chatto.

Searle, J. 1995. *The construction of social reality*. New York: Free Press.

Stegenga, J. 2009. Robustness, discordance, and relevance. *Philosophy of Science* 76: 650–661.

Tal, E. 2011. How accurate is the Standard Second? *Philosophy of Science* 78: 1082–1096.

Timmermans, S., and M. Berg. 2003. *The Gold Standard. The challenge of evidence-based medicine and standardization in health care*. Philadelphia: Temple University Press.

Ullmann-Margalit, E. 1977. *The emergence of norms*. Oxford/London/Glasgow: Oxford University Press.

Part IV
Philosophy of Mathematics and Formal Modeling

Chapter 12
Constitutive Inference and the Problem of a Complete Variation of Factors

Jens Harbecke

Abstract This paper aims to solve a potential problem for the methodology of constitutive inference offered by Harbecke (Stud Hist Philos Sci C: Stud Hist Philos Biol Biomed Sci 54:10–19, 2015b). The methodology is ultimately based on Mill's "method of difference", which requires a complete variation of factors in a given frame. In constitutive contexts, such a complete variation is often impossible. The offered solution utilizes the notion of a "mechanism slice". In a first step, an example of a currently accepted explanation in neuroscience is reconstructed, which serves as a reference point of the subsequent discussion. The regularity theory of mechanistic constitution and the corresponding methodology of constitutive inference are then introduced. Eventually, it is argued that the proposed solution accommodates well all schematic situations in which the impossibility of varying all test factors could be expected either to lead to false inferences or to preclude the establishment of correct constitutive claims.

Keywords Mechanistic explanation · Constitutive inference · Constitutive explanation · Scientific models · Regularity theory of constitution

12.1 Introduction

An adequate explanation in neurobiology, says the "mechanistic approach", demands the identification, location, and analysis of the mechanisms underlying a to-be-explained phenomenon on several levels (cf. Bechtel and Richardson 1993;

This research has been supported by the German-Israeli Foundation for Scientific Research and Development, grant No. G-199-116.2-2013.

J. Harbecke (✉)
Theoretical Philosophy and Philosophy of the Social Sciences, Witten/Herdecke University, Witten, Germany
e-mail: jens.harbecke@uni-wh.de; www.jensharbecke.com

A. Christian et al. (eds.), *Philosophy of Science*, European Studies in Philosophy of Science 9, https://doi.org/10.1007/978-3-319-72577-2_12

Machamer et al. 2000; Craver 2002, 2007). A "mechanism" should be thought of as a set of "...entities and activities organized such that they are productive of regular changes from start or set-up to finish or termination conditions." (Machamer et al. 2000, 3). A slightly different definition has been offered by Bechtel and Abrahamsen who describe a mechanism as "...a structure performing a function in virtue of its component parts, component operations, and their organization. The orchestrated functioning of the mechanism is responsible for one or more phenomena." (Bechtel and Abrahamsen 2005, 423)

A central challenge for the mechanistic approach consists in offering an account of how such mechanistic explanations can be established.[1] On what basis is it justified to characterize a mechanism, or perhaps a part of a mechanism, as "underlying" a given phenomenon? Many authors have observed that what Machamer et al. (2000, 3) call "being *productive* of regular changes" and what Bechtel and Abrahamsen (2005, 423) characterize as being "*responsible* for one or more phenomena" (*emphases added*) is not the same as *causing* a phenomenon or event. This is because a causal relationship is commonly considered to hold only between non-overlapping events (cf. Lewis 1973, 562). However, a mechanism is believed to overlap with the phenomenon in space and time. This non-causal and synchronous relation between the phenomena and their mechanisms is now usually referred to as "constitution", "composition", or "constitutive relevance".[2] The problem seems to be that even when all causal relationships among mechanisms or parts of mechanism have been identified, it remains unclear whether all constitutive relationships among mechanisms and phenomena have been established thereby as well.

In Harbecke (2015b), I have argued that this is not the case. It is possible to explicate a methodology for the establishment of constitutive explanations, but the latter differs substantially from methodologies establishing causal relationships. The regularity theory of mechanistic constitution (cf. Harbecke 2010, 2013, 2014, 2015a; Couch 2011) has formed the background of my methodological proposal.

In this paper, my aim is to discuss a potential problem for the methodology of constitutive inference offered in Harbecke (2015b). The basis of this proposal is an ideal discovery context as envisaged in Mill's method of difference. Mill's inferential methods for causal claims work if the factors in question are independently manipulable. However, such a full variation of test factors is impossible whenever the factors in question are themselves causally connected. In particular, if factors ϕ and ψ are parts of a causal chain within a mechanism $\mathcal{M}$ underlying a

[1]I presuppose the "epistemic" rather than the "ontic" interpretation of mechanistic explanations here. According to this interpretation, mechanistic explanations are models that represent physical mechanisms in the world. This might put me in opposition to Craver's understanding of a mechanistic explanation (cf. Craver 2007, 27).

[2]"Composition" has been used by Machamer et al. (2000, 13), Bechtel and Abrahamsen (2005, 426), and Craver (2007, 164); "constitution" occurs in Craver (2007, 153); "constitutive relevance" is found in Craver (2007, 139). As far as I can see, the authors intend these terms widely synonymously, despite the fact that they offer different conceptual analyses for the targeted notion.

phenomenon $\mathcal{P}$, then the relevance of ϕ and ψ with respect to $\mathcal{P}$ cannot be detected by Mill's method, because the causal connection between ϕ and ψ precludes their independent manipulation. The sections below offer some amendments to the original theory of constitutive inference developed in Harbecke (2015b) that solve the full variation problem.

The paper is structured as follows. In Sect. 12.2, I briefly summarize an example of a currently accepted explanation in neuroscience, which will serve as a reference point of the subsequent discussion. I present the regularity theory of mechanistic constitution in Sect. 12.3. Section 12.4 offers a brief introduction to the theory of constitutive inference. Section 12.5 discusses the problem of full variation for the proposed methodology of constitutive inference, and it offers a solution to it. Section 12.6 summarizes the results and points to some puzzles in the context of the methodology of constitutive inference that will have to be left for future research.

12.2 Explanation in Neuroscience

The adequacy of philosophical theories of mechanistic constitution and constitutive inference can only be assessed by taking into account successful and accepted research results in neuroscience. One such result that has become a standard case in the debate is the mechanistic explanation of spatial memory acquisition in rats (cf. Churchland and Sejnowski 1992, ch. 5; Craver and Darden 2001, 115–119; Craver 2002, sec. 2; Bickle 2003, chs. 3–5; Craver 2007, 165–170).

According to this theory, the phenomenon of long-term potentiation (LTP) at neural synapses within the rat's hippocampus is a central mechanism underlying spatial memory and learning in rats (cf. Bliss and Lømo 1973; Lømo 2003, 618; Morris et al. 1982). Hippocampal LTP in turn has been demonstrated to involve the activation of *N*-methyl-D-aspartat (NMDA) receptors on CA1 pyramidal cells (cf. also Harris et al. 1984; Morris 1984; Morris et al. 1986; Davis et al. 1992).

The mechanism underlying NMDA-receptor activation is highly complex. It is now believed that the NMDA-receptor channels of pyramidal cells are blocked by Mg^{+} ions during the rest potential phase (cf. Churchland and Sejnowski 1992, 255–270). If the postsynaptic membrane is strongly depolarized through a train of high-frequency stimuli and through activity of other receptors, the Mg^{+} ions are repelled whereby the blockade of NMDA-receptors is lifted. As a result, an increased influx of Na^{+}, K^{+}, and Ca^{2+} ions occurs. The resulting Ca^{2+} rise within the dendrite then activates calcium-dependent kinases (Ca^{2+}/Calmodulin-kinase and proteinkinase C). These processes add new channels to the postsynaptic dendrite, which requires in turn a modification of the cell genes expression (cf. Bourtchouladze et al. 1998), they alter the channels' sensitivity to glutamate, or they increase the channels' transmission capacity of Ca^{2+} ions (cf. Toni et al. 1999). Through all these paths an increase in sensitivity of the postsynaptic receptors is attained which can last for a period of up to several hours.

With these results in the background, the neurobiological explanation of spatial memory has been described as involving at least the following central phenomena and mechanisms (cf. Craver and Darden 2001, 118; Craver 2007, 166):

1. The development of spatial memory in the rat
2. The generating of a spatial map within the rat's hippocampus
3. The long-term potentiation of synapses of CA1 pyramidal cells
4. The activation of NMDA-receptors at the synapses of CA1 pyramidal cells

The overall explanation then consists in a conjunction of claims about the *constitutive relationships* holding between these phenomena and mechanisms. Arguably, it is this relation of constitution that scientists generally have in mind when they say that a mechanism "is responsible for" (Bliss and Lømo 1973, 331), "gives rise to" (Morris et al. 1986, 776), "plays a crucial role in" (Davis et al. 1992, 32), "contributes to", "forms the basis of" (both Bliss and Collingridge 1993, 38), "underlies" (Lømo 2003, 619; Frey et al. 1996, 703), or "is constitutively active in" (Malenka et al. 1989, 556) a phenomenon. To offer a transparent analysis of these natural language terms, and to provide an adequate analysis of the relation referred to, is the aim of the regularity theory of mechanistic constitution.

Along with this conjunctive claim stating various constitutive relationships comes a more or less explicit distinction of levels. If this reconstruction of the explanation is adequate, and if the example summarized above is paradigmatic for theories in neurobiology, it is evident that the notion of constitution plays a central role in standard neurobiological explanations. The following section offers an analysis of this notion.[3]

12.3 Regularity Constitution

As mentioned in Sect. 12.1, mechanistic constitution is a relation that differs from causation in important ways. The *relata* of mechanistic constitution do not have distinct instances, whereas causes are usually taken to be distinct from their effects. But if it is not causation, it becomes important to ask what people have in mind when they characterize a mechanism as "responsible for", "giving rise to", "playing a crucial role in", "contributing to", "forming the basis of", "underlying", or "being constitutively active in" a phenomenon. In Harbecke (2010, 2013, 2014, 2015a), I have argued that a regularity-based view reconstructs best what scientists have in mind here and what, in fact, we should take mechanistic constitution to be (cf. also Couch 2011).

The regularity-based analysis of mechanistic constitution focuses on general constitution, i.e. constitution between types of mechanistic events. Its formal statement

[3]Readers already familiar with Harbecke (2010, 2013, 2014, 2015a) and Couch (2011) may want to skip Sect. 12.3 and continue with Sect. 12.4.

comes with some ontological assumptions about what mechanistic "entities and activities" are (cf. Machamer et al. 2000, 3; as quoted in Sect. 12.1). In particular, it denies that there is an important distinction to be made between entities and activities. Rather, regions should simply be thought of as "hosting" active entities. A type such as "... is activated" that is instantiated, among other things, by NMDA-receptors then simply becomes the type "... hosts an active NMDA-receptor" that is instantiated by space-time regions.[4] Only space-time regions should then be considered individuals in the overall ontology of the framework.

For the formal definition, specific mechanistic type variables are expressed by Greek letters 'ϕ' and 'ψ'. Capital letters '$\mathbf{X}$', '$\mathbf{X}_1$', '$\mathbf{X}_2$',...,'$\mathbf{X}_n$' express conjunctions of types that can be co-instantiated (either in the same individual or in "co-located" individuals). The formulation goes as follows (cf. Harbecke 2010, 275–278; to improve legibility, type conjunctions such as "$\phi \wedge \mathbf{X}_1$" are always abbreviated to "$\phi\mathbf{X}_1$"):

> [**Constitution**]: A mechanistic type ϕ constitutes another mechanistic type ψ (written as "$\mathbf{C}\phi\psi$") if, and only if:
>
> (i) ϕ is contained in a minimally sufficient condition $\phi\mathbf{X}_1$ of ψ, such that...
> (ii) $\phi\mathbf{X}_1$ is a disjunct in a disjunction $\phi\mathbf{X}_1 \vee \mathbf{X}_2 \vee \ldots \vee \mathbf{X}_n$ of type conjunctions individually minimally sufficient for ψ, such that the disjunction is minimally necessary for ψ, and such that...
> (iii) if ϕ and $\mathbf{X}_1$ are co-instantiated, then (a) their instances are a mereological part of an instance of ψ, and (b) this instance of ψ is a mereological part of the mentioned fused instances of ϕ and $\mathbf{X}_1$.

In a nutshell, the idea is that mechanistic constitution is a relation between mechanistic types that are regularly, but not redundantly, co-instantiated such that their instances are mereologically related.[5] Since often no single mechanism is sufficient for the occurrence of a given phenomenon, the definition makes reference to complex mechanisms involving a range of mechanistic properties. Additionally, since sometimes more than one mechanism can secure the occurrence of the given phenomenon, the definition also allows for alternative constitutive conditions. The mereological requirement is introduced in order to ensure that the phenomenon must occur (at least partially) in the same place and time as the mechanisms that constitute it. All of these ideas are expressed by conditions (i)–(iii)(a).

Condition (iii)(b) additionally secures that a given phenomenon occupies no less space and time than its mechanism. Moreover, it excludes that the mechanisms underlying a given phenomenon $\mathcal{P}$ also constitute many of $\mathcal{P}$'s effects. Moreover, the condition turns **Constitution** into a criterion for reduction. According to **GEM** – the theory of parthood presupposed by the definition –, mutual parthood implies

[4] Compare this approach to the notion of specific types as proposed in Spohn (2006).

[5] The mereological theory presupposed here is *General Extensional Mereology* (**GEM**) as explicated by Varzi (2009).

identity. As a consequence, mutual constitution implies a co-extensiveness of the mechanism and the phenomenon. Lawful co-extensiveness is reasonably considered as sufficient for type identity and reduction.[6]

Constitution intentionally defines *partial* constitution primarily: A given mechanistic type constitutes a phenomenon if, and only if, it partially determines, and, in conjunction with some other mechanistic properties, entirely determines, the phenomenon. What is usually called *complete* constitution is captured by the notion of a conjunction of mechanistic types that is minimally sufficient for a given phenomenon. It follows that complete constitution is a borderline case of partial constitution, and identity is a borderline case of complete constitution.

The definition of **Constitution** now allows to reconstruct the logical form of actual explanations offered in neurobiology. Consider again the case of the explanation of spatial memory in rats reconstructed in Sect. 12.2. I use the second-order operator '$\Rightarrow_c$' to summarize the criteria specified by **Constitution** (from now on, we will sometimes speak of a type conjunction being "*c*-minimally sufficient" for another type in this sense). Moreover, '$\mathbf{Y}_1$', '$\mathbf{Y}_2$', ..., '$\mathbf{Y}_n$' are used to express disjunctions of conjunctions of properties all of which are minimally sufficient for the type on the right-hand side of the conditional. Then the above-mentioned research results can be expressed by the following proposition ("*SL*" abbreviates "spatial learning")...

$$\mathbf{P}_{SL}:\ (F\mathbf{X}_1' \vee \mathbf{Y}_1 \Rightarrow_c G) \wedge (G\mathbf{X}_2' \vee \mathbf{Y}_2 \Rightarrow_c H) \wedge (H\mathbf{X}_3' \vee \mathbf{Y}_3 \Rightarrow_c I),$$

...where the type symbols are intended as follows:

I : The development of spatial memory in the rat
H : The generating of a spatial map within the rat's hippocampus
G : The long-term potentiation of pyramidal cells in the hippocampus
F : The activation of NMDA-receptors at the synapses of CA1 pyramidal cells

Proposition $\mathbf{P}_{SL}$ says: "(If NMDA receptor activation at the synapses of certain neurons within a rat's hippocampus is instantiated together with certain other properties in an appropriate way, then a long-term potentiation of CA1 pyramidal cells is instantiated in the same place at the same time) and (If a long-term potentiation of CA1 pyramidal cells is instantiated together with certain other properties, then...) and..." and so on. The central hypothesis associated with the regularity theory of constitution is that this is essentially the relationship that scientists have in mind when they say that a mechanism "is responsible for", "gives rise to", "plays a crucial role in", "contributes to", "forms the basis of", "underlies", or "is constitutively active in" a phenomenon (cf. Sect. 12.2).

One question the regularity theory faces is its general adequacy. Even if it is adequate for explanations of spatial memory, this does not yet prove that it factually applies to all, or at least most, models developed in neurobiological research. Moreover, it is not clear yet whether models in this field *should* be constitutive

[6]For an argument for this position, cf. Mellor (1977, 308–309). Note that this position differs from the position defended in Lewis (1986, 50–51), according to which only metaphysically necessary co-extensiveness ensures type identity.

explanations in the sense of **Constitution**. More future research on the topic will be needed to offer satisfactory answers to these questions.

Yet another question concerns the methodology for the establishment of explanations in the sense of **Constitution**. What are the inference rules required to discover and explicate constitutive relationships? In Harbecke (2015b), I have attempted an answer by offering a methodology for constitutive inference. The following section summarizes the main points of this theory.[7]

12.4 Constitutive Inference

With Mackie's (1974) theory of regularity causation as the initial idea, May, Graßhoff, and Baumgartner developed a theory of causal inference and discovery in the 1990s and early 2000s (cf. Graßhoff and May 1995; May 1999; Baumgartner and Graßhoff 2004). The general method was based on Mill's "method of difference" (Mill 1843, 280), which the authors transformed into a strategy that can cope with multiple factors in order to establish more complex causal hypotheses.

My theory of constitutive inference and discovery shares some important ideas with these works. However, a number of adjustments are required in order to make the theory applicable in constitutive contexts. Among other things, the statement of the homogeneity assumption presupposed by causal inferences needs to be reformulated substantially. Similarly to inferences for causal regularities, constitutive inferences are always based on three general premises:

1. The occurrence of a complete constitutive condition, or of a constitutive mechanism, is sufficient for the occurrence of the constituted phenomenon (determination); moreover, if no such complete constitutive condition or mechanism occurs, the phenomenon also does not occur (dependence).
2. The testing situations obey constitutive homogeneity.
3. A constitutive regularity relationship of the form $\mathbf{X} \vee \mathbf{Y} \Rightarrow_c \phi$ and in the sense of **Constitution** is hypothetically true, where $\mathbf{X}$ stands for a conjunction of types, $\mathbf{Y}$ stands for a disjunction of conjunctions of types, and ϕ stands for a particular phenomenon type whose underlying mechanisms are to be identified.

The first premise states a determinism and a dependence within the constitutive hierarchy reminiscent of supervenience: If the complete complex underlying mechanism is present, then the phenomenon is present as well in a nomologically necessary way. However, if no such underlying mechanism occurs, it is impossible that a higher-level phenomenon such as the acquisition of spatial memory is present. The second claim states that the situations compared for the inference

[7]Readers already familiar with Harbecke (2015b) may want to skip Sect. 12.4 and continue with Sect. 12.5.

must be homogeneous in a specific sense. The third premise mentions the material hypothesis from which the investigation starts.

The following example illustrates the function of premises 1.–3. In the case of the investigation of spatial memory and learning, research in the time of Kimble (1963) initially had to start with a completely blank assumption containing no explicit information about constituting mechanisms of the described phenomenon (cf. premise 3):

$$\mathbf{P}^{I}_{SL} : \mathbf{X}_3 \vee \mathbf{Y}_3 \Rightarrow_c I$$

$\mathbf{P}^{I}_{SL}$ states that if a c-minimally sufficient condition $\mathbf{X}_3$ is instantiated (or another c-minimally sufficient condition contained in $\mathbf{Y}_3$) then an acquisition of spatial memory is instantiated in the same place and time as well.[8] Note that $\mathbf{P}^{I}_{SL}$ still leaves open the possibility that $\mathbf{X}_3 \vee \mathbf{Y}_3$ is empty. A difference test in Mill's sense has the sole goal of transforming a non-informative hypothesis such as $\mathbf{P}^{I}_{SL}$ into a more informative claim. An example for such an attempt to generate a more informative hypothesis is the series of experiments carried out by Kimble (1963) and later refined by Morris et al. (1982) who thereby discovered that "[d]eficits in spatial learning and memory [in rats] are found after lesions of the hippocampus [are brought about]." (Morris et al. 1982, 681) Their series of experiments had the general structure illustrated by Table 12.1.

The experiments aimed at a comparison of two kinds of situations: One (situation 1) in which rats had their hippocampal structure intact, and one (situation 2), in which the hippocampus was removed. The researchers then determined on the basis of certain instrumental variables such as the time required and distance travelled to complete an object-searching task whether the cognitive phenomenon ("spatial memory") was present ("1") or not ("0"). Based on their observation that "...the [spatial memory] deficit produced by hippocampal lesions was total" (Morris et al. 1982, 681), and under the presupposition of premises 1. and 2., they were able to transform hypothesis $\mathbf{P}^{I}_{SL}$ into $\mathbf{P}^{I'}_{SL}$.

$$\mathbf{P}^{I'}_{SL} : H\mathbf{X}'_3 \vee \mathbf{Y}_3 \Rightarrow_c I$$

$\mathbf{P}^{I'}_{SL}$ says that, if the generating of a spatial map within a rat's hippocampus is instantiated together with certain other properties in an appropriate way, then the

Table 12.1 Constitutive difference tests

	Situation 1	Situation 2
Presence of mechanistic test type	H	$\neg H$
Presence of phenomenon?	1	0

[8]The same-place-and-time implication follows from the definition of **Constitution**. See the comments on conditions (iii)(a) and (iii)(b) in Sect. 12.3.

acquisition of spatial memory is instantiated in the same place at the same time ("in the rat"). With respect to the formal versions of $\mathbf{P}^{I}_{SL}$ and $\mathbf{P}^{I'}_{SL}$, expressions "$\mathbf{X}_3$" and "$H\mathbf{X}'_3$" are intended as equivalent, but "$H\mathbf{X}'_3$" makes more constituting mechanistic types explicit than "$\mathbf{X}_3$" (this is why "$\mathbf{X}_3$" is transformed into its notational variant "$\mathbf{X}'_3$").

Arguably, the same general inference procedure was applied when Bliss and Lømo (1973) discovered the role of LTP in the formation of spatial representations in the hippocampus, and when Morris et al. (1986) blocked NMDA receptors and observed a breakdown of LTP. These tests and observations established the other two conjuncts of the full theory $\mathbf{P}_{SL}$:

$$\mathbf{P}^{H'}_{SL} : G\mathbf{X}'_2 \vee \mathbf{Y}_2 \Rightarrow_c H$$

$$\mathbf{P}^{G'}_{SL} : F\mathbf{X}'_1 \vee \mathbf{Y}_1 \Rightarrow_c G$$

A crucial, but usually implicit, premise in the inferences establishing $\mathbf{P}^{I'}_{SL}$, $\mathbf{P}^{H'}_{SL}$, and $\mathbf{P}^{G'}_{SL}$ is, of course, the homogeneity claim captured by premise 2. As Baumgartner and Graßhoff (2004, ch. 9) have shown already for causal contexts, Mill's (1843, 280) original homogeneity condition sets a too-strong requirement for any real-life experiment. Their weakened formulation (Baumgartner and Graßhoff 2004, 210) is more likely to be satisfied by actual testing procedures. This formulation can be adjusted for contexts of constitutive discovery as follows:

> [**Homogeneity** (HO):] In a difference test investigating whether a mechanistic type ϕ (resp. $\neg\phi$) constitutes another mechanistic type ψ (a phenomenon), two test situations S_1 and S_2 are homogeneous if, and only if, for all minimally sufficient constitutive conditions $\mathbf{X}_i$ of ψ (resp. $\neg\psi$), at least one conjunct of $\mathbf{X}_i$ is not instantiated in S_1 precisely if at least one conjunct of $\mathbf{X}_i$ is not instantiated in S_2, where the following holds for $\mathbf{X}_i$:
>
> (i) ϕ, $\neg\phi$, ψ, and $\neg\psi$ are not contained in $\mathbf{X}_i$,
> (ii) no type contained in $\mathbf{X}_i$ is located in between ϕ and ψ in the constitutive hierarchy,
> (iii) all types contained in $\mathbf{X}_i$ constitute ψ (resp. $\neg\psi$).

HO ensures that, for instance, the occurrence of a phenomenon ψ in S_1 and its non-occurrence in S_2 is not due to a change in a complex constituting mechanism of ψ that has nothing to do with ϕ. Consequently, false inferences to constitutive claims about ϕ relative to ψ are excluded. In order to ensure that what is tested by a constitutive difference test is not a causal relationship, the following condition must be satisfied by the mechanistic types investigated:

> [**Constitutive Types** (CT):] In a difference test investigating whether a mechanistic type ϕ (resp. $\neg\phi$) constitutes another mechanistic type ψ (a phenomenon), in all relevant test situations S_i, the instance of ϕ must be a mereological part of the instance of ψ.

If the instances of the mechanism and phenomenon overlap spatiotemporally, it is clear that no causal relationship is investigated as, according to the traditional understanding mentioned above, cause and effect never overlap spatiotemporally.

As a general fact about constitutive difference tests, it is important to note that the failure to generate a 1 in the first situation and a 0 in the second situation does *not* imply that the type considered does not constitute the phenomenon in question. Nor does such failure imply that the mechanistic type does constitute the phenomenon. Results of two 1's or two 0's simply render the difference test inconclusive. At the same time, a single 1–0 or 0–1 result that satisfies HO and CT is sufficient for an inference to a constitutive relationship (under the presupposition of premises 1.–3.). The reason is that premise 1. stated a determinism for the constitution relation: The occurrence of a complete complex constitutive mechanism is sufficient for the occurrence of the constituted phenomenon; and if no such complex constitutive condition or mechanism occurs, the phenomenon also does not occur. Hence, a 1–0 result can only mean that in the first situation there was a sufficient condition of the phenomenon present, whilst in the second there was no such condition. Since, according to the homogeneity premise 2., only the initially considered factor was changed in the relevant way, it must have been the case that precisely this factor was constitutively relevant for the phenomenon.

With the notion of a difference test and HO and CT in the background, it is now possible to formulate more complex tests, called "tests-of-four" to determine whether a type constitutes another one and how.[9] In such tests, four kinds of situations are brought about either simultaneously or subsequently, such that the presences and absence of exactly two factors are systematically alternated. Two examples of such tests, both of which take $\mathbf{P}^{I'}_{SL}$: $H\mathbf{X}'_3 \vee \mathbf{Y}_3 \Rightarrow_c I$ as their initial hypothesis, are the following ones:

Table 12.2 shows that H and J must be part of the same minimally sufficient condition for I, since when they both occur, I is instantiated as well, but removing one of the two leads to the non-occurrence of I. In contrast, Table 12.3 proves that J

Table 12.2 Explicating a conjunct

	J	$\neg J$
H	1	0
$\neg H$	0	0

$\models \mathbf{P}^{I''}_{SL}$: $HJ\mathbf{X}''_3 \vee \mathbf{Y}_3 \Rightarrow_c I$

Table 12.3 Explicating a disjunct

	J	$\neg J$
H	1	0
$\neg H$	1	0

$\models \mathbf{P}^{I''}_{SL}$: $H\mathbf{X}'_3 \vee J\mathbf{X}_{3'} \vee \mathbf{Y}'_3 \Rightarrow_c I'$

[9]To be precise, HO and CT have to be slightly reformulated to recognize the more complex structure of the test. For the corresponding case of causal inference, cf. Baumgartner and Graßhoff (2004, 216).

constitutes I because removing it always leads to a non-occurrence of the instance of I. However, it cannot be part of the minimally sufficient condition containing H (which was previously proven to constitute I), because removing H in the tested situations has no effect on the occurrence of I. This shows that there must be another c-minimally sufficient condition all of whose members constitute I.

For a real-world experimental case corresponding to Table 12.3, consider Grover and Teyler (1992) who demonstrated the existence of NMDA-independent Hebbian LTP. On the basis of their experiments the researchers were able to transform a hypothesis of the form $\mathbf{P}^{G'}_{SL}$ into a new theory $\mathbf{P}^{G''}_{SL}$: $F\mathbf{X}'_1 \vee C\mathbf{X}_{1'} \vee \mathbf{Y}'_1 \Rightarrow_c G$, which characterizes voltage-dependent calcium channel activation ("C") as constituting Hebbian LTP ("G") in conjunction with a set of mechanistic types not containing NMDA-receptor activation ("F").

Many other distributions of 1's and 0's in a test-of-four allow for different kinds of inferences, but not all support an inference to an extension of the initially assumed theory (cf. Harbecke 2015b).

12.5 The Problem of a Full Variation of Factors

The tests-of-four methodology presupposes that the testing factors are instantiable in all logically possible combinations. Otherwise, the test tables cannot be filled out and no inference is possible, because the inference rules do not apply. The problem is, however, that constituents of macrophenomena are often connected by causal relationships. As a consequence, they typically cannot be instantiated in all logically possible combinations. and tests-of-four are not applicable.

Consider, for instance, a mechanistic-constitutive structure as the one illustrated by Fig. 12.1. A (cognitive) phenomenon or capacity G is factually constituted by a complex (neural) mechanism with a complex inner causal structure involving seven mechanistic types or factors, namely the set: $\{F_1, F_2, F_3, F_4, F_5, F_6, F_7\}$. The

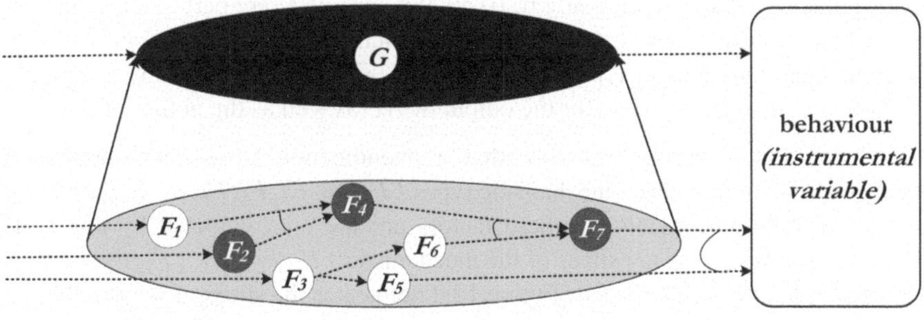

Fig. 12.1 Mechanistic causal chains; connected arrows, such as the arrows leading from F_4 and F_5 into F_7, represent causal minimal sufficiency & necessity; if only one arrow leads into another factor, such as the one leading from F_3 a into F_5, a single factor is characterized as causally sufficient & necessary for another factor

causal chain between factors F_2, F_4, and F_7 is part of the mechanism (highlighted in Fig. 12.1). Type F_7 is directly causally relevant for the behavioural variable serving as a proxy for G.

As a consequence of the causal connection, a situation in which F_7 is present but F_2 is inhibited through an intervention is only possible if further changes are made in the background of the mechanism. This is problematic due to two reasons: First, it means that, at least in some cases, the homogeneity condition HO is violated and no inference is possible. For instance, realizing F_7 whilst inhibiting F_2 may require activating a factor F_k that, in conjunction with other factors $\mathbf{X}_i$, forms a minimally sufficient constitutive condition of G. This is precisely what HO declares as not admissible. As a consequence, no constitutive inference is possible.

Secondly, if the causal chain $F_2 \rightarrow F_4 \rightarrow F_7$ is "repaired" in this way whilst HO is satisfied, the factual constitutive relationship of F_2 to the phenomenon G becomes invisible as no change will occur in the instrumental variable used as a proxy for G. In other words, even if F_2 is part of the mechanism underlying G, this will never be detected, since the only way to determine the presence or absence of G is through the instrumental variable described in Fig. 12.1 as the "behaviour". Again, no inference onto the relevance of F_2 is possible.

Of course, there are already methods available allowing for suboptimal data and incomplete data tables (cf. Qualitative Comparative Analysis (QCA) and Coincidence Analysis (CNA)). However, when there are systematic causal relationships among the investigated factors, also these methods are not applicable in a straightforward way.

What is required is an adjustment of the theory of constitutive inference. In Harbecke (2015b), I did not make the necessity of this adjustment explicit. However, as will become clear in a second, the difficulty of systematically varying all factors in the frame due to the causal interconnections among the factors is not a threat to the methodology as originally proposed.

The key notion of this adjustment is that of a "mechanism slice". It can be defined as follows:

> [**Mechanism Slice:**] A set of mechanistic types $\mathcal{N}$ is a mechanism slice relative to a phenomenon ϕ if, and only if, (i) all elements of $\mathcal{N}$ are part of a mechanism $\mathcal{M}$ that fully constitutes ϕ, and (ii) no element of $\mathcal{N}$ is downstream causally connected with another type in $\mathcal{M}$, but the elements of $\mathcal{N}$ are jointly minimally sufficient in a causal sense for the output of $\mathcal{M}$ (as well as the output of ϕ).

For illustration, suppose as before that a phenomenon G is constituted by a mechanism involving seven mechanistic types $\{F_1, F_2, F_3, F_4, F_5, F_6, F_7\}$, and that the causal relationships among these factors are the ones depicted by Fig. 12.1. Then the following sets are slices of the mechanism: $\{F_1, F_2, F_3\}$, $\{F_1, F_2, F_5, F_6\}$, $\{F_3, F_4\}$, $\{F_4, F_5, F_6\}$, $\{F_5, F_7\}$. Figure 12.2 delineates the slices of G visually.

The elements of each of the slices are instantiable in all logically possible combinations, as they are not directly causally connected. Moreover, *ex hypothesis*,

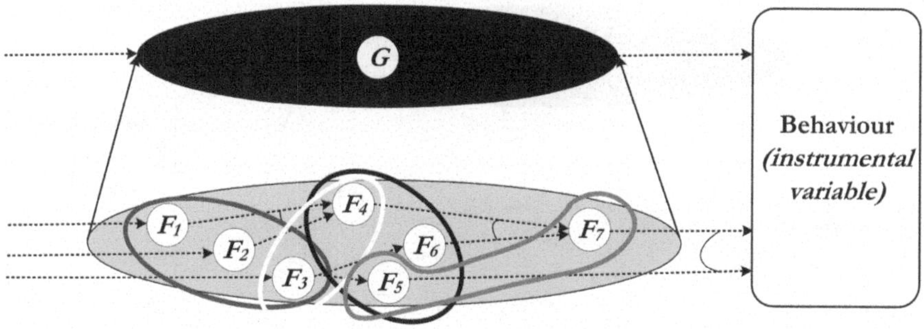

Fig. 12.2 Mechanistic slices

it will be the case that, for each set, eliminating one of its elements leads to a malperformance of the mechanism and, thereby, to the non-occurrence of the phenomenon's effect. The absence of the phenomenon's effect will serve as a proxy for the absence of the phenomenon itself. Consequently, the methodology will yield the conclusion that the following conjunctions of mechanistic types, i.e. the mechanism slices plus further background factors, are all minimally sufficient constituting conditions of G:

$$F_1F_2F_3\mathbf{X}_1,\ F_1F_2F_5F_6\mathbf{X}_2,\ F_3F_4\mathbf{X}_3,\ F_4F_5F_6\mathbf{X}_4,\ F_5F_7\mathbf{X}_5$$

This is how it should be, since no false constitutive claims are being made, and since the factually relevant factors are characterized as constitutively relevant. Note, though, that from these separate constitution hypotheses, it does **not** follow that $F_1F_2F_3F_4F_5F_6F_7\mathbf{X}' \vee \mathbf{Y}' \Rightarrow_c G$. Or in other words, the complex conjunction of mechanistic types $F_1F_2F_3F_4F_5F_6F_7\mathbf{X}'$ does not constitute G. The reason is that the antecedent of this conditional is sufficient, but not minimally sufficient for G. Moreover, notice that none of $\mathbf{X}_1, \mathbf{X}_2, \ldots, \mathbf{X}_5$ can be presupposed as identical to any $\mathbf{X}_i$ from the list except itself. The reason is that, at each slice, very different background factors may play the role of completing the minimally sufficient condition.

Nevertheless, with the list of constitutive conditions of G and the initial information about the internal causal structure of the mechanism, a complete and non-redundant description of the mechanism constituting G is possible. It will consist in a list of constitutive conditionals "$F_1F_2F_3\mathbf{X}_1 \Rightarrow_c G$", "$F_1F_2F_5F_6\mathbf{X}_2 \Rightarrow_c G$", "$F_3F_4\mathbf{X}_3 \Rightarrow_c G$", "$F_4F_5F_6\mathbf{X}_4 \Rightarrow_c G$", "$F_5F_7\mathbf{X}_5 \Rightarrow_c G$" along with a specification of the direct causal relationships among the factors in the set $\{F_1, F_2, F_3, F_4, F_5, F_6, F_7\}$. From these pieces of information a list of constitutive claims $\mathbf{C}F_1G$", "$\mathbf{C}F_2G$", "$\mathbf{C}F_3G$", "$\mathbf{C}F_4G$", "$\mathbf{C}F_5G$", "$\mathbf{C}F_6G$", "$\mathbf{C}F_7G$" can be inferred as well. Such a listing of the causal and constutive relationships between

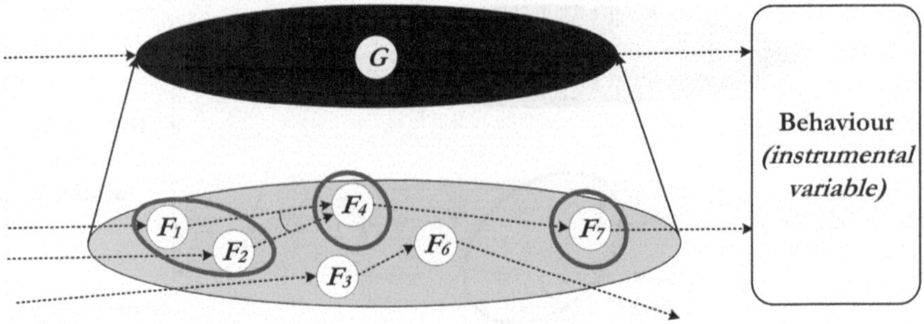

Fig. 12.3 Irrelevant mechanisms

the factors and the phenomenon G will form a satisfactory mechanistic explanation of G *on a single mechanistic level*.[10]

Now consider a case, in which not all mechanistic types regularly instantiated in the same space-time region as G's instantiation constitute G because their variation has no consequences for G's performance as a phenomenon. This might be the case in which there are two independent causal chains present within the spatiotemporal region of G's instantiation. For instance, one can imagine a case in which an independent projection passes through parts of the hippocampal structures whilst a rat builds up memory of a given environment.

A hypothetical case of this kind is illustrated by Fig. 12.3, as the causal chain $\ldots \rightarrow F_3 \rightarrow F_6 \rightarrow \ldots$ has no connection to the phenomenon's effect, which serves as an instrumental variable of the phenomenon's presence or absence. In such a mechanistic structure, the following sets are slices of the first mechanism $\{F_1, F_2\}$, $\{F_4\}$, $\{F_7\}$, and the following sets are slices of the second mechanism $\{F_3\}$, $\{F_6\}$.

An analysis of the mechanistic structure in line with the methodology of constitutive inference will reveal that only the following constitutive conditionals hold: $F_1F_2\mathbf{X}_1 \vee \mathbf{Y}_1 \Rightarrow_c G$, $F_4\mathbf{X}_2 \vee \mathbf{Y}_2 \Rightarrow_c G$; $F_7\mathbf{X}_3 \vee \mathbf{Y}_3 \Rightarrow_c G$. Again, this is just as desired. No false claims about the constitution of certain factors and the phenomenon are being made.

The same procedure works if a mechanism has a "dead end". Suppose a mechanistic structure as the one illustrated by Fig. 12.4 underlies a given phenomenon G. Factor F_5 is not relevant for the instrumental variable taken to represent the occurrence or non-occurrence of the phenomenon G. Then the following sets are slices of the mechanism underlying G: $\{F_1, F_2, F_3\}$, $\{F_1, F_2, F_6\}$, $\{F_3, F_4\}$, $\{F_4, F_6\}$, $\{F_7\}$. The constitutive inference rules based on extended tests-of-four would yield the following constitutive relationships. $F_1F_2F_3\mathbf{X}_1 \vee \mathbf{Y}_1 \Rightarrow_c G$, $F_1F_2F_6\mathbf{X}_2 \vee \mathbf{Y}_2 \Rightarrow_c G$; $F_3F_4\mathbf{X}_3 \vee \mathbf{Y}_3 \Rightarrow_c G$; $F_4F_6\mathbf{X}_4 \vee \mathbf{Y}_4 \Rightarrow_c G$; $F_7\mathbf{X}_5 \vee \mathbf{Y}_5 \Rightarrow_c G$.

[10]To be fair, it should be mentioned that, according to the mechanistic approach, a satisfactory description of the underlying mechanism at one level is not yet a complete explanation. To attain the latter, one would have to offer satisfactory descriptions of the mechanism at different levels (cf. Machamer et al. 2000, 13/14).

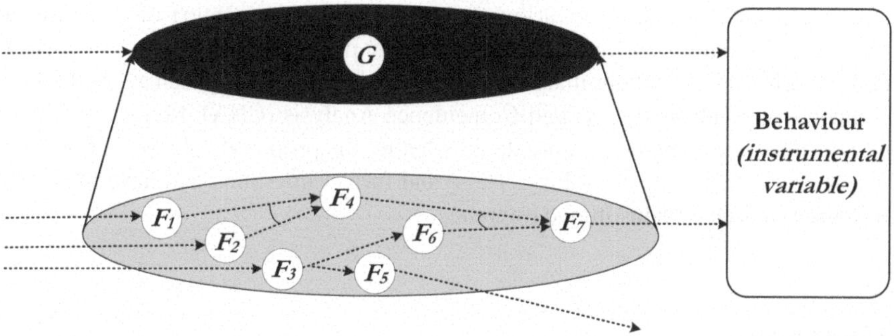

Fig. 12.4 Mechanistic dead ends

This as well is the desired result. Factor F_5 does not appear in the list of constituent mechanistic factors of G.

The situations illustrated by Figs. 12.2, 12.3, and 12.4 are actually the only general situations that could challenge a methodology of constitutive inference. The general challenge consists in avoiding false constitutive inferences and in allowing for the actual identification of relevant factors under the right conditions. Whenever mechanistic factors are exclusively instantiated outside the spatio-temporal regions of the phenomenon's instantiations, condition CT immediately excludes their potential relevance. The remaining cases are those in which (i) the factors are in fact relevant, or (ii) the factors' instantiations are part of an independent causal chain, or (iii) a factor's instantiation is part of a "dead end" of an otherwise constitutively relevant causal chain. All of these cases have been successfully analyzed with the help of the notion of a mechanism slice. Hence, the presented strategy solves the full variation problem characterized above as a potential threat to the methodology of constitutive inference.

12.6 Conclusion

The aim of this paper was to solve a potential problem for the methodology of constitutive inference as I had offered it in Harbecke (2015b). The solution consisted in an adjustment to the original theory that was essentially based on the notion of a "mechanism slice". I began by reconstructing an example of a currently accepted explanation in neuroscience, which served as a reference point of the subsequent discussion. I then presented the regularity theory of mechanistic constitution, and I gave an introduction to the theory of constitutive inference. I argued that the proposed solution accommodates well all schematic situations in which the impossibility of varying all test factors could be expected either to lead to false inferences or to preclude the establishment of correct constitutive claims.

The paper did not discuss various further topics in the context of constitutive inference. Among these are how the solution presented relates to established methods allowing for suboptimal data and incomplete data tables such as Qualitative Comparative Analysis (QCA) and Coincidence Analysis (CNA). Moreover, it was not possible to investigate in more detail whether the proposed solution is reflected by actual research in neurobiology. These and further questions will have to be left to future research on constitutive inference.

References

Baumgartner, M., and G. Graßhoff. 2004. *Kausalität und kausales Schliessen: eine Einführung mit interaktiven Übungen*. Bern: Bern Studies in the History and Philosophy of Science.

Bechtel, W., and A. Abrahamsen. 2005. Explanation: A mechanist alternative. *Studies in History and Philosophy of Biological and Biomedical Sciences* 36(2): 421–441.

Bechtel, W., and R. Richardson. 1993. *Discovering complexity: Decomposition and localization as scientific research strategies*. New York: Princeton University Press.

Bickle, J. 2003. *Philosophy and neuroscience: A ruthlessly reductive account*. Dordrecht: Kluwer.

Bliss, T., and G. Collingridge. 1993. A synaptic model of memory: Long-term potentiation in the hippocampus. *Nature* 361(6407): 31–39.

Bliss, T., and T. Lømo. 1973. Long-lasting potentiation of synaptic transmission in the dentate area of the anaesthetized rabbit following stimulation of the perforant path. *The Journal of Physiology* 232(2): 331–356.

Bourtchouladze, R., T. Abel, N. Berman, R. Gordon, K. Lapidus, and E.R. Kandel. 1998. Different training procedures recruit either one or two critical periods for contextual memory consolidation, each of which requires protein synthesis and PKA. *Learning & Memory* 5(4): 365–374.

Churchland, P.S., and T.J. Sejnowski. 1992. *The computational brain*. Cambridge: MIT Press.

Couch, M. 2011. Mechanisms and constitutive relevance. *Synthese* 183(3): 375–388.

Craver, C. 2002. Interlevel experiments and multilevel mechanisms in the neuroscience of memory. *Philosophy of Science* 69(3): 83–97.

Craver, C. 2007. *Explaining the brain*. New York: Oxford University Press.

Craver, C., and L. Darden. 2001. Discovering mechanisms in neurobiology. In *Theory and method in the neurosciences*, ed. P. Machamer, R. Grush, and P. McLaughlin, 112–137. Pittsburgh: University of Pittsburgh Press.

Davis, S., S. Butcher, and R. Morris. 1992. The NMDA receptor antagonist D-2-amino-5-phosphonopentanoate (D-AP5) impairs spatial learning and LTP in vivo at intracerebral concentrations comparable to those that block LTP in vitro. *Journal of Neuroscience* 12(1): 21–34.

Frey, U., S. Frey, F. Schollmeier, and M. Krug. 1996. Influence of actinomycin D, a RNA synthesis inhibitor, on long-term potentiation in rat hippocampal neurons in vivo and in vitro. *The Journal of Physiology* 490(Pt 3): 703.

Graßhoff, G., and M. May. 1995. Methodische analyse wissenschaftlichen entdeckens. *Kognitionswissenschaft* 5: 51–67.

Grover, L., and T. Teyler. 1992. N-methyl-D-aspartate receptor-independent long-term potentiation in area CA1 of rat hippocampus: Input-specific induction and preclusion in a non-tetanized pathway. *Neuroscience* 49(1): 7–11.

Harbecke, J. 2010. Mechanistic constitution in neurobiological explanations. *International Studies in the Philosophy of Science* 24(3): 267–285.

Harbecke, J. 2013. Regularity theories of mechanistic constitution in comparison. In *GAP 8 Was d?rfen wir glauben? Was sollen wir tun?*, ed. M. Hoeltje, T. Spitzley, and W. Spohn, 126–134. Duisburg-Essen: DuEPublico.

Harbecke, J. 2014. The role of supervenience and constitution in neuroscientific research. *Synthese* 191(5): 725–743.

Harbecke, J. 2015a. Regularity constitution and the location of levels. *Foundations of Science* 20(3), 323–338.

Harbecke, J. 2015b. The regularity theory of mechanistic constitution and a methodology for constitutive inference. *Studies in History and Philosophy of Science Part C: Studies in History and Philosophy of Biological and Biomedical Sciences* 54: 10–19.

Harris, E., A. Ganong, and C. Cotman. 1984. Long-term potentiation in the hippocampus involves activation of N-methyl-D-aspartate receptors. *Brain Research* 323(1): 132–137.

Kimble, D. (1963). The effects of bilateral hippocampal lesions in rats. *Journal of Comparative and Physiological Psychology* 56(2), 273.

Lewis, D. 1973. Causation. *The Journal of Philosophy* 70(17): 556–567.

Lewis, D. 1986. *On the plurality of worlds*, vol. 322. Oxford: Basil Blackwell.

Lømo, T. 2003. The discovery of long-term potentiation. *Philosophical Transactions of the Royal Society. London B. Biological Science* 358: 617–620.

Machamer, P., L. Darden, and C. Craver. 2000. Thinking about mechanisms. *Philosophy of Science* 67(1): 1–25.

Mackie, J. 1974. *The cement of the Universe*. Oxford: Clarendon Press.

Malenka, R., J. Kauer, D. Perkel, M. Mauk, P. Kelly, R. Nicoll, and M. Waxham. 1989. An essential role for postsynaptic calmodulin and protein kinase activity in long-term potentiation. *Nature* 340(6234): 554–557.

May, M. 1999. *Kausales Schliessen. Eine Untersuchung ?ber kausale Erkl?rungen und Theorienbildung*. Ph.D. thesis, University of Hamburg, Germany.

Mellor, D.H. 1977. Natural kinds. *British Journal for the Philosophy of Science* 28(4): 299–312.

Mill, J.S. 1882/1843. *A system of logic: Ratiocinative and inductive, Book III*, 8th ed. London: Harper and Brothers.

Morris, R. 1984. Developments of a water-maze procedure for studying spatial learning in the rat. *Journal of Neuroscience Methods* 11(1): 47–60.

Morris, R., P. Garrud, J. Rawlins, and J. O'Keefe. 1982. Place navigation impaired in rats with hippocampal lesions. *Nature* 297(5868): 681–683.

Morris, R., E. Anderson, G. Lynch, and M. Baudry. 1986. Selective impairment of learning and blockade of long-term potentiation by an N-methyl-D-aspartate receptor antagonist, AP5. *Nature* 319: 774–776.

Spohn, W. 2006. Causation: An alternative. *The British Journal for the Philosophy of Science* 57(1): 93–119.

Toni, N., P.-A. Buchs, I. Nikonenko, C. Bron, and D. Muller. 1999. LTP promotes formation of multiple spine synapses between a single axon terminal and a dendrite. *Nature* 402(6760): 421–425.

Varzi, A. 2009. Mereology. In *Stanford Encyclopedia of Philosophy*, ed. E.N. Zalta (Summer 2009 ed.). http://plato.stanford.edu/entries/mereology/.

Chapter 13
A Partial Calculus for Dag Prawitz's Theory of Grounds and a Decidability Issue

Antonio Piccolomini d'Aragona

Abstract We consider Dag Prawitz's theory of grounds and suggest a formal apparatus for it. On its basis we show that the account faces a decidability issue similar to one that is often associated with BHK semantics, and propose different readings of the issue that involve different consequences.

Keywords Decidability · Grounding · Operation · Proof · Typing

13.1 Introduction

Dag Prawitz's *Towards a foundation of a general proof theory* (Prawitz 1973) is famously 'interested in understanding the very proofs in themselves, i.e., in understanding not only *what* deductive connections hold but also *how* they are established' (Prawitz 1973, 225). A seminal example of what we currently designate as *general proof theory* comes from Gentzen's *Untersuchungen über das logische Schließen*, where sequent calculi and natural deduction systems stem from 'insights about the general structure of proofs' (Prawitz 1973, 225).

Prawitz's normalization theorems play in natural deduction systems the role that Gentzen's cut-elimination plays in sequent calculi. This purely syntactic result is almost immediately extended to semantics through what Schroeder-Heister calls the '*fundamental corollary* of normalization theory', stating that 'every closed derivation in intuitionistic logic can be reduced to one using an introduction rule in its last step' (Schroeder-Heister 2006, 531). On the one hand, the framework is inspired by Gentzen's well-known notion that 'the introductions represent, as it were, the *definitions* of the symbols concerned, and the eliminations are no more, in

A. Piccolomini d'Aragona (✉)
CNRS, CEPERC, Aix Marseille University, Aix en Provence, France

"La Sapienza" University of Rome, Rome, Italy
e-mail: antonio.piccolomini-daragona@etu.univ-amu.fr

A. Christian et al. (eds.), *Philosophy of Science*, European Studies in Philosophy of Science 9, https://doi.org/10.1007/978-3-319-72577-2_13

the final analysis, than the consequences of these definitions' (Gentzen 1934–1935, 80), and on the other, it matches Dummett's requirements for adequate explanations of meaning.

However, this is not the only possible approach to general proof theory, since one could also 'try to give a direct characterization of different kinds of proofs, where a *proof* is understood as the abstract process by which a proposition is established' (Prawitz 1973, 227). From this standpoint, the influence of Gentzen and Dummett is quite clearly accompanied by references to the Brouwer-Heyting-Kolmogorov (BHK) clauses, and by an analysis of the problems they pose to a verificationist theory of meaning. Already in 1977, in *Meaning and proofs: on the conflict between classical and intuitionistic logic*, Prawitz addressed, for example, the non-decidable character of the BHK proofs.

Since 2006 Prawitz has turned to what he calls *theory of grounds*. The new perspective diverges significantly from the earlier one, first and foremost in relation to the notion of valid inference and how it intertwines with the notions of valid argument and proof. At variance with proof theoretic semantics, valid inferences are now prior; they are used to define valid arguments and proofs, with the intention to capture the epistemically compelling power with which correct reasoning is endowed to confer evidence on its conclusion under assumed evidence for its premises. In his more recent papers, Prawitz outlines the general form of a language of grounds and provides indications on how the ground-theoretic framework should be developed. However, the overall project still seems to be in an embryonic stage.

In this work, we address a threefold task. First, we show how the decidability problem frames within the BHK approach (Sect. 13.2). Next, we propose a partial calculus for Prawitz's theory of grounds; after introducing a core calculus for Gentzen's introductions, we define two expansions of it, one for full first-order minimal logic, and another for a kind of "meta-language" of grounds (Sect. 13.3). Apart from their intrinsic interest, the expansions help understand the final task, a ground-theoretic reformulation of the BHK decidability issue. In this connection, we indicate some ways in which the issue may be further framed, leading to different results with respect to some epistemic concerns (Sect. 13.4).

13.2 Prawitz on BHK Proofs in 1977

In *Meaning and proofs: on the conflict between classical and intuitionistic logic* (Prawitz 1977), Prawitz analysed and developed Dummett's semantic argument 'in favour of the claim that intuitionistic rather than classical logic describes the correct forms of reasoning within mathematics' (Prawitz 1977, 2; Crocco and Audureau 2003). Here, the basic tenet that knowledge of meaning should manifest itself in relevant practices is further explained by the specification 'that quite generally, there are two aspects of the use of an (assertive) sentence: (1) the conditions under which it can be correctly asserted, and (2) the commitments made by asserting it', namely the direct means or 'rules for inferring a sentence', and the indirect means or 'rules for drawing consequences from sentences' (Prawitz 1977, 7). Dummett famously disregards a bivalent truth-conditional theory of meaning, and proposes an explanation in terms of correct assertability.

At a first glance, the BHK intuitionistic description of the notion of proof seems to be a perfect candidate for the requirements of a verificationist account. The clauses usually refer to a first-order language with names $\|k\|$ for each element k in a range domain D – and possibly symbols for functions $D^i \to D$ and relations in D^i ($i \in \mathbb{N}$). The definition proceeds by induction on the complexity of formulae with $\wedge$, $\vee$, $\to$, $\forall$ or $\exists$ as main sign – under the assumption that there is no proof of the atomic absurdity symbol $\bot$ (see also Heyting 1956).

($\wedge$) A proof of $\alpha \wedge \beta$ is an ordered pair where the first element is a proof of α and the second element is a proof of β.
($\vee$) A proof of $\alpha_1 \vee \alpha_2$ is a proof of α_i, with an indication that α_i has been proved ($i = 1, 2$).
($\to$) A proof of $\alpha \to \beta$ is an effective procedure that, when applied to any proof of α, produces a proof of β.
($\forall$) A proof of $\forall x \alpha(x)$ is an effective procedure that, when applied to any $\|k\|$, produces a proof of $\alpha(\|k\|)$.
($\exists$) A proof of $\exists x \alpha(x)$ is an ordered pair where the first element is a term t and the second element is a proof of $\alpha(t)$.

However, the BHK framework provides sufficient but not necessary conditions, for 'it is not true even intuitionistically that the condition for asserting a sentence is that we know a proof of it in this sense' (Prawitz 1977, 21). Although, for example, a BHK proof of a disjunction should be a proof of one of the disjuncts, with the indication of which disjunct is proved, we may 'even intuitionistically assert that $\alpha(n) \vee \beta(n)$ for some numeral n without knowing a proof of $\alpha(n)$ or of $\beta(n)$; it would be sufficient, e.g., if we know a proof of $\alpha(0) \vee \beta(0)$ and of $\forall x(\alpha(x) \vee \beta(x) \to \alpha(x+1) \vee \beta(x+1))$' (Prawitz 1977, 21). The impasse can be overcome by asking more generally that 'the condition for asserting a sentence is that we *either* know a proof of the sentence of the kind mentioned in the intuitionistic definition of proof exemplified above *or* know a procedure for obtaining such a proof' (Prawitz 1977, 22), thereby outlining a distinction between *canonical* and *non-canonical* proofs. The BHK approach can thus be said to depict the notion of canonical proof, rather than the notion of proof in its generality.

It is usually acknowledged that proofs epistemically force us to accept their conclusion. Yet, our framework so far only involves abstract mathematical objects that do not compel one in any way, unless one recognizes relevant properties they have in relation to expected results. In this connection, Prawitz stresses that 'we should not simply say that a canonical proof of, e.g., $\alpha \wedge \beta$ consists of a canonical proof of α and a canonical proof of β. [...] we must also be aware of the fact that these two proofs form a sufficient ground to go one step further and assert $\alpha \wedge \beta$. Or, more precisely, one should grant the existence of an operation which yields a canonical proof of $\alpha \wedge \beta$ when performed on canonical proofs of α and β' (Prawitz 1977, 25–26). Now, 'in the case when α is atomic or is built up of atomic sentences by $\wedge$, $\vee$ and $\exists$, the knowledge of a canonical proof [...] can be taken to consist of just the construction of the proof' (Prawitz 1977, 27). In contrast, as regards implication, universal quantification and non-canonical proofs, things are not easy, and it seems that the required recognition must be postulated quite strongly.

Indeed, 'when α is an implication or a universal sentence and in the cases when we know only a procedure for obtaining a canonical proof, we must require not only a construction or description of an appropriate procedure but also an understanding' (Prawitz 1977, 27) of the fact that when applied to arguments in its domain the procedure always yields certain results. The understanding may be straightforward, but this is in general not guaranteed. In fact, the procedure could be unfeasibly complex and impossible to calculate practically. Moreover, in order to establish that a specific result is produced for given arguments, we should in principle perform infinitely many substitutions from domains that cannot be easily regimented, or cannot be regimented at all. It is then perfectly conceivable that one has built constructions with significant epistemic features without realizing that they are so (Díez 2000).

Prawitz therefore admitted that 'the nature of this understanding is the problem' (Prawitz 1977, 27), because 'it is not clear in what sense, if any, one could decide the question whether this condition obtains in a certain situation' (Prawitz 1977, 29). For example, Kreisel's proposal (Kreisel 1962) that one should prove that a procedure has certain properties 'would lead to an infinite regress and would defeat the whole project of a theory of meaning' (Prawitz 1977, 27) based on the notion of proof. Nor can the recognition come from algorithms of the kind of which we dispose within specific formal systems, which enable us to establish whether a certain construction is well-formed with respect to fixed axioms or rules; the conditions for asserting a sentence are beyond the bounds of any formal system since, in the light of Gödel's theorems, 'there is no formal system generating all the procedures' (Prawitz 1977, 29) that we need.

13.3 The Theory of Grounds

13.3.1 Context

Prawitz's theory of grounds was first formulated in 2006, in *Validity of inferences* (now Prawitz 2013), and further developed in *Inference and knowledge* (Prawitz 2009) and in *The epistemic significance of valid inference* (Prawitz 2012). It partially connects to *An approach to general proof theory and a conjecture of a kind of completeness of intuitionistic logic revisited* (Prawitz 2014). Finally, *Explaining deductive inference*, the paper to which we will refer most, widely articulates and contextualizes the new approach.

When dealing with valid inferences, a preliminary issue is of course what an inference is. However, there cannot be a definitive answer because, as noted by Cozzo (2016), the notion varies considerably in relation to several different factors. Fortunately, Prawitz states that he intends to focus exclusively on reflective, deductive inferences, which leaves out intuitive, abductive or inductive inferences. In a first approximation, an inference is therefore said to be made of 'at least a number of assertions or judgments made in the belief that one of them, the conclusion, say β, is supported by the other ones, the premises, say $\alpha_1, \ldots, \alpha_n$. An inference [...] is not an assertion or judgment to the effect that β *follows* from $\alpha_1, \ldots, \alpha_n$, but is first of all a transition from some assertions (or judgments) to another one. [...] it is a compound act that contains the [...] assertions $\alpha_1, \ldots, \alpha_n$ and β, and in addition, the claim that the latter is supported by the former' (Prawitz 2015, 67). In turn, judgements and assertions are understood respectively as mental or linguistic acts by which one claims the truth of a propositional or sentential truth-bearer. Furthermore, the account also manages inferences where premises and conclusion are *hypothetical judgements* or *assertions* $\alpha_1, \ldots, \alpha_n \vdash \beta$, with β depending on the assumptions $\alpha_1, \ldots, \alpha_n$, and *open judgements* or *assertions*, since open formulae may occur.

An inference act can be described on different levels of abstraction, ranging from concrete individual inferences, where 'we pay attention to the agent that performs an inference and the occasion at which it is performed', to generic inferences, obtained 'by abstracting from the agent and the occasion', up to inference forms, which regards only 'how the logical forms of the sentences involved are related' (Prawitz 2015, 70). From a very reductive standpoint, such acts are merely transitions from some premises to a certain conclusion, which are optimally described by figures like

$$\frac{\Gamma_1 \vdash \alpha_1 \qquad \ldots \qquad \Gamma_n \vdash \alpha_n}{\Delta \vdash \alpha_{n+1}}$$

where, in the case of individual or generic transitions, $\Gamma_1, \ldots, \Gamma_n$ and Δ are (possibly empty) sets of open or closed sentences or else, in the case of forms of transitions, they are parameters, $\Delta \subseteq \bigcup\{\Gamma_1, \ldots, \Gamma_n\}$, and $\alpha_1, \ldots, \alpha_{n+1}$ are, in the case of individual or generic transitions, open or closed sentences or else, in the case of forms of transitions, they are parameters. Prawitz considers a 'crucial issue' of his 2015 paper whether such a narrow transitional picture is reasonable, namely whether an inference contains 'something more than an inferential transition, and if so, what' (Prawitz 2015, 68).

The main ground-theoretic question is 'why do certain inferences have the epistemic power to confer evidence on the conclusion when applied to premises for which there is evidence already?' (Prawitz 2015, 73). Prawitz's earlier proof-theoretic semantics describes valid inferences in terms of validity-preservation throughout arguments (Prawitz 1973), or in terms of provability-preservation throughout proofs (Prawitz 1977). However, for some reasons this account proves to be unsatisfactory. The theory of grounds is thus meant to provide an analysis of valid inferences that captures their epistemically compelling power in a way that reverses the order of explanation between valid inferences and valid arguments or proofs, defining the latter by means of the former (although the two accounts obviously intersect, see for example Tranchini 2014; Došen 2015).

13.3.2 Grounds and Their Language

If one needs to explain why and how valid inferences produce evidence for their conclusion, a more accurate characterization of evidence is first required. In this connection, Prawitz introduces a notion of *ground* that, in Cozzo's words, is 'multifaceted' (Cozzo 2015, 107). Grounds are 'abstract entities' of epistemic kind, since 'to be in possession of a ground is a mental state' (Cozzo 2015, 109). Actually, Prawitz says that the term "ground" is used 'in connection with judgments or assertions to have a name on what a person needs to be in possession of in order that her judgment is to be justified or count as knowledge' (Prawitz 2009, 187), or that 'it is convenient to think of evidence states as states where the subject is in possession of certain objects. I shall call these objects *grounds*' (Prawitz 2015, 88). However, in spite of their abstractness, grounds have a strong epistemic import, due to the fact that 'one finds something to be evident by performing a mental act' (Prawitz 2015, 88); thus, 'grounds on this conception will be something that one gets in possession of by doing certain things' (Prawitz 2012, 894), namely, by applying epistemic operations of a certain kind.

To speak about grounds, Prawitz develops a formal *language of grounds*, which is referred to what could be called a *background language*. The terms of the language of grounds are employed to describe grounds for judgements or assertions performed using the formulae of the background language. In turn, like in the Curry-Howard isomorphism (Howard 1980), such formulae provide types for the corresponding terms. Hereafter, our background language has a chosen range domain D.

Definition 1 The *alphabet* of L consists of the auxiliary symbols parentheses and comma, plus

- Individual variables x_i $(i \in \mathbb{N})$
- A set of individual constants $\mathtt{K} = \{\|k\|\|$ forevery $k \in D\}$
- A set of function symbols $\mathtt{F} = \{\|\phi^i\|\|$ forsome $\phi : D^i \to D$ $(i \in \mathbb{N})\}$
- A set of relation symbols $\mathtt{R} = \{\|r^i\|\|$ forsome $r \subseteq D^i$ $(i \in \mathbb{N})\}$
- Logical constants $\wedge, \vee, \rightarrow, \forall, \exists$
- Absurdity symbol $\bot$

The *set* $\mathtt{Term}_L$ *of the terms of* L is the smallest set X such that

- $x_i, \|k\| \in X$
- $t_1, \ldots, t_i \in X \Rightarrow \|\phi^i\|(t_1, \ldots, t_i) \in X$

The *set* $\mathtt{Form}_L$ *of the formulae of* L is the smallest set X such that

- $\bot \in X$
- $t_1, \ldots, t_i \in \mathtt{Term}_L \Rightarrow \|r^i\|(t_1, \ldots, t_i) \in X$
- $\alpha, \beta \in X \Rightarrow (\alpha \odot \beta) \in X (\odot = \wedge, \vee, \rightarrow$ and $\neg\alpha \stackrel{def}{=} (\alpha \rightarrow \bot))$
- $\alpha \in X \Rightarrow (\odot\, x_i\, \alpha) \in X$ $(\odot = \forall, \exists)$

An operational type is an expression

$$(\Gamma_1 \hookrightarrow \alpha_1), \ldots, (\Gamma_n \hookrightarrow \alpha_n) \hookrightarrow (\Gamma_{n+1} \hookrightarrow \alpha_{n+1}) \;\; (n \in \mathbb{N})$$

for finite $\Gamma_1, \ldots, \Gamma_{n+1} \subset \texttt{Form}_L$ and $\alpha_1, \ldots, \alpha_{n+1} \in \texttt{Form}_L$. In the case when $\Gamma = \emptyset$, we write $\Gamma \hookrightarrow \alpha$ simply as α. As we shall see, an operational type ω describes the behavior of an operation on grounds. Some of them are also used to type operational symbols F in the language of grounds for L – written $F\langle\omega\rangle$.

Definition 2 The *alphabet of* $G(L)$ consists of the auxiliary symbols parentheses and comma, plus

- every $t \in \texttt{Term}_L$
- A set of constants $\texttt{G}$ denoting grounds for closed atomic $\alpha \in \texttt{Form}_L$
- Variables ξ_i^α for $\alpha \in \texttt{Form}_L (i \in \mathbb{N})$
- Primitive operational symbols

  $\wedge I\langle\alpha, \beta \hookrightarrow \alpha \wedge \beta\rangle, \textit{for } \alpha, \beta \in \texttt{Form}_L$
  $\vee I_i\langle\alpha_i \hookrightarrow \alpha_1 \vee \alpha_2\rangle, \textit{for } \alpha_1, \alpha_2 \in \texttt{Form}_L \; (i = 1, 2)$
  $\rightarrow I\langle\beta \hookrightarrow \alpha \rightarrow \beta\rangle, \textit{for } \alpha, \beta \in \texttt{Form}_L$
  $\forall I_i\langle(\alpha(x_i) \hookrightarrow \forall x_i \alpha(x_i))\rangle, \textit{for } x_i \in \texttt{Term}_L, \alpha(x_i) \in \texttt{Form}_L$
  $\exists I_i\langle\alpha(t/x_i) \hookrightarrow \exists x_i \alpha(x_i))\rangle, \textit{for } t, x_i \in \texttt{Term}_L, \alpha(x_i) \in \texttt{Form}_L$

- Identity predicate $=$

Let $T : \alpha$ mean that T has type α. The *set* $\texttt{Term}_{G(L)}$ *of the terms of* $G(L)$ is the smallest set X such that

- $\texttt{G} \subset X$ *[*each constant denoting a ground for atomic α having type α*]*
- $\xi^\alpha : \alpha \in X$
- $T : \alpha, U : \beta \in X \Rightarrow \wedge I(T, U) : \alpha \wedge \beta \in X$
- $T : \alpha_i \in X \Rightarrow \vee I_i\langle\alpha_i \hookrightarrow \alpha_1 \vee \alpha_2\rangle(T) : \alpha_1 \vee \alpha_2 \in X$
- $T : \beta \in X \Rightarrow \rightarrow I\xi^\alpha(T) : \alpha \rightarrow \beta \in X$
- $T : \alpha(x) \in X \Rightarrow \forall Ix(T) : \forall x\alpha(x) \in X$
- $T : \alpha(t/x) \in X \Rightarrow \exists I(t, T) : \exists x\alpha(x) \in X$

$\rightarrow I$ and $\forall I$ respectively bind ξ^α and x. A term is open if it contains free variables, otherwise it is closed. In $\forall Ix(T)$, x must not occur free in β for ξ^β free in T.
The *set* $\texttt{Form}_{G(L)}$ *of the formulae of* $G(L)$ is the smallest set X such that

- $T, U \in \texttt{TERM}_{G(L)} \Rightarrow T = U \in X$

The *rules for the identity predicate* are

$(=_1)$ $T = T$
$(=_2)$ $T = U \Rightarrow U = T$
$(=_3)$ $T = U, U = Z \Rightarrow T = Z$
$(=_4^P)$ $T_i = U_i \Leftrightarrow F\,\vec{\xi}\,\vec{x}\,(T_1, \ldots, T_n) = F\,\vec{\xi}\,\vec{x}\,(U_1, \ldots, U_n)$ for F primitive defined on T_i, U_i, with each time respective arity of arguments and respective binding of variables.

We say that $B = \{\mathtt{K}, \mathtt{F}, \mathtt{R}, \mathtt{G}\}$ is our *base*. A ground relative to B – in short, a B-ground – for $\vdash \alpha$, with α closed, is an operation on grounds of type α. An operation on grounds relative to B – in short, a B-operation on grounds – of type $\alpha(x_1, \ldots, x_n)$, for $\alpha(x_1, \ldots, x_n)$ open with free variables $x_1, \ldots, x_n$, is an effective function f such that, for every $\|k_i\|$ $(i \leq n)$, $f(\|k_1\|, \ldots, \|k_n\|)$ is a B-ground for $\vdash \alpha(\|k_1\|/x_1, \ldots, \|k_n\|/x_n)$. Thus, f can be looked upon as a B-ground for $\vdash \alpha(x_1, \ldots, x_n)$. A B-operation on grounds of type $\alpha_1, \ldots, \alpha_n \hookrightarrow \alpha_{n+1}$ is an effective function f such that, for every g_i B-ground for $\vdash \alpha_i$ $(i \leq n)$, $f(g_1, \ldots, g_n)$ is a B-ground for $\vdash \alpha_{n+1}$. Of course, f is in this case a B-ground for $\alpha_1, \ldots, \alpha_n \vdash \alpha_{n+1}$. Finally, we can say that a B-operation on grounds of type $(\Gamma_1 \hookrightarrow \alpha_1), \ldots, (\Gamma_n \hookrightarrow \alpha_n) \hookrightarrow (\Gamma_{n+1} \hookrightarrow \alpha_{n+1})$ is an effective function f such that, for every g_i B-ground for $\Gamma_i \vdash \alpha_i$ $(i \leq n)$, $f(g_1, \ldots, g_n)$ is a B-ground for $\Gamma_{n+1} \vdash \alpha_{n+1}$.

Indeed, $G(L)$ is already sufficient to state denotational clauses concerning given closed atomic formulae of L, and closed formulae of L with $\wedge$, $\vee$ and $\exists$ as main sign. The elements of $\mathtt{G}$ denote B-grounds for asserting their respective type formulae and, for α, β, $\exists x\alpha(x)$ closed and t closed,

($\wedge$) if T denotes a B-ground for $\vdash \alpha$ and U denotes a B-ground for $\vdash \beta$, $\wedge I(T, U)$ denotes a B-ground for $\vdash \alpha \wedge \beta$.

($\vee$) If T denotes a B-ground for α_i, $\vee I_i\langle \alpha_i \hookrightarrow \alpha_1 \vee \alpha_2\rangle(T)$ denotes a B-ground for $\vdash \alpha_1 \vee \alpha_2$ $(i = 1, 2)$.

($\exists$) If T denotes a B-ground for $\vdash \alpha(t/x)$, $\exists I(t, T)$ denotes a B-ground for $\exists x\alpha(x)$.

Anyway, we are not yet able to do the same with $\rightarrow$ and $\forall$. In fact, a B-ground for an implication or a universal quantification could be obtained by applying the primitive operations represented in $G(L)$ by $\rightarrow I$ or $\forall I$ to B-operations on grounds that may be impossible to express in $G(L)$. Prawitz observes that 'knowing the meaning of conjunction, one also knows that there is an operation, namely $\wedge I$, such that when one is in possession of grounds for the assumptions of the closed hypothetical assertion $\alpha, \beta \vdash \alpha \wedge \beta$, one can get in possession of a ground for the categorical assertion $\vdash \alpha \wedge \beta$ by applying the operation' (Prawitz 2015, 91). This notwithstanding, primitive operations are not a satisfactory set, since 'this simple way of getting grounds for hypothetical assertions does not go very far [...] and in general, one has to define new operations' (Prawitz 2015, 92). From this point of view, $G(L)$ should be viewed as open to appropriate expansions.

Now, there are at least two ways to do this. One is attained by enriching L with wider $\mathtt{K}^+$, $\mathtt{F}^+$ and $\mathtt{R}^+$, or with new logical constants, so as to obtain a wider background language, say L^+. Accordingly, $G(L)$ should be expanded towards a wider $G(L^+)$ for L^+ with a new base containing $\mathtt{K}^+$, $\mathtt{F}^+$ and $\mathtt{R}^+$, a wider $\mathtt{G}^+$, and new primitive operational symbols. However, the above passage mentions the *definitions* of new operations, which entails keeping L fixed and passing from $G(L)$ to a wider $G(L)^+$ by adding *non-primitive* operations to $G(L)$. As we have seen, an operation on grounds is specified 'by stating the types of its domain and range' (Prawitz 2015, 92). Therefore, expanding $G(L)$ towards a wider $G(L)^+$, to deal with a certain non-primitive operation f, requires performing the following procedure: (1) expand the alphabet of $G(L)$ by adding a symbol F for f with alleged operational type; (2) provide clauses for forming new terms from F that comply with the intended type of the domain of f; (3) provide clauses for forming new formulae from the new terms; (4) provide one or more defining schemes of equations for F that comply with the intended type of the range of f. At this point, it is important to note that steps (1), (2) and (3) are not problematic. In contrast, step (4) is far from trivial. The new symbol should be fixed in such a way that it can be said to represent an effective function yielding specific results when it is applied to specific arguments. Nonetheless, the intended schemes of equations may be very complex or involve hardly detectable errors, thus making it very difficult to establish that the definition has been provided in a suitable way.

Definition 3 An *expansion* $G(L)^+$ of $G(L)$ is a language of grounds such that

- the alphabet of $G(L)^+$ is the alphabet of $G(L)$ plus new non-primitive operational symbols $F_1, \ldots, F_n$ with alleged operational types ($n \in \mathbb{N}$, so $G(L)$ is an expansion of itself)
- the clauses for $\mathtt{Term}_{G(L)^+}$ are the clauses for $\mathtt{Term}_{G(L)}$ plus clauses for forming terms built up from F_i according to the type of F_i ($i \leq n$)
- the types for $\mathtt{Term}_{G(L)^+}$ are the types for $\mathtt{Term}_{G(L)}$ plus types for terms built up from F_i according to the type of F_i ($i \leq n$)
- the clauses for $\mathtt{Form}_{G(L)^+}$ are the clauses for $\mathtt{Form}_{G(L)}$
- the rules for the identity predicate of $G(L)^+$ are the rules for the identity predicate of $G(L)$ plus schemes of equations for F_i ($i \leq n$) and

($=^N_4$) $T_i = U_i \Rightarrow F\,\vec{\xi}\,\vec{x}\,(T_1, \ldots, T_n) = F\,\vec{\xi}\,\vec{x}\,(U_1, \ldots, U_n)$ for F non-primitive defined on T_i, U_i, with each time respective arity of arguments and respective binding of variables, and T_i, U_i such that there is no instance of a scheme of equations for F whose left-hand side is $F\,\vec{\xi}\,\vec{x}\,(T_1, \ldots, T_n)$ or $F\,\vec{\xi}\,\vec{x}\,(U_1, \ldots, U_n)$.

Expansions are clearly closed under inclusion, so $(G(L)^+)^+$ is again an expansion of $G(L)$. We indicate with $\mathfrak{I}$ an arbitrary expansion of $G(L)$, and furthermore assume that we have defined a substitution function in $\mathfrak{I}$ and in its expansions.

Definition 4 Let R be the set of rules for the identity predicate of $\mathfrak{I}$. $A \in \mathtt{Form}_{\mathfrak{I}}$ is *correct* iff it can be obtained by a finite application of rules of R beginning with only applications of ($=_1$) or of schemes of equations for non-primitive operational symbols of $\mathfrak{I}$.

Definition 5 $T \in \mathtt{Term}_{\mathfrak{I}}$ is *canonical* iff it begins with a primitive operational symbol of $\mathfrak{I}$.

Definition 6 $T \in \mathtt{Term}_{\mathfrak{I}}$ closed (of type α for α closed) *denotes a B-ground for* $\vdash \alpha$ iff there is a closed canonical $U : \alpha \in \mathtt{Term}_{\mathfrak{I}}$ denoting a B-ground for $\vdash \alpha$, and such that $T = U$ is correct. $T(x_1, \ldots, x_n) \in \mathtt{Term}_{\mathfrak{I}}$ with free variables $x_1, \ldots, x_n$ (of type $\alpha(x_1, \ldots, x_n)$ for $\alpha(x_1, \ldots, x_n)$ with free variables $x_1, \ldots, x_n$) ($n \in \mathbb{N}$) *denotes a B-operation on grounds of type* $\alpha(x_1, \ldots, x_n)$ iff, for every $\|k_i\|$ ($i \leq n$), $T(\|k_1\|/x_1, \ldots, \|k_n\|/x_n)$ denotes a B-ground for

$\vdash \alpha(\|k_1\|/x_1,\ldots,\|k_n\|/x_n)$ – we also say that $T(x_1,\ldots,x_n)$ denotes a B-ground for $\vdash \alpha(x_1,\ldots,x_n)$. $T(\xi^{\alpha_1},\ldots,\xi^{\alpha_n}) \in \mathtt{Term}_{\Im}$ (of type α_{n+1}) with free variables $\xi^{\alpha_1},\ldots,\xi^{\alpha_n}$ ($n \in \mathbb{N}$) *denotes a B-operation on grounds of type* $\alpha_1,\ldots,\alpha_n \hookrightarrow \alpha_{n+1}$ iff, for every $\Im^+$, if $U_i \in \mathtt{Term}_{\Im^+}$ denotes a B-ground for $\vdash \alpha_i$ ($i \leq n$), $T(U_1/\xi^{\alpha_1},\ldots,U_n/\xi^{\alpha_n})$ denotes a B-ground for $\vdash \alpha_{n+1}$ – we also say that $T(\xi^{\alpha_1},\ldots,\xi^{\alpha_n})$ denotes a B-ground for $\alpha_1,\ldots,\alpha_n \vdash \alpha_{n+1}$ *[it might be objected that denotation of closed terms depends to an excessive degree on certain features of the underlying deductive system, e.g. convergence of reductions and Church-Rosser; if these constraints were not met, denotation would not be adequately captured. Although this is true, we nonetheless can – and will – keep the proposed definition, because the property that it requires to be satisfied should hold in any adequate account of denotation.]*

In the case of B-operations on grounds of type $(\Gamma_1 \hookrightarrow \alpha_1),\ldots,(\Gamma_n \hookrightarrow \alpha_n) \hookrightarrow (\Gamma_{n+1} \hookrightarrow \alpha_{n+1})$ there is clearly no denotation by terms, 'since the language of grounds has no variables of type $\Gamma \hookrightarrow \alpha$' (Prawitz 2015, 93). Now, for α, β, $\forall x\alpha(x)$ closed, we say that

($\rightarrow$) if T denotes a B-ground for $\alpha \vdash \beta$, $\rightarrow I\xi^{\alpha}(T)$ denotes a B-ground for $\vdash \alpha \rightarrow \beta$.
($\forall$) If T denotes a B-ground for $\vdash \alpha(x)$, $\forall Ix(T)$ denotes a B-ground for $\vdash \forall x\alpha(x)$.

Expansion 1 A simple expansion of $G(L)$ suggested by Prawitz.

Definition 7 The *alphabet of* $G(L)^1$ is the alphabet of $G(L)$ plus

- Non-primitive operational symbols

$\wedge E_i\langle\alpha_1 \wedge \alpha_2 \hookrightarrow \alpha_i\rangle$, for $\alpha_1, \alpha_2 \in \mathtt{Form}_L$ ($i = 1, 2$)
$\vee E\langle\alpha \vee \beta, \gamma, \gamma \hookrightarrow \gamma\rangle$, for α, β, $\gamma \in \mathtt{Form}_L$
$\rightarrow E\langle\alpha \rightarrow \beta, \alpha \hookrightarrow \beta\rangle$, for α, $\beta \in \mathtt{Form}_L$
$\forall E_k\langle\forall x_i\alpha(x_i) \hookrightarrow \alpha(\|k\|/x_i)\rangle$, for $\|k\|$, $x_i \in \mathtt{Term}_L$, $\alpha(x_i) \in \mathtt{Form}_L$
$\exists E\langle\exists x_i\alpha(x_i), \beta \hookrightarrow \beta\rangle$, for $x_i \in \mathtt{Term}_L$, $\alpha(x_i)$, $\beta \in \mathtt{Form}_L$

Let $T : \alpha$ mean that T has type α. The *set* $\mathtt{Term}_{G(L)^1}$ *of the terms of* $G(L)^1$ is the smallest set X such that

- $\mathtt{Term}_{G(L)} \subset X$
- $T : \alpha \wedge \beta \in X \Rightarrow \wedge E_i(T) : \alpha_i \in X (i = 1, 2)$
- $T : \alpha \vee \beta, U : \gamma, Z : \gamma \in X \Rightarrow \vee E\xi^{\alpha}\xi^{\beta}(T, U, Z) : \gamma \in X$
- $T : \alpha \rightarrow \beta, U : \alpha \in X \Rightarrow \rightarrow E(T, U) : \beta \in X$
- $T : \forall x\alpha(x) \in X \Rightarrow \forall E_k(T) : \alpha(\|k\|/x) \in X$
- $T : \exists x\alpha(x), U : \beta \in X \Rightarrow \exists E\xi^{\alpha(x)}x(T, U) : \beta \in X$.

$\vee E$ binds the variables ξ^{α}, ξ^{β}, while $\exists E$ binds the variables $\xi^{\alpha(x)}$, x. In $\exists E\xi^{\alpha(x)}x(T, U)$, x must not occur free in γ for ξ^{γ} free in U. *The set* $\mathtt{Form}_{G(L)^1}$ *of the formulae of* $G(L)^1$ is defined in the same way as $\mathtt{Form}_{G(L)}$. The *rules for the identity predicate* are($=_1$), ($=_2$), ($=_3$), ($=_4^P$), ($=_4^N$) plus

($\wedge_=$) $\wedge E_i(\wedge I(T_1, T_2)) = T_i$
($\vee_=$) $\vee E\xi^{\alpha_1}\xi^{\alpha_2}(\vee I_i\langle\alpha_i \hookrightarrow \alpha_1 \vee \alpha_2\rangle(T), U_1(\xi^{\alpha_1}), U_2(\xi^{\alpha_2})) = U_i(T/\xi^{\alpha_i})$
($\rightarrow_=$) $\rightarrow E(\rightarrow I\xi^\alpha(T(\xi^\alpha)), U) = T(U/\xi^\alpha)$
($\forall_=$) $\forall E_k(\forall I(T(x))) = T(\|k\|/x)$
($\exists_=$) $\exists E\xi^{\alpha(x)}x\,(\exists I(t, T), U(\xi^{\alpha(x)})) = U(T/\xi^{\alpha(x)})$

In the following definition, we try to circumvent the difficulty that, due to a lack of variables of appropriate type, certain operations on grounds of a certain operational type cannot be denoted by any term of our language of grounds. An operational symbol F captures an operational type $(\Gamma_1 \hookrightarrow \alpha_1), \ldots, (\Gamma_n \hookrightarrow \alpha_n) \hookrightarrow (\Gamma_{n+1} \hookrightarrow \alpha_{n+1})$ when two requirements are met: that F can be considered as a function, giving at most one value for each argument in its domain, and that, when applied to terms denoting operations on grounds of the operational type in the understood domain, it denotes an operation on grounds of the desired operational type.

Definition 8 Let F be an operational symbol of $\mathfrak{I}$ binding the sequences of variables $\vec{\xi}, \vec{x}$ and such that, for every $T_i \in \mathtt{Term}_{\mathfrak{I}}$ $(i \leq n)$, $T_i : \alpha_i$ iff $F\,\vec{\xi}\,\vec{x}\,(T_1, \ldots, T_n) \in \mathtt{Term}_{\mathfrak{I}}$. *$F$ fits with the operational type* $(\Gamma_h \hookrightarrow \alpha_h), \ldots, (\Gamma_k \hookrightarrow \alpha_k) \hookrightarrow (\Gamma_{n+1} \hookrightarrow \alpha_{n+1})$ iff, for every $\mathfrak{I}^+$,

1. for every $T_i, U_i : \alpha_i \in \mathtt{TERM}_{\mathfrak{I}^+}$ $(i \leq n)$, from the assumptions $T_i = U_i$ it is possible to derive in $\mathfrak{I}^+$ $F\,\vec{\xi}\,\vec{x}\,(T_1, \ldots, T_n) = F\,\vec{\xi}\,\vec{x}\,(U_1, \ldots, U_n)$ *[meant to ensure that F actually behaves functionally, yielding equal values on equal arguments. However, one should observe that this is required to be really provable only when F is applied to arguments that define it according to its schemes of equations, being instead assumed as an axiom in the remaining cases according to* $(=^N_4)$. *E.g.,* $\wedge E_i(\wedge I(T_1, T_2)) = \wedge E_i(\wedge I(U_1, U_2))$ *can be proved from* $\wedge I(T_1, T_2) = \wedge I(U_1, U_2)$ *by appealing to* $(=^P_4)$*, while that* $\xi^{\alpha\wedge\beta} = \wedge I(T_1, T_2) \Rightarrow \wedge E_i(\xi^{\alpha\wedge\beta}) = \wedge E_i(\wedge I(T_1, T_2))$ *is an instance of* $(=^N_4)$*.]*
2. for $\alpha_h, \ldots, \alpha_k$ types not occurring among those of the variables bound by F, for every $T_j \in \mathtt{Term}_{\mathfrak{I}^+}$ with $\alpha_j \in \{\alpha_h, \ldots, \alpha_k\}$, if T_j denotes a B-ground for $\Gamma_j \vdash \alpha_j$, $F(U_1, \ldots, U_n)$ denotes a B-ground for $\Gamma_{n+1} \vdash \alpha_{n+1}$, where U_i is T_i if $\alpha_i \in \{\alpha_h, \ldots, \alpha_k\}$, ξ^{α_i} otherwise $(i \leq n)$.

Proposition 9 *We mention some specific fittings:*

1. *$\rightarrow I$ fits with* $(\alpha \hookrightarrow \beta) \hookrightarrow \alpha \rightarrow \beta$
2. *$\wedge E_i$ fits with* $\alpha_1 \wedge \alpha_2 \hookrightarrow \alpha_i$
3. *$\vee E$ fits with* $\alpha \vee \beta, (\alpha \hookrightarrow \gamma), (\beta \hookrightarrow \gamma) \hookrightarrow \gamma$
4. *$\rightarrow E$ fits with* $\alpha \rightarrow \beta, \alpha \hookrightarrow \beta$
5. *$\forall E_k$ fits with* $\forall x\alpha(x) \hookrightarrow \alpha(\|k\|/x)$
6. *$\exists E$ fits with* $\exists x\alpha(x), (\alpha(x) \hookrightarrow \beta) \hookrightarrow \beta$*, of course, for x not free in* β.

In a sense, Definition 8 enables us to check that Definitions 2 and 7 attach appropriate operational types to the operational symbols of $G(L)$ and $G(L)^1$. More importantly, however, 1, 3 and 6 show that it is now possible to express wider properties of such symbols. Of course, a definition analogous to Definition 8 can be given for terms. Thus, 1–6 can be reformulated accordingly. It is now easy to see that we have a kind of Curry-Howard isomorphism between minimal logic – written $\mathtt{ML}$ – and $G(L)^1$. Moreover, indicating ground-theoretic validity by $\models_G$, we also have a soundness result that will be clearer after the definitions of ground-theoretic valid inferences and proofs in Sect. 13.3.3.

Proposition 10 *Each* $\mathtt{ML}$*-derivation can be translated into an isomorphic term of* $G(L)^1$ *and vice versa.*

Theorem 11 $\vdash_{ML} \alpha \implies \models_G \alpha$

Expansion 2 Another example for later purposes.

Definition 12 The *alphabet* of $G(L)^{1\mathcal{Q}}$ is the alphabet of $G(L)^1$ plus

- absurdity symbol $\bot^G$
- logical constants $\rightarrow^G$, $\forall^G$, $\exists^G$

The *set* $\mathtt{Term}_{G(L)^{1\mathcal{Q}}}$ *of the terms of* $G(L)^{1\mathcal{Q}}$ is $\mathtt{Term}_{G(L)^1}$. The *set* $\mathtt{Form}_{G(L)^{1\mathcal{Q}}}$ *of the formulae of* $G(L)^{1\mathcal{Q}}$ is the smallest set X such that

- $\mathtt{FORM}_{G(L)^1} \subset X$
- $\bot^G \in X$
- $A, B \in X \Rightarrow A \rightarrow^G B \in X$ $(\neg A \stackrel{def}{=} (A \rightarrow^G \bot^G))$
- $A \in X \Rightarrow \odot\, \xi^\alpha A \in X$ $(\odot = \forall^G, \exists^G)$

The *rules for the identity predicate* are the rules for the identity predicate of $G(L)^1$.

We now develop a language of grounds for $G(L)^{1\mathcal{Q}}$. First of all, the operational types with the formulae of $G(G(L)^{1\mathcal{Q}})$ can be taken to have the same shape as the operational types with the formulae of L. Then, let us call $\mathtt{K}^*$ the set of the $T \in \mathtt{Term}_{G(L)^{1\mathcal{Q}}}$ such that T denotes a B-ground for $\vdash \alpha$ with $\alpha \in \mathtt{Form}_L$. The base B^G of our language is $\{\mathtt{K}^*, \{\wedge I, \wedge E_i, \vee I, \vee E, \rightarrow I, \rightarrow E, \forall I_i, \forall E_k, \exists I_i, \exists E\}, \{=\}\}$.

Definition 13 The *alphabet* of $G(G(L)^{1\mathcal{Q}})$ consists of the auxiliary symbols parentheses and comma, plus

- every $T \in \mathtt{Term}_{G(L)^{1\mathcal{Q}}}$
- Variables θ_i^A for $A \in \mathtt{Form}_{G(L)^{1\mathcal{Q}}}$ $(i \in \mathbb{N})$
- Primitive operational symbols
  - one for each application of reflexivity, symmetry and transitivity. E.g., in the case of $(=_3)$ we will have
    $A_3\langle T = U, U = Z \hookrightarrow T = Z\rangle$, for $T, U, Z \in \mathtt{Term}_{G(L)^{1\mathcal{Q}}}$
  - one for applications of $(=_4^P)$ in both directions to primitive operational symbols of $G(L)^{1\mathcal{Q}}$, and one for each application of $(=_4^N)$ to non-primitive operational symbols of $G(L)^{1\mathcal{Q}}$. E.g., in the case of applications of $(=_4^N)$ to $\wedge E_i$, we will have
    $A^i_{4N,\wedge}\langle T = U \hookrightarrow \wedge E_i(T) = \wedge E_i(U)\rangle$, for $T, U : \alpha \wedge \beta \in \mathtt{Term}_{G(L)^{1\mathcal{Q}}}$ of the kind stated in Definition 3 with $\alpha \wedge \beta \in \mathtt{Form}_L$ $(i = 1, 2)$

- one for schemes of equations of each non-primitive operational symbol of $G(L)^{1Q}$. E.g., in the case of $(\wedge_=)$, we will have $A^i_\wedge \langle \wedge E_i(\wedge I(T_1, T_2)) = T_i \rangle$, for $T_1, T_2 \in \texttt{Term}_{G(L)^{1Q}}$ $(i = 1, 2)$
- $\rightarrow^G I \langle B \hookrightarrow A \rightarrow^G B \rangle$ for $A, B \in \texttt{Form}_{G(L)^{1Q}}$
- $\forall^G I_\alpha \langle A(\xi^\alpha) \hookrightarrow \forall^G \xi^\alpha A(\xi^\alpha) \rangle$ for $\alpha \in \texttt{Form}_L$, $A(\xi^\alpha) \in \texttt{Form}_{G(L)^{1Q}}$
- $\exists^G I \langle A(T/\xi^\alpha) \hookrightarrow \exists^G \xi^\alpha A(\xi^\alpha) \rangle$ for $T : \alpha \in \texttt{K}^*$ or where T is ξ^α, $\alpha \in \texttt{FORM}_L$, $A(\xi^\alpha) \in \texttt{FORM}_{G(L)^{1Q}}$

- Non-primitive operational symbols
  - $C^i \langle \exists^G \xi^{\alpha_i} \xi^{\alpha_h} (T = \wedge I(\xi^{\alpha_1}, \xi^{\alpha_2})) \rangle$, where $T : \alpha_1 \wedge \alpha_2 \in \texttt{K}^*$ or T is $\xi^{\alpha_1 \wedge \alpha_2}$, for α_1, $\alpha_2 \in \texttt{Form}_L$ $(i, h = 1, 2, i \neq h)$ *[every term denoting a ground for $\vdash \alpha_1 \wedge \alpha_2$ must reduce to a canonical term $\wedge I(T_1, T_2)$ where T_1 denotes a ground for $\vdash \alpha_1$ and T_2 denotes a ground for $\vdash \alpha_2$]*
  - $\rightarrow^G E \langle A \rightarrow^G B, A \hookrightarrow B \rangle$ for $A, B \in \texttt{Form}_{G(L)^{1Q}}$
  - $\forall^G E_T \langle \forall^G \xi^\alpha A(\xi^\alpha) \hookrightarrow A(T/\xi^\alpha) \rangle$, for $T : \alpha \in \texttt{K}^*$ or where T is ξ^α, $\alpha \in \texttt{Form}_L$, $A(\xi^\alpha) \in \texttt{Form}_{G(L)^{1Q}}$
  - $\exists^G E \langle \exists^G \xi^\alpha A(\xi^\alpha), B \hookrightarrow B \rangle$, for $\alpha \in \texttt{FORM}_L$, $A(\xi^\alpha), B \in \texttt{FORM}_{G(L)^{1Q}}$

- Identity predicate $\cong$

Let $\mathbf{R} : A$ mean that $\mathbf{R}$ has type A. The *set* $\texttt{Term}_{G(G(L)^{1Q})}$ *of the terms of* $G(G(L)^{1Q})$ is the smallest set X such that

- $\theta^A : A \in X$
- $A^i_\wedge : \wedge E_i(\wedge I(T_1, T_2)) = T_i \in X$ *[and similarly for other operations representing schemes of equations of non-primitive operational symbols of $G(L)^{1Q}$]*
- $C^i : \exists^G \xi^{\alpha_i} \xi^{\alpha_h} (T = \wedge I(\xi^{\alpha_1}, \xi^{\alpha_2})) \in X$
- $\mathbf{R} : T = U, \mathbf{S} : U = Z \in X \Rightarrow A_3(\mathbf{R}, \mathbf{S}) : T = Z \in X$ *[and similarly for other operations representing reflexivity and symmetry]*
- $\mathbf{R} : T = U \in X \Rightarrow A^i_{4N,\wedge}(\mathbf{R}) : \wedge E_i(T) = \wedge E_i(U) \in X$ $(i = 1, 2)$ *[and similarly for other operations representing applications of $(=^P_4)$ and of $(=^N_4)$ to operational symbols of $G(L)^{1Q}$]*
- $\mathbf{R} : B \in X \Rightarrow \rightarrow^G I \theta^A(\mathbf{R}) : A \rightarrow^G B \in X$
- $\mathbf{R} : A(\xi^\alpha) \in X \Rightarrow \forall^G I \xi^\alpha(\mathbf{R}) : \forall^G \xi^\alpha A(\xi^\alpha) \in X$
- $\mathbf{R} : A(T/\xi^\alpha) \in X \Rightarrow \exists^G I(T, \mathbf{R}) : \exists^G \xi^\alpha A(\xi^\alpha) \in X$
- $\mathbf{R} : A \rightarrow^G B, \mathbf{S} : A \in X \Rightarrow E(\mathbf{R}, \mathbf{S}) : B \in X$
- $\mathbf{R} : \forall \xi^\alpha A(\xi^\alpha) \in X \Rightarrow \forall^G E_T(\mathbf{R}) : A(T/\xi^\alpha) \in X$
- $\mathbf{R} : \exists^G \xi^\alpha A(\xi^\alpha), \mathbf{S} : B \in X \Rightarrow \exists^G E \theta^{A(\xi^\alpha)} \xi^\alpha(\mathbf{R}, \mathbf{S}) : B \in X$

$\forall^G I$ binds the variable ξ^α. $\exists^G E$ binds the variables $\theta^{A(\xi^\alpha)}$ and ξ^α. A term is open if it contains free variables, otherwise it is closed. In $\forall^G I(\mathbf{R})$, ξ^α must not occur free in B for θ^B free in $\mathbf{R}$. In $\exists^G \theta^{A(\xi^\alpha)} \xi^\alpha(\mathbf{R}, \mathbf{S})$, ξ^α must not occur free in C for θ^C free in $\mathbf{S}$.
The *set* $\texttt{Form}_{G(G(L)^{1Q})}$ *of the formulae of* $G(G(L)^{1Q})$ is the smallest set X such that

- $\mathbf{R}, \mathbf{S} \in \texttt{Term}_{G(G(L)^{1Q})} \Rightarrow \mathbf{R} \cong \mathbf{S} \in X$

The *rules for the identity predicate* are – we assume to have defined a substitution function –

($\cong_1$) $\mathbf{R} \cong \mathbf{R}$
($\cong_2$) $\mathbf{R} \cong \mathbf{S} \Rightarrow \mathbf{S} \cong \mathbf{R}$
($\cong_3$) $\mathbf{R} \cong \mathbf{S}, \mathbf{S} \cong \mathbf{W} \Rightarrow \mathbf{R} \cong \mathbf{W}$
($\cong_4^P$) $\mathbf{R}_i \cong \mathbf{S}_i\ (i \leq n) \Leftrightarrow \mathbf{F}\,\vec{\theta}\,\vec{\xi}\,(\mathbf{R}_1, \ldots, \mathbf{R}_n) \cong \mathbf{F}\,\vec{\theta}\,\vec{\xi}\,(\mathbf{S}_1, \ldots, \mathbf{S}_n)$ for F primitive defined on $\mathbf{R}_i$, $\mathbf{S}_i$, with each time respective arity of arguments and respective binding of variables.
($C^i_{\cong}$) $C^i(T) \cong \exists^G I(T_i, \exists^G I(T_h, \texttt{red}(T = \wedge I(T_1, T_2))))$ for $T \in \texttt{K}^*$ [$\texttt{red}(T = U)$ *is explained below]*
($\rightarrow^G_{\cong}$) $\rightarrow^G E(\rightarrow^G I\theta^A(\mathbf{R}(\theta^A)), \mathbf{S}) \cong \mathbf{R}(\mathbf{S}/\theta^A)$
($\forall^G_{\cong}$) $\forall^G E_T(\forall^G I\xi^\alpha(\mathbf{R}(\xi^\alpha))) \cong \mathbf{R}(T/\xi^\alpha)$
($\exists^G_{\cong}$) $\exists^G E\theta^{A(\xi^\alpha)}\xi^\alpha(\exists^G I(T, \mathbf{R}), \mathbf{S}(\theta^{A(\xi^\alpha)})) \cong \mathbf{S}(\mathbf{R}/\theta^{A(\xi^\alpha)})$
($\cong_4^N$) $\mathbf{R}_i \cong \mathbf{S}_i\ (i \leq n) \Rightarrow \mathbf{F}\,\vec{\theta}\,\vec{\xi}\,(\mathbf{R}_1, \ldots, \mathbf{R}_n) \cong \mathbf{F}\,\vec{\theta}\,\vec{\xi}\,(\mathbf{S}_1, \ldots, \mathbf{S}_n)$ for F non-primitive defined on $\mathbf{R}_i$, $\mathbf{S}_i$, with each time respective arity of arguments and respective binding of variables, and $\mathbf{R}_i$, $\mathbf{S}_i$ such that there is no instance of a scheme of equations for $\mathbf{F}$ whose left-hand side is $\mathbf{F}\,\vec{\theta}\,\vec{\xi}\,(\mathbf{R}_1, \ldots, \mathbf{R}_n)$ or $\mathbf{F}\,\vec{\theta}\,\vec{\xi}\,(\mathbf{S}_1, \ldots, \mathbf{S}_n)$

With $\texttt{kerK}^*$ we indicate the subset of $\texttt{K}^*$ whose elements T are such that, if $T = U$ is correct, then either U is T or U contains more non-primitive operational symbols than T. One should now observe that we have the following results.

Proposition 14 *Each application of the rules for the identity predicate of* $G(L)^{1Q}$ *can be translated into an isomorphich term of* $G(G(L)^{1Q})$*, and vice versa.*

Proposition 15 *Given* $T : \alpha \in \texttt{K}^*$*, there is a unique canonical* $U : \alpha \in \texttt{kerK}^*$ *such that* $T = U$ *is correct.* [by normalization and Church-Rosser]

Hence, for every $T : \alpha \in \texttt{K}^*$, we choose a term of $G(G(L)^{1Q})$ showing that $T = U$ is correct, for some unique canonical $U : \alpha \in \texttt{kerK}^*$. We call this chosen term $\texttt{red}(T = U)$. [A precise definition should give choice parameters. Note that $\texttt{K}^*$ is actually the set of the closed terms: every term of type $\alpha_{n+1}(\vec{x}_{n+1})$ with free variables $\xi^{\alpha_1(\vec{x}_1)}, \ldots, \xi^{\alpha_n(\vec{x}_n)}$, $\vec{x}$ ($\vec{x}_i \subseteq \vec{x}$ for every $i \leq n$) denotes a B-ground for $\alpha_1(\vec{x}_1) \ldots, \alpha_n(\vec{x}_n) \vdash \alpha_{n+1}(\vec{x}_{n+1})$, and, if a term is closed, every reduction ends in a unique non-reducible canonical form – by strong normalization and Church-Rosser. Enumerate all the closed terms and, at each step, apply in any order the rules for the non-primitive operational symbols. When reduction stops, apply transitivity up to the term you begun with. Translate the whole in a term of the higher language, and choose what you obtain as $\texttt{red}$.]

With respect to B^G, we can define different notions of B^G-operation on grounds – in short, B^G-grounds – in exactly the same way as we have done for $G(L)^{1Q}$ and B. Also, we can adapt Definitions 4, 5, 6 and 8 to $G(G(L)^{1Q})$ and its expansions. More specifically, an operation on grounds relative to B^G – in short, a B^G-operation on grounds – of type $A(\xi^{\alpha_1}, \ldots, \xi^{\alpha_n})$, for $A(\xi^{\alpha_1}, \ldots, \xi^{\alpha_n})$ open with free variables $\xi^{\alpha_1}, \ldots, \xi^{\alpha_n}$, will be an effective function f such that, for every $T_i : \alpha_i \in \texttt{K}^*$

$(i \leq n)$, $f(T_1, \ldots, T_n)$ is a B^G-ground for $\vdash A(T_1/\xi^{\alpha_1}, \ldots, T_n/\xi^{\alpha_n})$, in which case we speak of a B^G-ground for $\vdash A(\xi^{\alpha_1}, \ldots, \xi^{\alpha_n})$. Likewise, $\mathbf{R}(\xi^{\alpha_1}, \ldots, \xi^{\alpha_n})$ with free variables $\xi^{\alpha_1}, \ldots, \xi^{\alpha_n}$ (of type $A(\xi^{\alpha_1}, \ldots, \xi^{\alpha_n})$, for $A(\xi^{\alpha_1}, \ldots, \xi^{\alpha_n})$ with free variables $\xi^{\alpha_1}, \ldots, \xi^{\alpha_n}$) $(n \in \mathbb{N})$ will denote a B^G-operation on grounds of type $A(\xi^{\alpha_1}, \ldots, \xi^{\alpha_n})$ iff, for every $T_i : \alpha_i \in \mathbb{K}^*$ $(i \leq n)$, $\mathbf{R}(T_1/\xi^{\alpha_1}, \ldots, T_n/\xi^{\alpha_n})$ denotes a B^G-ground for $\vdash A(T_1/\xi^{\alpha_1}, \ldots, A_n/\xi^{\alpha_n})$, in which case we say that $\mathbf{R}(\xi^{\alpha_1}, \ldots, \xi^{\alpha_n})$ denotes a B^G-ground for $\vdash A(\xi^{\alpha_1}, \ldots, \xi^{\alpha_n})$.

- $(A^i_\wedge)$ $A^i_\wedge(T_1, T_2)$ denotes a B^G-ground for $\vdash \wedge E_i(\wedge I(T_1, T_2)) = T_i$ $(i = 1, 2)$ [and similarly for other operations representing schemes of equations of non-primitive operational symbols of $G(L)^{1Q}$].
- (A_3) If $\mathbf{R}$ denotes a B^G-ground for $\vdash T = U$, and $\mathbf{S}$ denotes a B^G-ground for $\vdash U = Z$, $A_3(\mathbf{R}, \mathbf{S})$ denotes a B^G-ground for $\vdash T = Z$ [and similarly for other operations representing reflexivity and symmetry].
- $(A^i_{4N,\wedge})$ If $\mathbf{R}$ denotes a B^G ground for $\vdash T = U$ with T and U of the kind required in Definition 3, $A^i_{4N,\wedge}(\mathbf{R})$ denotes a B^G-ground for $\vdash \wedge E_i(T) = \wedge E_i(U)$ $(i = 1, 2)$ [and similarly for other operations representing applications of $(=^P_4)$ and of $(=^N_4)$ to operational symbols of $G(L)^{1Q}$].
- $(\rightarrow^G)$ If $\mathbf{R}$ denotes a B^G-ground for $A \vdash B$, $\rightarrow^G I\theta^A(\mathbf{R})$ denotes a B^G-ground for $\vdash A \rightarrow^G B$.
- $(\forall^G)$ If $\mathbf{R}$ denotes a B^G-ground for $\vdash A(\xi^\alpha)$, $\forall^G I\xi^\alpha(\mathbf{R})$ denotes a B^G-ground for $\vdash \forall^G \xi^\alpha A(\xi^\alpha)$.
- $(\exists^G)$ If $\mathbf{R}$ denotes a B^G-ground for $\vdash A(T/\xi^\alpha)$, $\exists^G I(T, \mathbf{R})$ denotes a B^G-ground for $\vdash \exists^G \xi^\alpha A(\xi^\alpha)$.

Proposition 16 *We mention some specific fittings:*

- $C^i(\xi^{\alpha_1 \wedge \alpha_2})$ *denotes a* B^G*-ground for* $\vdash \exists^G \xi^{\alpha_i} \xi^{\alpha_h}(\xi^{\alpha_1 \wedge \alpha_2} = \wedge I(\xi^{\alpha_1}, \xi^{\alpha_2}))$
- $\rightarrow^G$ *fits with the operational type* $(A \hookrightarrow B) \hookrightarrow B$
- $\forall^G E_T$ *fits with the operational type* $\forall^G \xi^\alpha A(\xi^\alpha) \hookrightarrow A(T/\xi^\alpha)$
- $\exists^G E$ *fits with the operational type* $\exists^G \xi^\alpha A(\xi^\alpha), (A(\xi^\alpha) \hookrightarrow B) \hookrightarrow B$*, of course, for* ξ^α *not free in* B*.*

$G(L)^{1Q}$ is a language of grounds with which one can say something about terms of $G(L)^1$; likewise, $G(G(L)^{1Q})$ is a kind of meta-language of grounds with which one can build grounds for some assertions about terms of $G(L)^1$. However, for full expressivity, both $G(L)^{1Q}$ and $G(G(L)^{1Q})$ should of course be enriched. $G(L)^{1Q}$ should at least contain a ground-theoretic analogue of all the usual first-order intuitionistic logical constants as well as other linguistic resources that allow speaking about types and denotation; similarly, $G(G(L)^{1Q})$ should have operations dealing with a ground-theoretic analogue of all the usual first-order intuitionistic logic as well as typing and denotational rules. Despite this, what we have done so far is sufficient for the discussion in Sect. 13.4.

13.3.3 Valid Inferences and a Decidability Issue

According to Prawitz, a merely transitional picture of inferences cannot answer the main ground-theoretic question, since 'it is hard to see how an act that consists of only an inferential transition [...] could be able to confer evidence on its conclusion' (Prawitz 2015, 94). However, operations on grounds take us a step further, because 'it suggests itself that an inference is essentially an operation on grounds that produces grounds' (Prawitz 2015, 94), so that 'to *perform a reflective inference* is, in addition to make an inferential transition, to apply an operation on grounds' (Prawitz 2015, 94).

An individual inference J will thus be individuated by premises $\Gamma_1 \vdash \alpha_1, \dots, \Gamma_n \vdash \alpha_n$ and conclusion $\Gamma_{n+1} \vdash \alpha_{n+1}$ in a language L, by an agent that performs J and the occasion where J is performed, by alleged grounds $g_1, \dots, g_n$ for the premises and, finally, by an operation on grounds f to be applied to $g_1, \dots, g_n$. Both $g_1, \dots, g_n$ and f should be described respectively by terms $T_1, \dots, T_n$ and an operational symbol F in an appropriate language of grounds $\Im$ with base B. Thus, J is B-valid iff $F(T_1, \dots, T_n)$ denotes a B-ground for $\Gamma_{n+1} \vdash \alpha_{n+1}$, and J is valid iff it is B-valid for every base B. A generic inference will be B-valid, or valid, essentially under the same conditions. Finally, an inference form J consists of a structural relation between premises and conclusion, and of an operation on grounds f. For the inference to be B-valid, it must hold that, given any instance of the structural relation with specific premises $\Gamma_1 \vdash \alpha_1, \dots, \Gamma_n \vdash \alpha_n$ and specific conclusion $\Gamma_{n+1} \vdash \alpha_{n+1}$ in a language L, the resulting generic inference is B-valid. Again, J is valid iff it is B-valid for every base B. Proofs and valid arguments can be defined as finite chains of valid inferences.

The epistemically compelling power of valid inferences is now easily explained by noting that 'whether an inference confers evidence on its conclusion depends on the nature of the operation f that is a constitutive part of an inference, as this notion has been reconstructed' (Prawitz 2015, 96). When f is meaning-constitutive, the fact that it always produces a ground for the conclusion when it is applied to grounds for the premises 'is just an expression of what the sentence asserted by the conclusion means'; when f is not meaning-constitutive, this fact will be 'a consequence of how the operation f has been defined' (Prawitz 2015, 96).

However, in the concluding remarks of *Explaining deductive inference*, Prawitz refers to an issue that may undermine the efficiency of the ground-theoretic setup. Indeed, grounds 'may have become known under a different description than the normal form', and one may ask whether 'a ground as now defined really amount[s] to evidence' (Prawitz 2015, 96). In fact, 'when a subject performs a valid inference and applies an operation f to what she holds to be grounds for the premises, she forms a term T that in fact denotes a ground for the drawn conclusion α, but it is not guaranteed [...] that she knows that T denotes a ground for α' (Prawitz 2015, 96). Such knowledge is achieved without problems when α is atomic and T complies with how the meaning of α has been fixed, or when α is compound and T is in canonical form. When T is non-canonical the agent, to see that T denotes a ground for the conclusion, must reflect on how the non-primitive operational symbol with which T begins is defined. Here, 'she can in fact carry out the operations that T is built up of and bring T to normal form in this way, but she may not know this fact' (Prawitz 2015, 96). Moreover, 'when T is an open term, it denotes a ground for an open assertion or an assertion under assumption, and it is first after appropriate substitutions for the free variables that one the previous two cases arises' (Prawitz 2015, 96–97).

In other words, a non-canonical, and possibly open term could be unfeasibly complex, or require infinitely many substitutions from expansions of the original language of grounds that need not be regimented at all. Prawitz's crucial question is therefore 'if it is decidable for an arbitrary definition of an operation, which we may contemplate to add to a given closed language of grounds, whether it always produces a ground of a particular type when applied to grounds in its domain' (Prawitz 2015, 98). Since non-canonical terms begin with non-primitive operational symbols, the question comes down to whether it is decidable that given schemes of equations make a non-primitive operational symbol fit with an intended operational type. Here we cannot make recourse to a solution *à la* Kreisel, by requiring the decision be made on the basis of a possessed ground for an assertion of the kind "F fits with the given operational type". Such ground would have 'to be delivered by a proof' that 'will in turn depend on its inferences giving evidence for their conclusion' (Prawitz 2015, 97), and a regressive explanation would result. Nor can the decision be made on algorithms of the kind of which we dispose in closed languages of grounds, which allow us to decide whether a certain term denotes a ground for a certain judgement or assertion. Because of Gödel's incompleteness, we know 'that already for first-order arithmetical assertions there is no closed language of grounds in which all grounds for them can be defined' (Prawitz 2015, 98).

13.4 General and Specific Decidability

What does "decidable" mean? Clearly, rejection of the algorithmic solution entails that decidability must indicate a wider notion than the one usually employed in mathematical logic. Here we propose two parameters through which the notion can vary according to the generality degree of its range. In particular, we make a distinction between *general* and *specific* decidability.

General decidability involves a kind of uniform procedure that allows to establish homogeneously whether an arbitrary operational symbol is well-defined with respect to an intended operational type. Even though this procedure cannot be a recognition algorithm, it would be unreasonable to employ it without being at least able to describe how it behaves. Therefore, in this case one should be able to provide a description – maybe even an approximate description – or to list its instructions. In contrast, specific decidability should be conceived of as a heterogeneous ability to state, case-by-case, whether given operational symbols fit with intended operational types. No uniformity or restrictions are imposed on the means that can be used to reach the desired decision, the idea rather being that, when presented with an operational symbol, one can reason on how it is defined and thereby decide whether it has been adequately fixed. Clearly, the same reasoning may not work for other symbols; therefore, while the general notion binds "decidability" to a precise procedure, specific decidability involves an unrestricted variety of epistemic acts.

Both the general and the specific interpretation appear to match epistemic concerns. Then, which should one choose? Far from being irrelevant, the choice significantly affects the way in which one can frame the idea that whether arbitrary operational symbols are well defined is decidable. Indeed, general and specific decidability amounts respectively to

(G) there is a procedure $\wp$ such that, for every operational symbol F, it is possible by means of $\wp$ to decide whether F fits with a given operational type.
(S) for every operational symbol F, there is a reasoning $\wp$ such that, by means of $\wp$, it is possible to decide whether F fits with a given operational type.

Clearly, (G) and (S) do not have the same theoretical weight. In particular, (S) seems to be reasonable, whereas (G) is open to several objections. Provided that terms and definitions in a certain language of grounds are epistemic objects, used to account formally for our deductive practice and the results it produces, then they must also be thought of as being endowed only with epistemic properties. In turn, epistemic properties are always in principle knowable; this holds in particular for the property of being well-defined with respect to a given operational type. Hence, a case-by-case understanding would be possible each time. Moreover, in a framework where *reduction ad absurdum* holds, (S) seems to be even more cogent, for its denial here would be tantamount to the existence of an operational symbol F such that, for every reasoning $\wp$, one cannot decide by means of $\wp$ whether F fits with a given operational type; in other words, there would be an epistemic object with an absolutely unknowable property. As regards the second option, it is clearly unreasonable to have recourse to a procedure of the kind envisaged by (G) without some awareness of what it is or how it behaves. Once one endorses (G), one should outline the procedure it involves, and it is as yet unclear how, and far from obvious that, this can be done. Actually, Prawitz is openly sceptical 'of the idea that the condition for something to be a proof or to constitute evidence is luminous' (Prawitz 2015, 98).

Why does the decidability problem play a crucial role in the ground-theoretic context? One should note that the theory of grounds describes valid inferences as applications of operations that, when actually performed on grounds for the premises, yield a ground for the conclusion. Now, if the operation is non-primitive and the ground obtained for the conclusion is to be denoted by a non-canonical term, it is possible that 'to have come in possession of a ground for the conclusion of a valid inference' is not also to know 'that what one is in possession of is such a ground' (Prawitz 2015, 97). The issue could therefore seriously affect the entire project of explaining the compelling power of valid inferences, since 'one may ask if to make a valid inference really gives the evidence that one should expect' (Prawitz 2015, 97).

To this Prawitz replies that 'to make an inference is not to assert that the inference is valid, nor is it to make an assertion about the grounds that one has found for the conclusion of the inference' (Prawitz 2015, 97). When one performs a ground-theoretically valid inference, what one is in possession of is a ground, not a term, and grounds are, so to speak, always canonical. The term only describes the act: it is non-canonical when the act does not end with an introduction, and denotes the obtained ground in an indirect way. Ground-theoretically valid inferences thus always provide the agent with the content of a mental state of justification, although the description of how the content has been obtained may amount to a very complex non-canonical term. Therefore, it might be wholly sufficient to make it 'a conceptual truth that a person who performs a valid inference is aware of making an operation that produces what she takes to be a ground for the conclusion, although the latter is not what she asserts' (Prawitz 2015, 98).

If one is not satisfied with Prawitz's solution, one should 'explicate the involved notions in another way so that the performance of a valid inference resulted in an even greater awareness and further knowledge' (Prawitz 2015, 98). Perhaps, one could impose 'greater restrictions on the operations that can be used to form grounds', in such a way that 'a valid inference produces not only evidence for its conclusion but also evidence for its own validity' (Prawitz 2015, 98). However, there seems to be reason to doubt that this could ever be accomplished. This stronger claim, when considered from a particular standpoint, may for example conflict even with the aforementioned tenability of (S).

Let us consider the idea that the mere possession of a ground is not sufficient to obtain evidence, and that one could be asked to have also evidence for the fact that what one is in possession of is such a ground. The possession of grounds is attained through certain epistemic acts, and the whole process of obtaining grounds through certain epistemic acts is coded by ground terms. In short, on this conception

(K) P is in a state of evidence for α iff P's act can be coded by a term T denoting a ground for $\vdash \alpha$ and, if T is non-canonical, P has evidence for "T denotes a ground for $\vdash \alpha$".

An interesting and important, but difficult question is clearly what are the reasonings that (S) involves. Since (S) is an epistemic claim, such reasonings must be either epistemic objects or epistemic acts. Furthermore, it seems legitimate to maintain that there must be a relation J between them and P such that P can enter J and that, by doing so, P can build the intended grounds and obtain evidence for the fact that these grounds have certain properties. The nature of our J will of course depend on the kind of epistemic objects or acts with which P is in relation. Here, we consider two options: (1) the $\wp$s are terms denoting grounds and J is a building relation, and (2) the $\wp$s are proofs and J is a carrying-out relation.

Once noted that decidability can easily be translated in terms of evidence – P can decide whether α is the case or not iff P can be in a state of evidence for α or $\neg\alpha$ respectively – then (S) can be further specified by describing, by means of (1) or (2), the reasonings it involves. If one accepts Prawitz's view on the decidability issue, these reformulations are both reasonable. If, however, one endorses (K), the reformulations may fail – note that this claim is closely reminiscent of the one according to which solving the decidability problem *à la* Kreisel would involve a regressive explanation.

In the first place, with (1) we have

(S_K) for every operational symbol *F*, there is a term *T* denoting a ground for asserting either that *F* fits or that it does not fit with a given operational type, and such that, if *P*'s act can be coded by *T*, then *P* has evidence for, respectively, "*F* fits with the given operational type" or "*F* does not fit with the given operational type".

Closer inspection of expansion 2 easily shows that terms denoting grounds for asserting that *F* fits, or does not fit, with a given operational type can be conceived of as being built in a language of grounds for a language of grounds with implication and quantification over ground variables. Suppose now that *F* does fit with a given operational type, and let *T* denote a ground for asserting this fact. According to (K), when *T* is non-canonical, and *P* seeks evidence for asserting that *F* fits with that type, it is not sufficient for *P* to have built *T*. In fact, *P* must also have evidence for "*T* denotes a ground for asserting that *F* fits with the given operational type", which may result in an infinite regress. In contrast, if *T* is canonical, the mere building of *T* is sufficient to have evidence for asserting that *F* fits with the intended operational type; the problem in this case is that there is no guarantee that *P* can enter a building relation with *T*. As shown by expansion 2, the primitive operational symbol with which *T* begins could range on a term *U* denoting an operation on grounds of a certain operational type, namely a ground for a certain hypothetical or open judgement or assertion. However, in order to build *T*, *P* must apply to *U* the primitive operational symbol with which *T* begins; since the only condition under which *P* can perform such an act is that *P* has evidence for asserting that the operational symbol with which *U* begins fits with an appropriate operational type, an infinite regress may again result.

Similarly, with (2) we have to distinguish between proofs ending with introduction inferences and proofs ending with inferences in non-introductory form. However, since the acts of which such inferences consists are described respectively by canonical or non-canonical terms, we will obtain a picture in every respect similar to the one described in the previous paragraph, and one that is undermined by possible infinite regresses.

13.5 Concluding Remarks

If one does not agree with Prawitz's answer to the decidability issue, and maintains that both (S) and (K) are reasonable, one should explain the reasonings involved in (S) in a different way than (1) or (2). In other words, one should find appropriate epistemic objects or acts that are neither terms denoting grounds nor proofs – or at least, not simply objects or acts of this kind – or else find a feasible relation that does not merely involve building a term or carrying out a proof – or at least, not simply a relation of this kind (a proposal in this direction can be found in Usberti 2015, 2017). If one is forced to adopt (1) or (2), the only viable solution seems that of singling out 'inferences that give evidence for their conclusions without it necessarily being known that they give such evidence' (Prawitz 2015, 97).

Whether the suggested solution is viable, depends of course on how Prawitz's theory of grounds is formalized; this work provides only *partial* calculus, and further refinements are of course required. More to the point, when setting up a full-scale system, one should specify the general form of a defining equation for a non-primitive operational symbol. Only two options seem to be available: either a defining equation has no restrictions apart from its being constructive, thus behaving *mutatis mutandis* like a reduction in the old proof-theoretic framework (Prawitz 1973), or its complexity is limited by certain parameters in a way that enables the decidability issue to be overcome.

Acknowledgements This work has been carried out thanks to the support of the A*MIDEX grant (n ANR-11-IDEX-0001-02) funded by the French Government "Investissement d'Avenir" program.

References

Cozzo, C. 2015. Necessity of thought. In *Dag Prawitz on proofs and meaning*, ed. H. Wansing, 101–120. Heidelberg: Springer.

Cozzo, C. 2016. Inference and compulsion. In *Second pisa colloquium in logic, language and epistemology*, ed. E. Moriconi. Pisa: ETS, forthcoming.

Crocco, G., and E. Audureau 2003. Intuitionisme et Constructivisme chez Brouwer. In *Calcul et formes de l'activité mathématique*, ed. J. Boniface, 190–217. Paris: Ellipses.

Díez, G.F. 2000. Five observations concerning the intended meaning of the intuitionistic logical constants. *Journal of Philosophical Logic* 29(4): 409–424.

Došen, K. 2015. Inferential semantics. In *Dag Prawitz on proofs and meaning*, ed. H. Wansing, 147–162. Heidelberg: Springer.

Gentzen, G. 1934–1935. Untersuchungen über das logische Schließen. *Matematische Zeitschrift* XXXIX: 176–210, 405–431.

Heyting, A. 1956. *Intuitionism. An Introduction*. Amsterdam: North-Holland.

Howard, W. 1980. The formula-as-types notion of construction. In *To H. B. Curry: Essays on combinatoriy logic, lambda calculus and formalism*, ed. J. R. Hindley and J. P. Seldin, 479–490. London: Academic.

Kreisel, G. 1962. Foundations of intuitionistic logic. In *Logic methodology and philosophy of science*, ed. E. Nagel, 198–212. Stanford: Stanford University Press.

Prawitz, D. 1973. Towards a foundation of a general proof-theory. In *Logic methodology and philosophy of science IV*, ed. P. Suppes, 225–307. Amsterdam: North-Holland.

Prawitz, D. 1977. Meaning and proofs: On the conflict between classical and intuitionistic logic. *Theoria* 43: 2–40.

Prawitz, D. 2006. *Natural deduction. A proof theoretical study*. New York: Dover Publications.

Prawitz, D. 2009. Inference and knowledge. In *The Logica Yearbook 2008*, ed. M. Pelis, 175–192. London: College Publications.

Prawitz, D. 2012. The epistemic significance of valid inference. *Synthese* 187(3): 887–898.

Prawitz, D. 2013. Validity of inferences. In M. Frauchiger, *Reference, rationality, and phenomenology: Themes from Føllesdal*, 179–204. Ontos Verlag: Dordrecht.

Prawitz, D. 2014. An approach to general proof theory and a conjecture of a kind of completeness of intuitionistic logic revisited. In *Advances in natural deduction*, ed. L. C. Pereira, E. H. Haeusler, and V. de Paiva, 269–279. Heidelberg: Springer.

Prawitz, D. 2015. Explaining deductive inference. In *Dag Prawitz on proofs and meaning*, ed. H. Wansing, 65–100. Heidelberg: Springer.

Schroeder-Heister, P. 2006. Validity concepts in proof-theoretic semantics. *Synthese* 148(3): 525–571.

Tranchini, L. 2014. Proof-theoretic semantics, proofs and the distinction between sense and denotation. In *Journal of Logic and Computation*. https://doi.org/10.1093/logcom/exu028.

Usberti, G. 2015. A notion of C-justification for empirical statements. In *Dag Prawitz on proofs and meaning*, ed. H. Wansing, 415–450. Heidelberg: Springer.

Usberti G. 2017. Inference and epistemic transparency. In "*Inferences and proofs*", special issue of TOPOI, ed. G. Crocco, and A. Piccolomini d'Aragona, https://doi.org/10.1007/s11245-017-9497-1.

Chapter 14
Models in Search of Targets: Exploratory Modelling and the Case of Turing Patterns

Axel Gelfert

Abstract Traditional frameworks for evaluating scientific models have tended to downplay their exploratory function; instead they emphasize how models are inherently intended for specific phenomena and are to be judged by their ability to predict, reproduce, or explain empirical observations. By contrast, this paper argues that exploration should stand alongside explanation, prediction, and representation as a core function of scientific models. Thus, models often serve as starting points for future inquiry, as proofs of principle, as sources of potential explanations, and as a tool for reassessing the suitability of the target system (and sometimes of whole research agendas). This is illustrated by a case study of the varied career of reaction-diffusion models in the study of biological pattern formation, which was initiated by Alan Turing in a classic 1952 paper. Initially regarded as mathematically elegant, but biologically irrelevant, demonstrations of how, in principle, spontaneous pattern formation could occur in an organism, such Turing models have only recently rebounded, thanks to advances in experimental techniques and computational methods. The long-delayed vindication of Turing's initial model, it is argued, is best explained by recognizing it as an *exploratory tool* (rather than as a purported representation of an actual target system).

Keywords Exploratory models · Scientific modelling · Models · Turing patterns · Reaction-diffusion systems

14.1 Introduction

It is a recurring feature of contemporary philosophical writing on scientific models that it begins by recognizing their heterogeneous and diverse nature. "Many different things", we are told, "can serve as models including physically constructed

A. Gelfert (✉)
Department of Philosophy, Literature, History of Science and Technology, Technical University of Berlin, Germany
e-mail: axel@gelfert.net

A. Christian et al. (eds.), *Philosophy of Science*, European Studies in Philosophy of Science 9, https://doi.org/10.1007/978-3-319-72577-2_14

scale models, model organisms, and mathematical objects such as sets of trajectories through a state-space" (Weisberg 2007, 216–217). It has been argued persuasively that, in spite of such diversity, models are instruments of inquiry in their own right, enjoying partial autonomy from both theory and data. Thus understood, models are neither mere approximations to an 'underlying' fundamental theory, nor are they simply a convenient 'shorthand' for aggregating observations. As Margaret Morrison and Mary Morgan put it, models, in a manner of speaking, are located "outside the theory–world axis" (Morrison and Morgan 1999, 17); their construction "involves a complex activity of integration" (Morrison 1999, 44). While attention to the specifics of case studies has vastly improved our understanding of what models are, and how they are being constructed, when it comes to their evaluation and testing, more traditional criteria are typically being invoked. Thus, models are thought to be "inherently intended for specific phenomena" (Suárez 1999, 75) and, importantly, are to be judged by their ability to predict, reproduce, or explain observational data. Indeed, in contexts that require selecting one of a set of candidate models, classical hypothesis-testing remains an important 'intuition pump' for how we ought to assess choices in modelling (e.g. Steele and Werndl forthcoming), thereby subsuming the assessment of models under well-understood – but ultimately limiting – frameworks for the evaluation of theoretical hypotheses.

In this chapter, I shall argue that traditional frameworks for evaluating models have tended to downplay their *exploratory* function; or, to give my thesis a more positive spin, I shall make the case that *exploration* should stand alongside explanation, prediction, and representation as a core function of scientific models. This is not to deny that, often, scientific models are intended for specific observed phenomena, nor is it to claim that all modelling must be inherently exploratory; rather, it is intended to draw attention to the – largely overlooked – role of models and modelling in the context of exploratory science. Given the complementary nature of the goal of establishing exploration *alongside* (rather than *in place of*) the more traditional functions of scientific models, the main focus of this chapter is to bring out with as much clarity as possible what we stand to gain from admitting exploration as a criterion for the evaluation of models. The overall thrust of the present paper, thus, is a constructive one, and little space will be devoted to the critique of more traditional criteria of assessment. (A comprehensive survey and critique of traditional accounts of modelling, along with a proposal of a more practice-oriented functional approach, is given in (Gelfert 2016).) When formulated at this level of generality, the goal of establishing models as – at least in part – exploratory tools in science might seem vague and unspecific. At the very least, it calls for significant clarification, which is what the next two sections will attempt. Sect. 14.2 ('Exploration, heuristics, and the (im)possibility of "theory-free" science') positions the present paper in relation to the – by now, well-established – philosophical debate about *exploratory experimentation*, and tries to demarcate the notion of 'exploration' from the concept of 'heuristics', which has been the topic of a long-standing debate about the place of heuristic reasoning in science. In particular, it is argued that exploratory inquiry exhibits structure and is characterized by recurring strategies, both of which lend themselves to analysis and assessment in terms of their adequacy and fruitfulness. Section 14.3 ('Functions and uses of exploratory models') follows recent analyses by distinguishing between four main exploratory uses of models: as starting points for future inquiry, as proofs of principle, as sources of potential explanations, and as a tool for reassessing the suitability of the target system.[1] These four aspects of exploratory modelling are neither intended to be exhaustive, nor are they mutually exclusive. Indeed, as I shall argue, the role of exploratory models as *proofs of principle* can be understood as addressing both methodological issues and explanatory desiderata. Sometimes what needs to be demonstrated is the feasability of a particular (e.g., mathematical) approach, whereas on other occasions the proof of principle consists in specifying, by way of example, a 'how-possibly' explanation.

Section 14.4 ('The case of reaction-diffusion models for biological pattern formation') undertakes a detailed case study of how the various functions of exploratory modelling interact, and work out, in practice. This is illustrated by the varied career of reaction-diffusion models in the study of biological pattern formation. First proposed by Alan Turing in 1952, the basic idea is that pattern formation in biological systems may arise from the interplay of two chemical substances ('morphogens'), one of which is locally activated, while the other gives rise to long-range inhibition, with both differing in their diffusion rates. As a result, such Turing models predict distinct spatial patterns of different 'chemical wavelengths', corresponding to the concentration of the two morphogens, which in turn may trigger the expression of different phenotypes. Initially regarded as a mathematically elegant, but biologically irrelevant, proof of how, in principle, spontaneous pattern formation could occur, the model was subsequently obscured by other, more intuitively representational and empirically well-supported models. Only recently, due in part to the development of better computer simulation techniques, has the Turing model rebounded and has come to be recognized as contributing to our understanding of real-world cases of biological pattern formation. Exploration, then, holds out the promise of, but neither requires nor guarantees, long-term applicability of a model to specific real-world target systems.

14.2 Exploration, Heuristics, and the (Im)possibility of 'Theory-Free' Science

On what is little more than a caricature view of the scientific method, science progresses through a series of bold hypotheses, from which predictions are deduced which are subsequently put to the test, leading us to give up our hypotheses when the predictions are found to be flawed, or to hold on to them – albeit only tentatively – for as long as our hypotheses pass whatever empirical tests

[1] Section 14.3 follows my discussion in (Gelfert 2016, 71–99).

we subject them to. The limitations, both descriptive and normative, of such extreme hypothetico-deductivism have, of course, long been a topic of philosophical discussion, and one would be hard-pressed, in this day and age, to find *bona fide* hypothetico-deductivists in large numbers. Yet, elements of this view have proved remarkably stubborn, not least the idea that the only way to assess science is by looking at its past track record of predictions and explanations (rather than, say, by developing a measure of what, following Thomas Nickles, one might call its "generative potential" (Nickles 1985, 184)). Furthermore, *testing* – that is, the practice of bringing about circumstances that allow researchers to compare reality against previously derived, empirically observable consequences of a theory – becomes the gold standard for evaluating hypotheses, thereby neglecting the fact that "scientists sometimes reason their way *towards* hypotheses" and not "*from* hypotheses" towards testable conclusions (Hanson 1960, 105).

The issue has traditionally – sometimes misleadingly – been characterized in terms of Hans Reichenbach's distinction between the "context of discovery" and the "context of justification", with only the latter being deemed capable of being rationally assessed. Research activities that precede the explicit formulation of hypotheses and theories, on this caricature view, would be considered to be of merely auxiliary value, subject to psychological, social, and other circumstantial factors which might perchance lead a scientist to stumble upon an improbable, yet ultimately fruitful, idea, but which could not easily be generalized to result in methodological guidelines for what constitutes good science. It goes (almost) without saying that this stark opposition between a (rationally assessable) "context of justification" and an (unruly and arbitrary) "context of discovery" is philosophically indefensible. As numerous authors have argued, it is based on conflating various aspects of scientific practice and its philosophical analysis. For one, it equates 'discovery' with the generation of novel ideas, as in the much-cited example of Friedrich August Kekulé's (possibly apocryphal) daydream of the snake eating its own tail, which led him to hypothesize the novel chemical structure of the benzene ring. Yet, as Theodore Arabatzis has forcefully argued, a mere novel hypothesis – even if it later turns out to be true – does not yet qualify as a discovery, since 'discovery' is a success term and, as such, needs to be grounded in justificatory activity: "The context of discovery is 'laden' with the context of justification because 'discovery' is a term which refers to an epistemic achievement: if one succeeds in discovering something then, no doubt, this something exists." (Arabatzis 2006, 217) Instead of a binary distinction between two contexts, what one finds, then, is a spectrum of perspectives, each highlighting different (normative and descriptive) aspects and desiderata; any attempt to tease these apart will lead not to a single, but to a multiplicity of distinctions.[2]

[2]Paul Hoyningen-Huene, in a much-cited paper, distinguishes between five different versions of how one might contrast discovery and justification in scientific practice. See (Hoyningen-Huene 2006).

An overly narrow view of the relation between discovery and justification would be especially restrictive when turning one's attention to the early stages of scientific inquiry, more specifically: to *exploratory* modes of inquiry. Labelling an episode of scientific research 'exploratory' is intended to convey more than just a sense of its priority in the chronological order of events. Rather, the label 'exploratory' pertains to a particular mode of doing science: one that aims at getting a grasp of a phenomenon or scientific problem in the absence of a well-understood and workable theory of the domain in question. For a theory to be well-understood and workable, it is not, of course, required that it be true, nor that it must aim for a *reductionist* account of whatever phenomenon is under investigation. Rather, the idea is that, in those cases that have traditionally received the most attention in philosophy of science, a significant prior body of theoretical knowledge can be assumed to be available, which in turn suggests – not by itself, of course, but in the hands of able scientists – a way of rendering the phenomenon theoretically tractable, at least in principle. By contrast, in exploratory research the existence of such a body of theoretical knowledge cannot be assumed, or is itself at issue. This means that, as analysts of the scientific process, we cannot readily turn to familiar patterns of making sense of what is going on in terms of applying theoretical knowledge to a particular case (or a particular class of cases).

Nowhere is this more obvious than in cases of *exploratory experimentation*, which have garnered considerable interest from historians and philosophers of science over the past 20 years or so. In 1997, Friedrich Steinle and Richard Burian, independently of one another, published papers arguing for 'exploratory experimentation' as a distinctive mode of experimental inquiry in science. Where Steinle drew on the example of nineteenth-century electromagnetism to characterize exploratory experimentation as a research activity driven by "the elementary desire to obtain empirical regularities and to find out proper concepts and classifications by means of which those regularities can be formulated" (Steinle 1997, S70), Burian cited work in mid-twentieth century histochemistry as aiming for a *stabilization* of phenomena, notably via "stabilization of the protocols for locating particular molecular species" and for rendering them "relevant to the experimental and theoretical analyses of such other investigative traditions" (Burian 1997, 42) as may be available. Exploratory experimentation, as Steinle puts it, cannot be divorced from "the process of forming and stabilizing the [new] conceptual framework" (Steinle 1997, S72), which – needless to say – is made all the more difficult by the fact that, typically, in exploratory science "no well-formed theory or even no conceptual framework is available or regarded as reliable" (Steinle 1997, S70). C. Kenneth Waters gives a rather succinct, yet insightful characterization of the relation between exploration and the absence (or at least inaccessibility) of fundamental theory, when he writes that "*the aim of exploratory experiments is to generate significant findings about phenomena without appealing to a theory about these phenomena for the purpose of focusing experimental attention on a limited range of possible findings*" (Waters 2007, 279) (italics original).

As this brief characterization makes plain, exploratory experimentation cannot easily be – and certainly should not be! – assimilated to a view of science that regards experiments as exclusively, or even just primarily, in the business of testing theoretical hypotheses. For one, as the various discussants cited above agree, no well-formed body of theoretical principles from which such hypotheses may be derived exists in exploratory settings, and sometimes even the conceptual frameworks themselves are part of what is at issue. *Testing*, then, can at best be a tentative and local affair, and will necessarily be heavily invested with contentious background assumptions, since it will not generally be the case that one can simply hold the "theoretical" part of one's exploratory investigations fixed. This is why, as Kevin Elliott puts it, exploratory experimentation is perhaps best understood as "an attempt to study a phenomenon using as many tools and techniques as possible so as to understand it more fully and to gain more solid epistemic access to it" (Elliott 2007, 328). Necessarily, this process of getting a grasp on a phenomenon will often require multiple attempts and approaches from different directions. Hans-Jörg Rheinberger captures the spirit of such exploratory inquiry well, albeit in somewhat metaphorical language, when he insists that an "experiment is not only a test or an examination", but more frequently "is an exploratory movement, a game in which one plays with possible positions, an open arrangement" (Rheinberger 2010, 247). However, unlike in a well-defined game, where the rules specify when the game is over and who the winner is, in exploratory settings there is no guarantee that, at the end of an episode of inquiry, we will find ourselves having obtained a "winning" (true) hypothesis – even if exploration is carried out in the hope of at least identifying a fruitful approach. This reflects the hope, long cultivated by proponents of the heuristic value of abductive reasoning in science, that in contexts where it is unclear what is required in order to show a hypothesis H to be true (or false), we may yet be able to gain a better sense of "what is required for deciding that H constitutes a plausible *kind* of conjecture" (Hanson 1960, 93).

At the same time, it would be misleading to think of exploration as devoid of *all* theory. Theory can enter at various levels, sometimes explicitly so, without thereby reducing exploratory experimentation to merely a means of 'testing' theoretical hypotheses. For one, significant background knowledge – including background theory – will often be required for devising experiments if the latter are to improve our understanding of the phenomena that stand in need of 'stabilization'. Furthermore, interpreting data and comparing experimental designs will likewise often be informed by theoretical considerations, even when the latter do not drive interpretation and design. Finally, as Elliott has noted, theory often "plays the role of a starting point of a 'foil' in the exploratory process" (Elliott 2007, 327); this observation is borne out by a growing number of historical case studies of exploratory experimentation across the physical and biological sciences. Thus, while exploratory research often proceeds in the absence of a fully developed theory, is not driven by pre-existing theoretical concerns, and does not primarily aim at testing theoretical predictions, it need not (and perhaps could not) be entirely theory-free either.

In order to avoid possible misunderstanding, it is perhaps worth drawing two further contrasts, with *abduction* and *heuristics*, respectively. Though closely related to *exploration* – insofar as both notions also aim to shed light on the interplay between discovery and justification – they highlight somewhat different aspects of the process of scientific inquiry and innovation. In its most generic form, abduction may be described as the process of inferring hypotheses (purported facts or, in science, laws of nature) that, if true, would render certain sentences plausible. Thus, if the latter report observations that have been made, abduction is of *explanatory* use – the inferred hypotheses are (potential) explanations of what has been observed – whereas if they concern predictions, it may be a powerful tool for unearthing new *discoveries*. Though the term 'abduction' is sometimes reserved for the process of *generating* plausible hypotheses (irrespective of their relative merits), more often than not it is equated to the process of *inference to the best explanation*, which has an obvious evaluative component. 'Abduction', then, may refer both to the generation of hypotheses concerning a given empirical phenomenon and to the process of assessing their relative merit. Yet note that both of these established usages take the existence of the explanandum for granted: what is at issue is not whether a stable phenomenon exists, but what explains it. By contrast, in contexts of exploratory research, it is often an open question whether or not a series of observations exhibits sufficient unity to count as a stable research object in the first place; exploration, in such a situation, may consist in probing the (experimental and/or theoretical) conditions under which one can hope to zero in on a purported phenomenon. In this sense, exploration may very well precede any explicit attempt at abductive inference.[3] Regarding the second contrast, one first needs to distinguish between the use of *heuristics* as mental shortcuts in judgment and decision-making – which, as a fact about human psychology, can be expected to play an important role in exploration as much as in any other domain of human cognition and behaviour – and the *heuristic appraisal* specific to science, which "evaluates the promise, the future potential (including what is at stake), the problem-solving capacity, or what we might call the 'opportunity profile' of a claim, technique, proposal, etc." (Nickles 2006, 161). To the extent that both may feature in science, they are concerned with (and are acknowledged by their proponents to be a matter of) "considerations relating to economy of research" (Nickles 2006, 159) and to the efficient allocation of (material or temporal) resources. As such, they figure in "all stages of research" (Nickles 2006, 165), not just in exploratory research. Though heuristic appraisal may *encourage* researchers to pursue exploratory strategies, since doing so may – at various points in scientific inquiry – be deemed the most promising allocation of resources, it is not itself identical to *exploration*. Exploration, in a nutshell, cannot be easily assimilated to either abductive inference or heuristic appraisal.

[3]Lorenzo Magnani has proposed the notion of "manipulative abduction" (Magnani 2004), which appears to have greater affinity to exploration as discussed in this chapter than the standard "theoretical abduction"; on this point see also (Gelfert 2016, 116, fn. 6).

14.3 Functions and Uses of Exploratory Models

While exploration precedes explicit hypothesis-testing and the emergence of a settled theory of the phenomena in question, it need not – as the discussion in the previous section makes plain – be shielded from rational analysis. Even as the criteria for what constitutes fruitfulness may vary across subdisciplines, agreement as to whether a given exploratory move has been more or less fruitful can often be reached, sometimes quite straightforwardly. Why is this so? In the case of exploratory experimentation, a strong case can be made that in spite of its independence from specific theoretical frameworks regarding the target phenomenon, "experimental activity may well be highly systematic and driven by typical guidelines". Such guidelines, as Steinle puts it, are of a "general methodological type" (Steinle 1997, S70). At a descriptive level, they will be instantiated by recurring exploratory strategies; that is, by typical moves an experimenter can make in an attempt to gain an epistemic foothold. Some of these may seem trivial – an experimenter might, for example, systematically vary the different experimental parameters to see which changes have an influence on the phenomenon under investigation – others may require considerable ingenuity, such as trying to isolate hidden factors in order to find out which are superfluous and which are indispensable. Crucially, which exploratory strategies are promising, and which constraints they are operating under, will depend on the type of inquiry – e.g., measuring, experimentation, modelling, computer simulation etc. – and on the particulars of a given case. For example, when performing a field experiment in ecology, isolating a putative causal factor, or varying experimental parameters, will arguably take great skill and require significant resources; by contrast, when numerically modelling a phenomenon using a polynomial equation, systematically varying the model parameters may border on the trivial. As in the case of experimentation, then, exploratory uses of models need to constantly (or at least regularly) be assessed for their fruitfulness and suitability to the task at hand.

A close affinity between modelling and exploration is only to be expected, given the origins of scientific models in what, at the time, were called 'mechanical analogies'. When, for example, James Clerk Maxwell set out to get a grasp of electromagnetic phenomena, he proposed his molecular vortex model which, in retrospect, may be characterized as an attempt to model the electromagnetic field in mechanical terms. Maxwell is explicit about the exploratory role of his vortex model: anyone who understands its "provisional and temporary character", he argues, "will find himself rather helped than hindered by it in his search after the true interpretation of the phenomena" (Maxwell 1890, 486). That his model was not merely a prop for the theoretical imagination, in the way that a sketch of a geometrical figure might be for a mathematician trying to construct a proof, is evident from the fact that Maxwell frequently moved back and forth between viewing the model as auxiliary and crediting it with new ontological content which allows it to guide our inquiry into the nature of electromagnetic phenomena. (See e.g. (Harman 1998).) In such cases, to adapt a passage written by Herman Meyer as early as 1951, "the construction of scientific models" itself is an instance of our "specifically human way of progressing from phenomenological description to *scientific knowledge*"; it is, Meyer argues, "a mental operation *sui generis*" (Meyer 1951, 118). As we shall see, this point – that the *construction* of a model, irrespective of its detailed empirical performance, can afford knowledge – can be made more precise by highlighting the role of exploratory models in constructing what one might call *proofs of principle*.

Before turning to a list of some of the distinct functions that exploratory models can serve, it is worth highlighting just one of the many strands in philosophical theorizing about what models are. (For a review of the ontology of models, see (Gelfert 2017).) The starting point for this way of thinking about scientific models is somewhat akin to Meyer's characterization of model construction as a "mental operation" in its own right, in that it ties scientific models closely to the cognitive processes involved in *mental modelling*. The recognition that much of human reasoning proceeds with 'mental models', by carrying out thought experiments on internal representations of scenarios (rather than by rigorously applying logical rules to propositional representations), has led some philosophers of scientific models, for example Nancy Nersessian, to the view that a mental model is "a structural analog of a real-world or imaginary situation, event, or process", where this is meant to convey that "it embodies a representation of the spatial and temporal relations among, and the causal structures connecting the events and entities depicted and whatever other information that is relevant to the problem-solving tasks" (Nersessian 1999, 11). There is, on this view, considerable continuity between the way we construct scientific models and the way we represent the world around us in thinking, whether we actively seek to represent real-world situations or, as in the quote above, imaginary situations, events, or processes. Given this symmetry between representing real and imagined, actual and possible scenarios, it should not come as a surprise that models – including, importantly, scientific models – allow us to extrapolate beyond the actual, thereby allowing us to also explore possible, e.g. counterfactual, scenarios. The use of models, then, is not restricted to the initial goal of representing actual target systems. Indeed, some classes of scientific models – such as toy models – typically only aim to provide *potential* explanations of general patterns, which happen to be *actually* instantiated, without thereby claiming to be able to identify the actually operative causal or explanatory factors.[4] For present purposes, I shall neither pursue different accounts of what, in general, scientific models are, nor limit myself to just one class of models; rather, I shall focus on some important *functions* of exploratory modelling, giving brief examples of each, before turning to an in-depth example, of models of biological pattern formation, in the next section.

It is important to preface the subsequent discussion with a disclaimer to the effect that any list of exploratory functions of scientific models is likely to be incomplete

[4]On this point, see (Reutlinger et al. forthcoming).

and tentative. In distinguishing between four general functions of model-based exploration, I am following recent analyses by emphasizing that these functions are neither exhaustive nor mutually exclusive.[5] Indeed, as we shall find, the same model may sometimes serve different exploratory goals at the same time. Another disclaimer concerns the fact that the models discussed here, for the most part, are formulated in the language of mathematics. While this is not by chance, it should not be taken to imply that exploration, in any of the senses to be discussed, is restricted to mathematical models only. A full explanation of why mathematical models lend themselves so well to exploratory uses is beyond the scope of this paper and would require close engagement with the literature concerning the applicability of mathematics to the natural sciences and with the role of well-developed symbolic formalisms in guiding, and constraining, the making of inferences.[6] For present purposes, and in order to defuse worries about the choice of examples, perhaps it will suffice to note that scientists themselves – including those from the non-exact sciences – frequently presuppose this connection indirectly, and sometimes note it quite explicitly. Thus, in one of the rare explicit discussions of the exploratory role of models, the ecologist Peter J. Taylor notes that, while many theoretical and conceptual tools in biology function as *schemas*, i.e. as way of highlighting relevant basic processes, constraints, or (e.g. reproductive) strategies, "if the schema can be expressed in a mathematical formulation, the model becomes what I call an *exploratory tool*". Mathematical investigation, Taylor notes, in a manner reminiscent of Steinle's exploratory strategy of varying experimental parameters, allows biologists to study how "the system's behaviour changes as its parameters change or become variables, as time lags are added, and so on" (Taylor 1989, 122). In doing so, it allows for the formulation of new research questions or may suggest new terms and classifications, e.g. of systems-level behaviours.

In distinguishing between different *exploratory functions* of models, it makes sense to proceed from 'weaker' to 'stronger' versions, where this does not imply that any such ordering is either necessary or unique. Rather, for reasons of exposition, it will be helpful to realize that there is a spectrum of exploratory strategies, some of which are more ambitious than others. At the 'weak' end of the spectrum, then, we find exploration being conducted in the hope of identifying fruitful ways of proceeding, in spite of the absence of a well-formed underlying theory. As William Wimsatt notes in relation to scientific models, an "oversimplified model may act as a starting point in a series of models of increasing complexity and realism" and may serve "to generate new predictive tests of or to give new significance to features of an alternative preferred model" (Wimsatt 2007, 104/127). This first, relatively weak function of exploratory modelling may thus be described as *aiming at starting points for future inquiry*. One might worry that, in the early stages of research, virtually any move by a scientist or modeller, in some shape or form, aims

[5]On this point, see also (Gelfert 2016, 95).

[6]I discuss these issues in more detail in (Gelfert 2016), (Gelfert 2014), (Gelfert 2011), and (Gelfert 2009).

at enabling future inquiry, so that – in the absence of additional criteria for what should count as exploration – even "back-of-the-envelope" sketches would count as 'exploratory models' in this first sense. This may be so, but simply means that we may need to exercise a certain amount of self-restraint in deploying the label. History and context will usually allow us to distinguish quite clearly between mere uninformed guesswork and fruitful, informed exploration that led to increasingly realistic and sophisticated models. Consider the case of models of traffic flow in sociodynamics. Early such models, which looked towards fluid dynamics for inspiration, were unsuccessful at explaining various features of the flow of human-operated vehicles. By the middle of the twentieth century, it had become clear that, for a model of car traffic to have any chance of being successful, it would need to include a variety of disparate influences, from physical quantities (such as velocity, acceleration, geometry of the vehicles) to psychological phenomena (e.g., reaction time of drivers). In 1953, the American engineer Louis Pipes proposed the first *car-following model* (Pipes 1953), which was based on the idea that traffic flow is nothing but the cumulative effect of each driver, n, responding to the car in front of her, $n - 1$. The corresponding positions can then be modelled, with d being the distance between the two vehicles at rest, l_{n-1}^{veh} the length of the car in front, and Tv_n the (velocity-dependent) 'legal distance', in the form of the following equation:

$$x_{n-1} = x_n + d + Tv_n + l_{n-1}^{veh}.$$

It is clear that this equation is neither fundamental nor complete, but – as the subsequent proliferation of more refined car-following models demonstrates (see Wageningen-Kessels et al. 2015) – it proved a fruitful starting point for thinking of traffic as the cumulative effect of how drivers respond to each other.

The second and third functions of exploratory modelling I wish to highlight – viz., providing *proofs of principle* and *potential explanations* – can, without much loss in terms of the sharpness of this distinction, be illustrated using the same example from population biology, the Lotka-Volterra model of predator-prey dynamics. This model has become the textbook explanation for one of the most intriguing phenomena in population biology, namely the correlated oscillations in the size of populations of predator and prey species. Mathematically, the model consists of a pair of first-order, non-linear, differential equations linking the population size of one species to that of the other, in such a way as to reflect that one species is feeding on the other. The rate of change in each population is thus dependent on the other, but not in exactly the same way. Since the prey is typically a fast-reproducing species, its population size is primarily determined by reproduction (proportionate to the existing population size) and mortality due to predation (proportionate to its own population size and to that of the predator species). By contrast, the predator species is modelled as having a constant death rate, so the total number of deaths is proportionate to its population size, while its population growth (=the total number of births) is assumed to be proportionate to both its own population size and to that of the prey which, after all, sustains the predator species. Mathematically, this is represented as follows (with x indicating the size of the prey population, y the

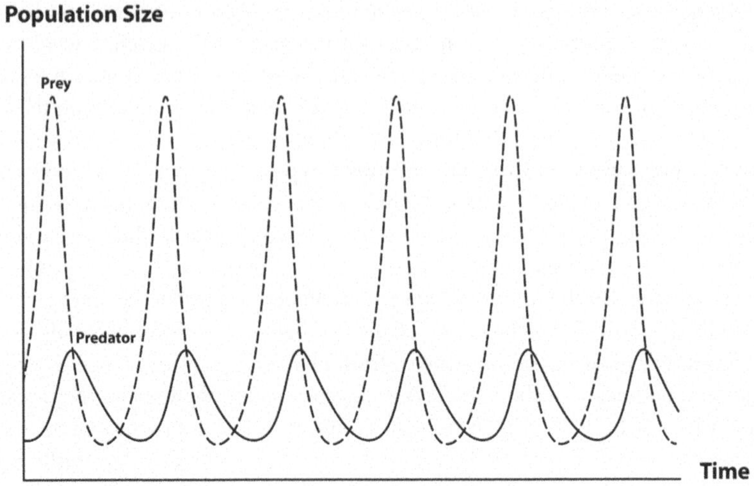

Fig. 14.1 Lotka-Volterra model

number of predators, t time, a the prey's birth rate, b the predator's death rate, and α, β positive coefficients representing the effect each population has on the other):

$$\frac{dx}{dt} = x\,(a - \alpha y)$$

$$\frac{dy}{dt} = y\,(\beta x - b)$$

Because of the way the variables representing the two population sizes depend on each other, they exhibit a remarkable behaviour in time: both populations will oscillate indefinitely, the predator population lagging behind slightly, while the prey population overshoots quite dramatically. (See Fig. 14.1) No stable equilibrium exists, only an unstable one that could not withstand the slightest (and, in any real-world situation, inevitable) perturbation. In other words, the Lotka-Volterra model exhibits a distinctive, cyclical pattern: When the number of predators is low, the prey population will increase rapidly, even as the predator population begins to recover, which in turn will grow and will eventually bring down the total number of prey below the number needed to sustain the (now increased) predator population.

What makes the Lotka-Volterra model significant as a tool of exploration is its dual character as a methodological *proof of principle* and a *potential explanation*. Consider the methodological aspect first. At any given point in time, the size of a population will be an integer number: there may be forty-five lynx roaming the forest, or four hundred twenty-two, but not 1.33 or $\sqrt{2}$. Changes in population size, likewise, come in units of one animal at a time: a live birth adds a member (or perhaps several) to the population, each death brings down the total number by one. Furthermore, births and deaths are discrete events in time, which raises the question of whether one should perhaps model population dynamics using discrete time steps. Yet, the Lotka-Volterra model demonstrates that it *is* possible to reproduce the qualitative behaviour of predator-prey systems using continuous differential equations; that is, it constitutes a *proof of principle* that the methodology of continuous differential equations is suitable for generating insight into the dynamics of (discrete) populations. But the Lotka-Volterra model does more than prove the point that we do not always need a discretized model to make sense of how populations behave, it also offers a *potential explanation* of an empirically salient phenomenon: the cyclical patterns of population growth, followed by near-collapse, in predator-prey systems. These patterns, observed across various species in different ecosystems – e.g., in lynx and snowshoe hares in the Hudson's Bay area (inferred from hunting data) as well as in sharks and their prey (as reflected in the catch of fishermen in the Adriatic Sea) – had long been regarded as puzzling and as standing in need of an explanation. The Lotka-Volterra model *explains* how it is possible that such stark cyclical patterns, of rapid population growth periodically followed by steep decline, can emerge spontaneously, from within the two-species system containing the predator and prey species, without the need to posit any external forcings such as famine or disease.

A perhaps less obvious, but nonetheless important function of exploratory modelling is the *search for*, or *revision of*, potential target systems. It is common to think of models as tools for representing actual target systems, or indeed as "inherently intended for specific phenomena" (Suárez 1999, 75). From this perspective it may seem puzzling why, when faced with a mismatch between model and target, we should ever consider reassessing *the target system*, rather than revising our model to better fit the target. But recall that in an ongoing episode of exploratory research, it is often not obvious whether one has even succeeded in properly delineating a stable target system. And even if one has, one may find that the model that resulted from aiming at a specific target turns out to be better suited to the study of a neighbouring phenomenon, or that one's model has unexpected applications outside the originally intended domain of phenomena. In other words, what the target of our modelling efforts will eventually be – or, in more extreme cases, whether we can even expect there to be actual target systems that fit – may well remain an open question for as long as exploratory research has not yet become settled science.[7] Sometimes, as the title of this chapter suggests, our models are *in search of* targets, and the next section will develop in more detail one such example of a well-developed set of model equations whose overall applicability to real target systems remains very much a matter of dispute.

[7]It is also important to keep in mind that there is value in modelling counterfactual situations, including scenarios that could not possibly be instantiated in the actual world (such as higher-dimensional systems in physics).

14.4 The Case of Reaction-Diffusion Models for Biological Pattern Formation

Only a couple of years before his untimely death, Alan Turing published an ambitious article, "The Chemical Basis of Morphogenesis" (Turing 1952), which set out "to account for the main phenomena of morphogenesis" by providing "a possible mechanism by which the genes of a zygote may determine the anatomical structure of the resulting organism". Turing was careful to emphasize that he did not wish to "make any new hypotheses" of a biologically substantive kind; rather, he aimed to demonstrate in general terms how it may be the case "that certain well-known physical laws are sufficient to account for many of the facts" of the development of biological form. Previous attempts to apply mathematics to the problem of biological form, notably D'Arcy Thompson's influential treatise *On Growth and Form* (1917), essentially tried to recreate various actual biological shapes in mathematical form and, via the method of transformed coordinates, showed how they can be transformed into one another. (Thompson 1917) Yet, while Thompson's text contains considerable discussion of the constraints on biological forms – arising from differences in the relative importance of the various forces, like gravitation and capillary forces, all operating at different scales – overall there is little concern with modelling potential causal mechanisms. By contrast, Turing explores biological pattern formation by mathematically modelling a *possible mechanism* how biological form develops from the zygote.

Turing's starting point is a general puzzle concerning the origins of biological form. All organisms begin as undifferentiated zygotes that, for all intents and purposes, exhibit no preferred directionality or spatial structure. On the one hand, an embryo in its early (blastula) stage, exhibits "spherical symmetry, or if there are any deviations from perfect symmetry, they cannot be regarded as of any particular importance, for the deviations vary greatly from embryo to embryo within a species, though the organisms developed from them are barely distinguishable" (Turing 1952, 41). On the other hand, fully grown organisms, such as a horse ("which is not spherically symmetrical", as Turing quite rightly notes; ibid.), have considerable internal structure, which defines their morphological characteristics. It might seem puzzling how, at some point during an organism's development, the symmetry and homogeneity of the early embryo can give way to distinctive, and formative, spatial patterns. Part of the puzzlement regarding this transition, however, arises from what Turing calls the "fallacy" of assuming "that the deviations from spherical symmetry in the blastula could be ignored because it makes no particular difference what form of asymmetry there is" (Turing 1952, 42). For, it is important that *some* such deviations will inevitably occur, and if a developing embryo responds by *amplifying* them in systematic ways, it may reach a new and stable equilibrium that lacks the original symmetry and brings determinate spatial patterns into existence. In his paper, Turing aims to specify just such a mechanism, by which small fluctuations can trigger, in a predictable and systematic fashion, pattern formation in what would otherwise be homogeneous systems.

The central idea behind Turing's model is that two (or more) diffusible substances (dubbed "morphogens") can interact with each other in such a way as to form "reaction-diffusion systems" which establish chemical gradients within an organism, which, in turn, give rise to self-organized biological patterns. For such pattern formation to emerge spontaneously in an otherwise homogeneous system, at minimum two linked morphogens are required. If only one morphogen were present, though it could be locally produced and diffuse across an organism, it could at best trigger the expression of a *preformed* pattern, e.g. by "switching on" a gene that was unevenly distributed across the organism (or responded to different concentrations of the *same* morphogen). It could not, however, bring about, say, periodic patterns in an otherwise homogeneous population of cells. Adding a second morphogen changes the situation fundamentally, provided we allow for the possibility of the rates of production and activation of the two substances being coupled to one another. In addition to diffusion, we then have a *reaction* that produces the morphogens, in ways that are not merely superpositions of two independent processes, but are dynamically coupled. Turing's model, thus, includes two morphogens, S and P, diffusing homogeneously (except for random fluctuations) within a certain space, with one being locally self-enhanced and the other being capable of long-range inhibition of the first. Because of the differential diffusion rates of the two molecules, and the way in which the more slowly diffusing molecule both stimulates the production of itself (self-enhancement) and the production of its own inhibitor, a range of possibilities for pattern formation opens up. To see how this may occur, consider a random distribution of the two morphogens at time t_1 and its evolution over time (see Fig. 14.2).

P enhances its own production (e.g. autocalytically) and diffuses slowly, so has a tendency to concentrate into peaks, even without the presence of S (time t_1). Moreover, since P also stimulates production of S, its own (quickly diffusing) inhibitor, these peaks will become more localized as P's concentration will fall quickly as one moves away from a given peak (time t_2). Of course, as S diffuses further in space, its concentration will eventually fall below the level needed to inhibit the formation of further peaks in the concentration of P. Where S is sufficiently diluted, new peaks of P can emerge in the same fashion. The result is a "standing wave" pattern, with localized peaks of P at more or less regular intervals in space (time t_3).

The discussion so far makes it plausible that the mutual interaction between two substances, a (self-)activating morphogen and a quickly diffusing inhibitor which together have the twin effects of local activation and long-range inhibition, can explain spontanenous pattern formation in what would otherwise, barring inevitable fluctuations, be a homogeneous medium. The specific pattern, of course, will depend on the various parameters that characterize the reaction and diffusion process: the rate of production of each substance (and its dependence on the concentration of the other), the rate of degradation and the differential speed with which each substance

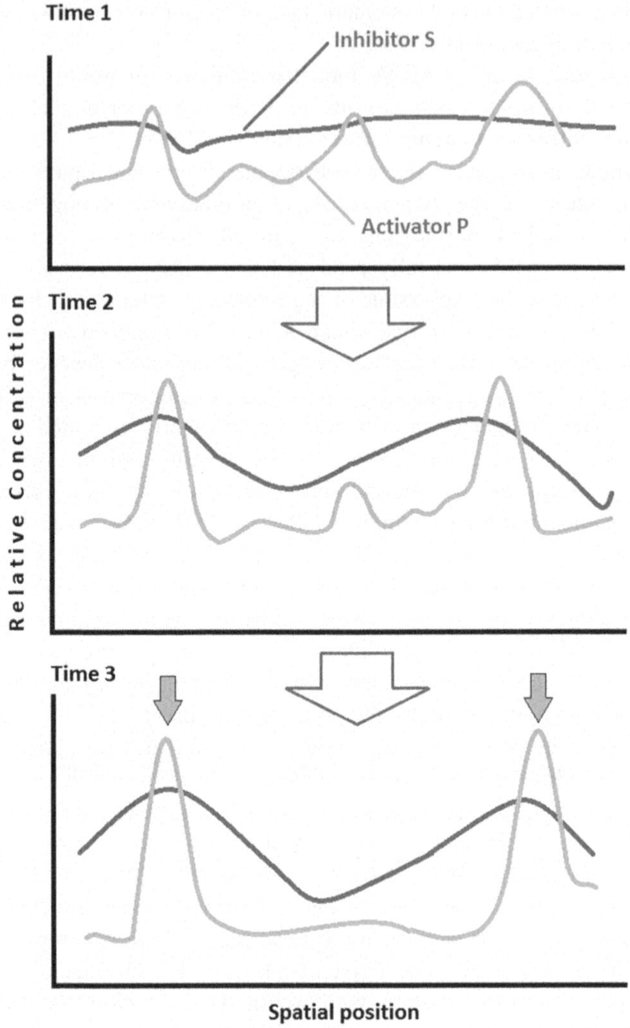

Fig. 14.2 Reaction-diffusion time evolution

diffuses. Mathematically, this is captured by the following set of equations, which express the rates of concentration change as the sum of reaction terms for the production, degradation, and diffusion of the two morphogens, respectively:

$$\frac{\partial u}{\partial t} = F(u, v) - d_u v + D_u \Delta u$$

$$\frac{\partial v}{\partial t} = G(u, v) - d_v v + D_v \Delta v$$

Fig. 14.3 Turing patterns (Image courtesy of Shigeru Kondo)

where u and v refer to the concentrations of the P and S, respectively. Not all mathematically stable solutions to the model equations are associated with the formation of salient and stationary patterns; some give rise to oscillatory waves or uniform distributions. Those solutions that essentially correspond to stationary waves with finite wavelength – or, *Turing patterns*, as they are now called – themselves exhibit great variety (see Fig. 14.3). Visual inspection shows that these basic solutions already bear significant resemblance to actual patterns found in various organisms, such as giraffes, the striped zebrafish, and the African leopard. Through minor modifications – e.g. by tweaking the boundary conditions, or 'adding on' another Turing-style system of morphogens – it is possible to recreate a vast number of different arrangements of spots, stripes, swirls, and splodges, such that, through clever choice of parameters, it should be possible to explain various types of patterns found in nature. Or so it would seem.

Turing wrote as a non-biologist at a time when little was known about the molecular basis of biological pattern formation. So it is hardly surprising that he was unable to specify any actual biological examples of his two-morphogen reaction-diffusion model. Neither was this his goal: as he makes clear throughout the paper, his main objective – in line with our third function of exploratory modelling discussed in the previous section – was to provide a *potential explanation* of how biological patterns form spontaneously as part of an organism's development. He is explicit in characterizing his model as merely "a possible mechanism" (Turing 1952, 37) and, in the concluding paragraph of his paper, weakens this goal further when he expresses his hope "that the imaginary biological systems which have been treated, and the principles which have been discussed, should be of some help in interpreting real biological forms" (Turing 1952, 72). Arguably, more is required than merely functioning as an interpretative tool if a model is to count as doing actual explanatory work. As Shigeru Kondo and Takashi Miura put it in a favourable review of half a century or so of work on the Turing model: "No matter how vividly or faithfully a mathematical simulation might replicate an actual biological pattern, this alone does not constitute proof that the simulated state reflects the reality." (Kondo and Miura 2010, 1617) It will be instructive, then, to look in a little more detail at the track record of Turing's model and how it has fared with respect to real-world cases.

Turing is not entirely silent on what kinds of pattern formation he thinks may potentially be explained by his model. First, he begins his paper by stating that he will describe "a mathematical model of the growing embryo" (Turing 1952, 37), thereby suggesting that one of the explananda he is interested is how the early embryo's spherical symmetry is broken and how organisms acquire basic spatial organization. In this sense, Turing's professed goal is *more* ambitious than simply providing a potential mechanism for how certain surface characteristics – such as animal coat patterns – may form. Second, towards the end of his paper, Turing refers to a number of botanical examples, such as the arrangement, number, and symmetries of whorls of leaves of certain plants such as Woodruff (*Asperula odorata*). (The choice of the latter example – leaves growing around a circular stem of a plant – may be partly driven by his mathematical choice of a ring-like structure as one of the boundary conditions for his mathematical model.) Both kinds of cases, through no fault of Turing's, turned out to be less than ideal as illustrations of his model. From the late 1970s onwards, one of the main model organisms for the study of general morphogenesis was the fruit fly (*Drosophila melanogaster*), largely due to its manageable size and quick generation time, which made fruit flies the preferred model of how spatial organization comes about. However, while early embryonic development in *Drosophila* does exhibit standing wave patterns that are reminiscent of what the Turing model might have predicted, further investigation showed development and spatial patterning to be far more complicated, with the sequence of segments in the fly's body being determined by a set of maternal, gap, pair-rule and segment polarity genes, each of which is responsible for a different aspect of the organism's orgniazation (e.g., anterior/posterior asymmetry, division into head, thorax, and abdominal regions, etc.). In other words, not only did the dominant model organism's morphogenetic development not lend itself to being modelled in line with Turing's proposal, the very complexity of the *actual* developmental process suggested that the simple design of the Turing model might simply not have been favoured by evolution.

The case of plant development may be seen as even less convincing. For one, plants, due to their fixed location in space, may be subject to external factors (such as the relative position to the sun) whose influence could, at least potentially, suffice to break the symmetry, which would render the example rather less relevant to the question of embryonic morphogenesis in (esp. higher) animals. For plant morphogenesis in general, there must likewise be other considerations and constraints, as Turing was well aware. Recall that stable Turing patterns may be thought of as 'standing waves' of sorts (i.e., patterns that are brought about by the 'standing wave'-like distribution of the relevant morphogens in the organism). Whorls of leaves are presumed, by Turing, to "originate in rings of active tissue" (Turing 1952, 68), with the number of leaves corresponding roughly to the circumference of the ring divided by the 'chemical wavelength' of whatever morphogen is responsible for leaf formation. Whereas in species such as *Asperula odorata* the arrangement of between five and nine leaves into a whorl fits reasonably well with Turing's idea, his model cannot easily be extended even to something as closely related as flower petals. For, as Turing himself notes, "when all species are taken into account one must expect that the diameters of the rings concerned will take on nearly all values within a considerable range" (Turing 1952, 69) and that there will also be some variation in chemical wavelength. By the logic of his own model, then, the number of petals on flowers should be expected to vary considerably, less so within the same species (though even there we should expect the variation across individual specimens to give rise to significant numerical variation in the number of petals and other elements), but most definitely across different species since, in principle, within a certain reasonable interval, any number should be as likely as any other to be the result of dividing the circumference by the (likewise variable) chemical wavelength. Yet, as Turing notes, this is not what one finds in nature: "The number five is extremely common, and the number seven rather rare." (Turing 1952, 69)

For the longest time, developmental biologists regarded Turing's model as mathematically elegant, but biologically irrelevant. There are many reasons why this was so, and a full discussion of the history of the Turing model and its reception is beyond the scope of this paper. Suffice it to say that Turing's mathematical approach and his formal presentation, together with the underdeveloped (and ultimately unconvincing) biological examples he had given, did not mesh well with the epistemic culture of developmental biology. Solving the equations of Turing's model even approximately required mathematical training and ingenuity that few developmental biologists had, and by the time computer simulation methods were gradually becoming more widely used, another – more intuitive – model, Lewis Wolpert's 'French flag model' (Wolpert 1969), had won over the majority of those developmental biologists interested in biological pattern formation.

In order to understand Wolpert's model, consider a layer of cells which, for ease of illustration, we can imagine to be in the shape of a rectangle. A group of cells along the left-hand edge of the tissue is continuously producing a single morphogen S which diffuses linearly, so as to create a smooth gradient from left (high concentration of S) to the right (low concentration). (Let us further posit that, on the right-hand edge, there is a 'sink', where S disappears, or perhaps decays, so that we need not worry about saturation due to the continuous production of S on the left-hand side.) Once a stable distribution has been reached throughout the system, the concentration of S at a given point in space effectively provides *positional information* regarding how far along the rectangular strip of tissue we are. In a second step, cells can then respond to the positional information that is available to them. In the simplest case, a cell might simply be programmed to react differently to different concentrations of S around it. If, for example, cells reacted to high concentrations of $c > c_B$ (where c_B is a threshold concentration required for certain genes to be 'switched on', say) by turning blue, to intermediate concentrations $c_B > c > c_R$ by turning white, and to low concentrations $c_R > c$ by turning red, what would emerge in the scenario described is a tri-color pattern, blue-white-red: hence, the designation of 'French flag model'. Empirical support came from a variety of experiments and observations. In a series of experiments in the late 1950s and early 1960s, researchers identified signal-producing structures of just the kind described in the thought experiment: in experiments with chicken embryos, it was found that the apical ectodermal ridge, a thick layer of ectoderm that rims the tip of the developing limb bud, is crucial to chondrogenesis. If removed, the formation of the limb would be truncated, whereas if a piece of thigh mesoderm was grafted on top of it, rather than developing into more thigh tissue, it would grow into (sometimes an additional) limb. (See Saunders et al. 1957). Additional support for the idea that it only takes a single morphogen to 'unlock' different genetic programmes in response to the positional information contained in its concentration gradient comes from the further observation that, in many organisms (e.g., salamanders), whole limbs can be regenerated when removed. Furthermore, the French flag model can explain why patterns are stable even as overall size varies: if, as in the thought experiment, cells respond in only three distinct ways – by turning blue, white, or red – then, whether the area is small or large (and, hence, the concentration gradient more or less steep), one will always find a tri-color pattern. By contrast, a Turing mechanism would likely predict a variable number of stripes, due to the interplay between the chemical wavelength associated with the reaction-diffusion process and the overall geometry of the tissue sample. In light of such seemingly clearcut empirical evidence, and given the intuitive appeal of the 'French flag model', interest in the Turing model diminished. When, in 1979, Stuart Newman and Harry Frisch put forward a mathematical model of chick limb that had "affinities to that of Turing" (Newman and Frisch 1979, 662), this generated some initial interest, but did not as a whole diminish the dominance of the 'French flag model'. Though Newman and Frisch included a (given the computer technology at the time, highly 'pixelated') sketch of how chondrogenesis might come about over time within their model, this does not seem to have tipped the balance of evidence in their favour either – not least since the development of the chick's distinctive limb with its three, morphologically very different fingers was seen as an especially convincing case of how, in the spirit of the 'French flag model', very different patternings needed to be 'unlocked' for each of the three fingers.

It was not until the mid-1990s that new experimental evidence and simulation techniques became available that led to a revival of interest in the Turing model. Importantly, it was the *interplay* between experimental manipulation and model-based simulation that, as we shall see, made all the difference. Recall that one of the perceived advantages of the 'French flag model' was its ability to explain how, in the words of a recent textbook, "the system could also regenerate the complete original pattern if it were cut in half" (Wolpert et al. 2015, 29) or otherwise disturbed. In a series of experiments on striped tropical fish (*Pomacanthus imperator*, or Emperor angelfish), Shigeru Kondo and Rihito Asai (Kondo Asai 1995) manipulated their distinctive striped pattern via laser ablation of pigment cells, which removed a small portion of a stripe. Instead of the parallel stripes simply regrowing to complete the original pattern, as would be expected from within the 'French flag model', the lower stripe changed direction and shifted upwards, partially completing the *upper* stripe, while maintaining the spatial interval between the two stripes. This led to a distinctive and reproducible pattern of regrowth, different from the original pattern, which was more in line with a dynamic response to the twin processes of local activation and long-range inhibition than with the expression of a preformed pattern or a simple 'triggering' of a response by a simple gradient. Computer simulations confirmed that a suitably adjusted Turing model, but not a 'French flag model', was able to account for the distinctive spatial patterning associated with regrowth. (See Fig. 14.4; for a review, see (Kondo and Miura 2010).)

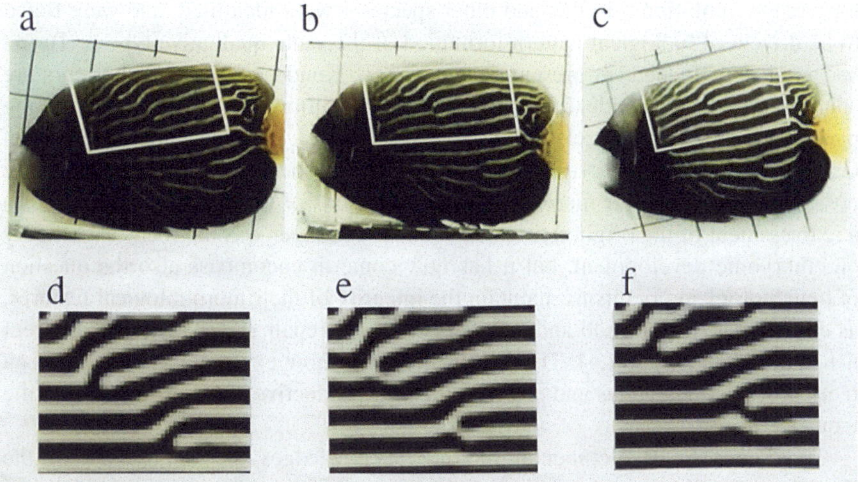

Fig. 14.4 Re-arrangement of stripe pattern of *Pomacanthus imperator*; observation (**a**–**c**) vs. Turing model simulations (**d**–**f**) (Image courtesy of Shigeru Kondo)

At least in part, then, it was due to the advent of better simulation techniques, which allowed for more subtle ways of exploring the phenomenon – for instance, by 'disturbing' (and otherwise manipulating) existing patterns and comparing the simulated response of the Turing system with observed patterns of re-growth – that Turing's approach gained new traction as a model of biological pattern formation. In addition, there has been the growing realization among researchers that, for the Turing model to be fruitful and afford insight into how biological pattern formation works, we need not actually discover two molecules whose chemical characteristics match those posited by Turing's account of the interaction between an activator and an inhibitor. Rather than treating Turing's model as a hypothetical representation of a two-morphogen system and making it our goal to discover an actual pair of substances that exhibit the posited chemical behaviour, we can consider the model as itself a hypothetical realization of the twin processes of *local activation* and *long-range inhibition*. After all, we should expect that what matters for self-organized pattern formation is not the precise number of different substances involved, or their specific chemical characteristics, but rather the fact that *some* underlying structure must realize the coupling of local activation with long-range inhibition. Interestingly, it was subsequently shown that in zebrafish the skin patterns are set up and maintained not directly in response to differential diffusion rates of two morphogens, but by a network of interactions between neighbouring pigment cells. While "the shape of the network is different from that of the original Turing model, it fits the short-range positive, long-range negative feedback description" (Kondo and Miura 2010, 1619). While it would thus be incorrect to consider the original Turing model a faithful *representation of* pattern formation in this particular species, it is in virtue of *exploration* on its basis that other realizations of local activation and long-range inhibition – in this and other species – were identified, and were found to be drivers of biological pattern formation of the same qualitative type as Turing had predicted. Indeed, in presenting their work, Kondo and his collaborators are adamant that they have identified "the essential conditions for generating the Turing pattern" in zebrafish and related species. (Cf. Osaka University Pattern Formation Group) By tweaking the original Turing model ever so slightly, it has thus become possible to open up ever more potential applications of the model. This has led to a reassessment of the original research agenda: no longer was the focus exclusively on embryonic development, but it has now come to encompass also the question of how biological organisms maintain the integrity of their morphological features, as demonstrated by Kondo and Asai's research on repair skin patterns in different fish species (see Fig. 14.4). This nicely illustrates how exploratory modelling can reorient research agendas and can help identify productive new targets of scientific inquiry.

Contemporary developmental biology acknowledges the importance of the Turing mechanism as one of several key elements in the formation of biological patterns. While it may be too early, and while the field of developmental biology may be too diverse, to speak of a consensus, there appears to be a growing sense that real-life cases of pattern formation in biology often involve both, a (basic) periodic Turing pattern and a more specific patterning for each of the periodic elements thus brought about. A paradigmatic case would be limb formation. (See (Sheth, et al. 2012).) Whereas in the past it was thought that each digit was encoded by a separate genetic programme, research involving the 'knocking out' of Hox genes has convinced many researchers that, early on in its development, the limb bud is structured by a wave-like Turing pattern, and that the specific differences between digits are the result of further 'Wolpert-style' patterning of each individual element: in other words, the initial "periodic patterning is normally subject to a type of positional signal that varies along the proximodistal axis of the limb" (Green and Sharpe 2015, 1210). The story of the Turing model as presented in this paper, then, is not one of an "underdog theory" displacing "the received view": both models, Turing's reaction-diffusion model and Wolpert's positional information model, remain relevant and are legitimate and important sources of insight. Rather, what this historical episode illustrates is how a model that did, in a very clear sense, constitute a proof of principle for self-organized pattern formation, can go out of fashion and subsequently rebound. Some of the reasons that likely contributed to the initial lack of uptake among developmental biologists have already been mentioned: sociological reasons having to do with different epistemic styles (in this case, the clash between mathematical methods and experimentally-oriented developmental biology); lack of sufficiently sophisticated computer simulation techniques that would have allowed for a more detailed rendering of the finer details of Turing patterns (and which might have lent more credibility to Newman and Frisch's calculation of the development of the chick limb); and the absence of empirical observations of the posited two-morphogen systems. Yet, once this began to change, notably through the study of tell-tale distortions of small-scale structures in angelfish and zebrafish (thanks to the availability of new experimental and simulation techniques), the model was able to rebound – not necessarily as a faithful representation of pattern formation in any one species in particular, but as a tool of exploration, which allowed scientists to converge upon fruitful avenues of research. Indeed, the very fact that researchers now consider the Turing model useful not only in the (recently more promising; see (Maini et al. 2006)) search for two-morphogen systems, but also in relation to other ways of realizing the twin processes of local activation and long-range inhibition – e.g., via networks of interaction between neighbouring cells – attests to the model's flexibility and exploratory potential.

14.5 Conclusion

At the beginning of this chapter, I wrote that exploration should stand alongside more traditional criteria for assessing models (such as explanatory, predictive, and representational success). The case of the Turing model illustrates why it may be worth valuing exploratory fruitfulness in its own right. At several crucial junctures in the Turing's model varied career, it would have been quite misguided to hold it to the standards of traditional hypothesis-testing, given the level of generality at which it was formulated and the inability of scientists, at the time, to identify the molecular basis of biological development. To put things another way, had Turing's model been treated exclusively as a way of representing real target systems, it would have had to be regarded, if not as a non-starter, then certainly as empirically unsuccessful. And, as we have seen, such doubts about the model's applicability to real-world systems were common among developmental biologists. Yet, in spite of its perceived empirical shortcomings, the model was never discarded completely, and rightly so. This is because, in the spirit of Turing's own professed aim – that is, to offer no more and no less than "a possible mechanism" by which spontaneous pattern formation in living beings may occur – a minority of researchers recognized that the model's primary function was (and, to a large extent, remains) *exploratory*. That is, it was not geared towards representing any one target system, or class of target systems, in particular, but towards 'standing in' (or, as one might put it in order to highlight its function as a proof of principle, 'filling in') for whatever it is that realizes the twin processes of local activation and long-range inhibition and thereby brings about spontaneous pattern formation in some biological systems. From the start, the same basic model equations were intended to be potentially applicable to biochemically quite different potential target systems, and, as discussed in the previous section, this exploratory potential of the model was subsequently vindicated by novel observations across a range of species. Without an acknowledgment of the fundamental role of exploration in scientific modelling, the longevity of Turing's reaction-diffusion model – which made its recent renaissance possible – would be little more than a historical anomaly, and its productive role in opening up new avenues of inquiry would have to be seen as a fortuitous, but ultimately puzzling accident.

Acknowledgments I would like to thank the conference participants at the GWP. 2016 meeting in Düsseldorf in March 2016 and at the workshop "Models in Science", held at Lingnan University (Hong Kong) in March 2017, for their discussion and feedback. An anonymous referee provided especially helpful and incisive comments, for which I am grateful. Professor Shigeru Kondo at Osaka University kindly gave permission to reproduce the two images labelled as Figs. 14.3 and 14.4 above.

References

Arabatzis, T. 2006. On the inextricability of the context of discovery and the context of justification. In *Revisiting discovery and justification*, ed. J. Schickore and F. Steinle, 215–230. Dordrecht: Springer.

Burian, R.M. 1997. Exploratory experimentation and the role of histochemical techniques in the work of Jean Brachet, 1938–1952. *History and Philosophy of the Life Sciences* 19 (1): 27–45.

Elliott, K.C. 2007. Varieties of exploratory experimentation in nanotoxicology. *History and Philosophy of the Life Sciences* 29 (3): 313–336.

Gelfert, A. 2009. Rigorous results, cross-model justification, and the transfer of empirical warrant. *Synthese* 169 (3): 497–519.

———. 2011. Mathematical formalisms in scientific practice: From denotation to model-based representation. *Studies in History and Philosophy of Science* 42 (2): 272–286.

———. 2014. Applicability, indispensability, and underdetermination: Puzzling over Wigner's 'unreasonable effectiveness of mathematics'. *Science & Education* 23 (5): 997–1009.

———. 2016. *How to do science with models: A philosophical primer*. New York: Springer.

———. 2017. The ontology of models. In *Springer handbook of model-based science*, ed. L. Magnani and T. Bertolotti, 5–23. Heidelberg: Springer.

Green, J.B., and J. Sharpe. 2015. Positional information and reaction-diffusion: Two big ideas in developmental biology combine. *Development* 142 (7): 1203–1211.

Hanson, N.R. 1960. Is there a logic of scientific discovery? *Australasian Journal of Philosophy* 38 (2): 91–106.

Harman, P.M. 1998. *The natural philosophy of James Clerk Maxwell*. Cambridge: Cambridge University Press.

Hoyningen-Huene, P. 2006. Context of discovery versus context of justification and Thomas Kuhn. In *Revisiting discovery and justification*, ed. J. Schickore and F. Steinle, 119–131. Dordrecht: Springer.

Kondo, S., and R. Asai. 1995. A reaction-diffusion wave on the skin of the marine angelfish Pomacanthus. *Nature* 376: 765–768.

Kondo, S., and T. Miura. 2010. Reaction-diffusion model as a framework for understanding biological pattern formation. *Science* 329 (5999): 1616–1620.

Magnani, L. 2004. Model-based and manipulative abduction in science. *Foundations of Science* 9 (3): 219–247.

Maini, P.K., R.E. Baker, and C.-M. Chuong. 2006. The Turing model comes of molecular age. *Science* 314 (5804): 1397–1398.

Maxwell, J.C. 1890. On physical lines of force. In *Scientific papers of James Clerk Maxwell*, ed. W.D. Niven, 451–513. Cambridge: Cambridge University Press.

Meyer, H. 1951. On the heuristic value of scientific models. *Philosophy of Science* 18 (2): 111–123.

Morrison, M. 1999. Models as autonomous agents. In *Models as mediators: Perspectives on natural and social science*, ed. M. Morgan and M. Morrison, 38–65. Cambridge: Cambridge University Press.

Morrison, M., and M. Morgan. 1999. Models as mediating instruments. In *Models as mediators*, ed. M.S. Morgan and M. Morrison, 10–37. Cambridge: Cambridge University Press.

Nersessian, N. 1999. Model-based reasoning in conceptual change. In *Model-based reasoning in scientific discovery*, ed. L. Magnani, N. Nersessian, and P. Thagard, 5–22. New York: Plenum Publishers.

Newman, S.A., and H.L. Frisch. 1979. Dynamics of skeletal pattern formation in developing chick limb. *Science* 205 (4407): 662–668.

Nickles, T. 1985. Beyond divorce: Current status of the discovery debate. *Philosophy of Science* 52 (2): 177–206.

———. 2006. Heuristic appraisal: Context of discovery or justification? In *Revisiting discovery and justification*, ed. J. Schickore and F. Steinle, 159–182. Dordrecht: Springer.

Osaka University Pattern Formation Group, n.d. *Papers of the Pattern Formation Group*. Retrieved March 31, 2017, from http://www.fbs.osaka-u.ac.jp/labs/skondo/publication.html

Pipes, L. 1953. An operational analysis of traffic dynamics. *Journal of Applied Physics* 24 (3): 274–281.

Reutlinger, A., D. Hangleiter, and S. Hartmann. forthcoming. Understanding (with) toy models. *British Journal for the Philosophy of Science*.

Rheinberger, H.-J. 2010. *An epistemology of the concrete: Twentieth-century histories of life*. Durham: Duke University Press.

Saunders, J.W., J.M. Cairns, and M.T. Gasseling. 1957. The role of the apical ridge of ectoderm in the differentiation of the morphological structure and inductive specificity of limb parts in the chick. *Journal of Morphology* 101 (1): 57–87.

Sheth, R., L. Marcon, M. Bastida, M. Junco, L. Quintana, R. Dahn, and M. Ros. 2012. Hox genes regulate digit patterning by controlling the wavelength of a turing-type mechanism. *Science* 338 (6113): 1476–1480.

Steele, K., and C. Werndl. forthcoming. Model-selection theory: The need for a more nuanced picture of use-novelty and double-counting. *The British Journal for the Philosophy of Science*

Steinle, F. 1997. Entering new fields: Exploratory uses of experimentation. *Philosophy of Science* 64(Proceedings of the PSA1996, Pt. II) S65–S74.

Suárez, M. 1999. The role of models in the application of scientific theories: Epistemological implications. In *Models as mediators: Perspectives on natural and social science*, ed. M.S. Morgan and M. Morrison, 168–195. Cambridge: Cambridge University Press.

Taylor, P. 1989. Revising models and generating theory. *Oikos* 54 (1): 121–126.

Thompson, D.W. 1917. *On growth and form*. Cambridge: Cambridge University Press.

Turing, A. 1952. The chemical basis of morphogenesis. *Philosophical Transactions of the Royal Society of London. Series B, Biological Sciences* (641): 37–72.

Wageningen-Kessels, F.V., H. van Lint, K. Vuik, and S. Hoogendoorn. 2015. A genealogy of traffic flow models. *EURO Journal on Transportation and Logistics* 4 (4): 445–473.

Waters, C.K. 2007. The nature and context of exploratory experimentation: An introduction to three case studies of exploratory research. *History and Philosophy of the Life Sciences* 29 (3): 275–284.

Weisberg, M. 2007. Who is a modeler? *British Journal for the Philosophy of Science* 58: 207–233.

Wimsatt, W.C. 2007. *Re-engineering philosophy for limited beings: Piecewise approximations to reality*. Cambridge, MA: Harvard University Press.

Wolpert, L. 1969. Positional information and the spatial pattern of cellular differentiation. *Journal of Theoretical Biology* 25 (1): 1–47.

Wolpert, L., C. Tickle, and A. Martinez Arias. 2015. *Principles of development*. 5th ed. Oxford: Oxford University Press.

Author Index

A. Christian et al. (eds.), *Philosophy of Science*, European Studies in Philosophy of Science 9, https://doi.org/10.1007/978-3-319-72577-2

Subject Index

A. Christian et al. (eds.), *Philosophy of Science*, European Studies in Philosophy of Science 9, https://doi.org/10.1007/978-3-319-72577-2

The manufacturer's authorised representative in the EU is Springer Nature Customer Service Centre GmbH, Europaplatz 3, 69115 Heidelberg, Germany. If you have any concerns regarding our products, please contact ProductSafety@springernature.com

Printed and bound by CPI Group (UK) Ltd, Croydon, CR0 4YY
05/10/2026
02228141-0007